ENVIRONMENTAL CHEMISTRY

COLIN BAIRD

University of Western Ontario

W. H. Freeman and Company
New York

Back cover image: Chlorine monoxide (ClO) and the Antarctic ozone hole. These maps, produced by the Microwave Limb Sounder aboard the Upper Atmosphere Research Satellite, show the amount of ClO (left) and ozone (right) in the stratosphere above 20 kilometers (66,000 feet). Very small abundances of ozone appear where there are large abundances of ClO, the dominant form of chlorine that destroys ozone. Data from September 21,1991 (top) are compared with those from September 20, 1992 (bottom). (Photos supplied for both front and back covers by National Aeronautics and Space Administration.)

Library of Congress Cataloging-in-Publication Data
Baird, Colin
    Environmental chemistry/Colin Baird.
        p.   cm.
    Includes index.
    ISBN 0-7167-2404-9
    1. Environmental chemistry.   I. Title.
TD192.B35   1995
628.5'01'54—dc20   94-23239
                                                          CIP

Printed in the United States of America

1 2 3 4 5 6 7 8 9 0   VB 9 9 8 7 6 5 4

To my daughter, Jenny, and all the other members of her generation: may these lessons of the past and present be a guide to the avoidance of environmental catastrophes in the future.

CONTENTS

CHAPTER 3

GROUND-LEVEL AIR CHEMISTRY AND AIR POLLUTION   77

CHAPTER 4

THE GREENHOUSE EFFECT
AND GLOBAL WARMING   149

CHAPTER 5

BACKGROUND ORGANIC CHEMISTRY 193

CHAPTER 6

CHAPTER 7

NATURAL WATERS: CONTAMINATION AND PURIFICATION   287

CHAPTER 8

NATURAL WATERS: ACID-BASE CHEMISTRY OF THE CARBONATE SYSTEM   325

CHAPTER 9

HEAVY METALS AND THE CHEMISTRY OF SOILS   347

CHAPTER 10

ENERGY PRODUCTION AND ITS ENVIRONMENTAL CONSEQUENCES   395

PREFACE

I am writing these introductory remarks while seated on the dock of a small island lying just off the North Atlantic coast. Some of the tiny islands in this bay will become permanently submerged in the twenty-first century if the predictions are accurate about sea-level rise from greenhouse warming (a phenomenon discussed in Chapter 4). Because the sun is shining brightly, I've applied some extra sunscreen to protect myself from the increased ultraviolet radiation arising from ozone depletion (Chapter 2). Further down the Nova Scotia coast, what were three of North America's best salmon rivers now require annual stocking because salmon will no longer travel up them to spawn in their acidic waters (Chapter 3). During its recent reconstruction, the public footbridge to the island was painted with so much creosote wood preservative that local residents have stopped harvesting mussels from the beds that lie below it, because they suspect PAH contamination of the sea life (Chapter 6). Before I bought my house on the island, the well water was tested for arsenic, a common heavy metal pollutant in this region of abandoned gold mines (Chapter 9).

I traveled to the island from my home in London, Ontario, which is Canada's capital for PCB storage and the site of an ongoing debate as to how to destroy these chemicals (Chapter 6). The city is an hour's drive north of Lake Erie, famous for nearly having "died" of phosphate pollution (Chapter 7). An hour to the west lies Lake Huron, site of several nuclear power reactors that generate plutonium and other radioactive wastes for which no one has yet devised a safe disposal plan (Chapter 10). Nearby, farmers are trying to convince governments to subsidize the production of gasohol from corn (Chapter 10).

These environmental problems probably have parallels to those where you live, and learning more about them may convince you that environmental chemistry is not just a topic of academic interest but that it touches your life every day in very practical ways.

WHO THIS BOOK IS WRITTEN FOR

There are many definitions of environmental chemistry. To some it is the chemistry of the Earth's natural processes in air, water, and soil. More commonly, as in this book, it is concerned principally with the chemical aspects of problems that human beings have created in the natural environment. Although the science underlying environmental problems is often highly complex, the central aspects of it can usually be understood and appreciated with only introductory chemistry as background. In this book, this is the only chemical knowledge that is assumed, although students who have not had organic chemistry as part of their introductory course should work through such material in Chapter 5 before tackling the toxic organics discussed in Chapter 6.

Students interested in pursuing topics in more detail should consult the "Suggestions for Further Reading" found at the end of each chapter. I have purposely refrained from giving in-text references to specific facts to avoid cluttering up what is an undergraduate text rather than a reference book or advanced treatise on the subject. Nevertheless, I am receptive to corrections or suggestions from students and their instructors for inclusion in potential future editions.

ACKNOWLEDGMENTS

I wish to express my gratitude and appreciation to a number of people who in various ways have contributed to this book:

To my students who used draft versions of the text and pointed out sections that needed clarifying.

To the principal reviewers of the text for their helpful comments and suggestions. These include Robert Haines (Sir Wilfred Grenfell College), Howard Lee McLean (Rose-Hulman Institute of Technology), Harry Pence (State University of New York, College at Oneonta), Michael Perona (California State University at Stanislaus), Wilmer Stratton (Earlham College), and Darrell Watson (University of Mary Hardin-Baylor).

To Dr. Myra Gordon for providing masses of "clippings" of environmental issues from the popular and technical literature.

To my editorial assistant Kim Grainger for tracking down references and checking answers to problems.

To my secretaries Shannon Woodhouse, Wendy Smith, and Diana Timmermans for their brave attempts to decipher my writing.

To Earl and Val Meister for their hospitality during the summertime writing stints in Stonehurst.

To the Freeman editors—Deborah Allen for her encouragement, ideas, insightful suggestions, and patience; Denis Cullinan for his careful copy editing and suggestions; and Christine Hastings for skillfully "quarterbacking" the book's production.

Colin Baird
Stonehurst, Nova Scotia
July 1994

CHAPTER

INTRODUCTION TO ENVIRONMENTAL CHEMISTRY

Chemistry plays a major role in our environment. Indeed, it is common for the public to blame synthetic chemicals and their creators for most current pollution problems. But it passes unrecognized that most of the environmental problems of past centuries and decades, such as the biological contamination of drinking water, were solved only when the methods of science in general—and chemistry in particular—were applied to them. The phenomenal rise in human life expectancy and in the material quality of life that has come about in recent decades is due in no small measure to chemicals and chemistry.

Nevertheless, it is true that chemicals, as defined in the broadest sense, lie at the heart of most of today's environmental problems. The byproducts of the substances used to improve our health and standard of living have in some instances returned to haunt us by degrading our health and that of plants and animals. In short, our conquest of widespread biological pollution and the increase in our standards of health and material wealth in developed countries have been achieved at the price of the widespread, low-level chemical pollution of the Earth.

There is as yet no consensus among scientists as to whether such low-level contamination by chemicals adversely affects the health of humans or other living organisms. Some scientists completely discount the danger of any deleterious effects due to synthetic chemical influences (especially

1

those involving cancer), by pointing to the much higher concentrations of "natural" toxic substances, such as pesticides produced by plants themselves, to which we all are exposed in our diets. At the opposite end of the spectrum are scientists who believe that environmental chemicals play a large role in inducing certain types of cancers and birth defects in both humans and wildlife.

In this book, we shall study the chemistry of our environment and the chemistry underlying our modern environmental problems. Understanding the science that underlies these problems is vital if they are to be solved and if as a society we wish to avoid their recurrence in new contexts. We shall also discuss what is currently known about the health effects associated with these substances and processes, making an attempt to be as objective as possible rather than adopting either extremist orientation.

We begin the discussion with a case study of current concern, the chemical contamination of the Great Lakes that illustrates many facets of environmental chemistry that will be taken up in more detail in subsequent chapters.

Toxic Chemicals and the Newborn

Public concern regarding environmental chemicals usually is centered upon their potential to cause cancer. However, researchers do not focus solely on possible cancerous effects. Scientists recognize that problems associated with reproduction, including those that induce birth defects, are equally characteristic in animals exposed in experiments to large concentrations of environmental chemicals. Recently, research has been initiated to discover whether humans who have been exposed to the same chemicals, albeit at lower levels, also are subject to reproductive problems. In order to detect the rather subtle effects likely to occur in humans, it is necessary to find and study human populations whose geographic location, employment, or diet may subject them to higher-than-average amounts of the chemicals of concern.

One such subpopulation lives along the shores of Lake Michigan and consists of family members who regularly fish its waters as a hobby. Indeed, the anglers and their families often eat salmon or trout from the lake two or three times a week. Of course, they've probably all heard that the waters of Lake Michigan are "full of chemicals," but since they've never suffered any overt ill effects from pollutants in the fish, they happily continue with their sport. Scientists would not have expected them to notice any deleterious health effects immediately, since chronic exposure to low levels of contaminants does not usually culminate in overt illness quickly; potential damage is more likely to include small increases in the rates of cancers and reproductive problems.

The research team of Sandra and Joseph Jacobson and their coworkers at Wayne State University in Detroit have spent more than a decade studying the offspring of people from this area, including children whose mothers eat lake fish as well as those whose mothers do not. They have discovered statistically significant differences in children born to women who have high levels of certain chemicals in their bodies; these differences are present not only at birth but persist to the age of four years and perhaps beyond. Of various chemicals present in the fish, there is one group whose concentration in the children correlates most significantly with the occurrence of the effects; these chemicals are called **PCBs,** which stands for **polychlorinated biphenyls.** These chemicals were used for many purposes in the last four decades, and their improper disposal has given rise to areas of great contamination, especially in some of the communities around Lake Michigan. The properties, chemistry, and use of PCBs will be discussed in further detail in Chapter 6. For the present, it is sufficient to know that they are widespread, long-lasting pollutants in the Great Lakes and elsewhere and that they eventually become concentrated in the fatty tissue of fish, particularly those from Lake Michigan.

The Jacobsons and their colleagues studied a group of several hundred children born in western Michigan hospitals during 1980–1981; many of the children's mothers had been heavy consumers of lake fish in the preceding six years. The mothers in the sample were predominantly urban, white, and middle class. The prenatal (i.e., prebirth) exposure of the infants to PCBs was determined by analyzing the blood of their umbilical cord for these chemicals after birth; the postnatal exposure of the children was assessed by analyzing their mother's breast milk and also by analyzing blood samples from the children at the age of four years.

The Jacobsons discovered that at birth the children born to mothers who had transmitted the highest amounts of PCBs to the children before birth had a slightly lower birth weight, slightly smaller head circumference, and were slightly more premature than those born to women who passed along lower amounts. The severity of these "deficits" was larger the greater the prenatal exposure to PCBs. When tested at the age of seven months, many affected children displayed small difficulties with visual recognition memory, again with the extent of the problem increasing with prebirth transmission of PCBs. At the age of four years, the lower body weight observed at birth for highly exposed infants still lingered. More serious was the observation that the four-year-olds displayed progressively lower scores on several tests of mental functioning—with respect to some verbal and memory abilities—the greater their prenatal PCB exposure (see Figures 1-1 and 1-2). Interestingly, at four years of age, the children's total body content of PCBs, which is determined almost entirely from the breast milk they consumed as infants rather than from any prenatal transmission of the chemicals, was not the relevant factor in determining these physical and mental deficits; rather, it was the smaller amounts of PCBs transmitted from mother to fetus that were important. Thus PCBs appear to interfere with the proper prebirth development of the brain and with the mechanisms that determine physical size.

The study by the Jacobsons on the Lake Michigan children is one of the clearest examples available concerning the influence of environmental chemicals on the health of human beings. It is important to realize

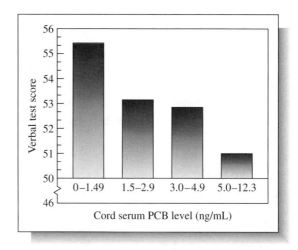

FIGURE 1-1

Test outcomes (McCarthy verbal test scores) of the 1990 Lake Michigan case study of four-year-old children. The children's scores are graphed versus the PCB concentrations in the umbilical cord serum at birth. (Source: Redrawn from J. L. Jacobson, S. W. Jacobson, and H. E. B. Humphrey (1990). Effects of in utero exposure to polychlorinated biphenyls and related contaminants on cognitive functioning in young children. *Journal of Pediatrics* 116(1): 38–44.)

that the highest levels of PCBs to which the children in this group were exposed before birth are not much greater than those to which the majority of unborn children in the general population are subjected. And while the chemicals did not produce gross birth defects in the children (they were not stunted in size or mentally retarded), they did produce small and consistent deficits of several kinds.

The PCB chemicals that gave rise to the reproductive problems around Lake Michigan are widespread in our environment, even though no new production of these substances has taken place since the late 1970s. PCBs are not solely a water pollution problem, or solely an air pollution problem, or solely a problem involving soil and sediments. They are present in the air that we breathe, the water we drink, the food we eat, and in the sediments at the bottom of our rivers and lakes. PCBs cycle and recycle among these physical media, and as a consequence they have been dispersed to locations far beyond the points at which they initially entered

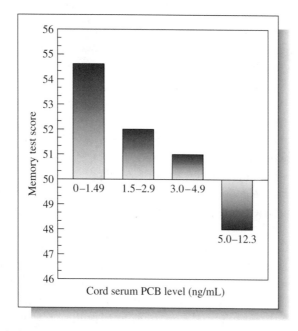

FIGURE 1-2

Test outcomes (McCarthy verbal test scores) of the 1990 Lake Michigan case study of four-year-old children. The children's scores are graphed versus the PCB concentrations in the umbilical cord serum at birth. (Source: Redrawn from J. L. Jacobson, S. W. Jacobson, and H. E. B. Humphrey (1990). Effects of in utero exposure to polychlorinated biphenyls and related contaminants on cognitive functioning in young children. *Journal of Pediatrics* 116(1): 38–44.)

the natural environment. Though not very volatile (i.e., easily evaporated), some PCBs do evaporate from bodies of water and from waste disposal sites on land. These airborne PCBs can travel hundreds or thousands of kilometers before they condense onto the land or onto another body of water. Once in a river or lake, they become incorporated into the sediments or into the plants and fish living in the water. Slowly, over a period of decades, the current environmental burden of PCBs will decline because the sediments will become covered and make the compounds inaccessible. However, we shall remain involuntarily exposed to them for decades. Additional quantities of PCBs could well be added to the environmental burden if care is not exercised when those still in use—particularly in electrical equipment—are eventually removed.

APPROACHES TO THE PREVENTION OF POLLUTION

Given the results in the case study above and the findings of many other research projects, it has become clear that human health and well-being is probably being affected in subtle but significant ways by small concentrations of chemicals in our environment. We shall spend much of Chapter 6 exploring the nature of PCBs and highly toxic chemicals such as "dioxins," as well as the mechanisms by which they become concentrated in the food that human beings consume.

However, before studying the specifics of different pollution processes, whether in air, in water, in our food supply, or in soil, it is instructive to consider the general philosophy that is now being implemented in developed countries to reduce human exposure to such chemicals in the future.

Historically, it was assumed implicitly or explicitly that chemicals emitted into the environment would be assimilated by nature: either the natural system would convert them into harmless, naturally occurring substances, or the chemicals would be diluted to such an extent they would pose no threat to life. ("The solution to pollution is dilution" was a well-known saying in the early and middle decades of the twentieth century.) It became clear in the 1960s and 1970s that many substances are not assimilated because they are **persistent,** that is, they are unaltered by the action of light, water, air, or microorganisms (which often serve to break down, or degrade, many pollutants) for very long periods of time.

Examples of these persistent substances include pesticides such as DDT, the refrigerants called CFCs, the gas carbon dioxide, and toxic forms of the element mercury. Due to their persistence and their continuous release, the environmental concentrations of such substances increased in the past to worrisome levels. Just as disturbing, it has been discovered that many persistent substances do not become dispersed evenly throughout the environment but rather concentrate within living organisms, thereby reaching levels in humans and other animals that in some cases have affected their health and even led to premature death.

The response to the problem of persistent, biologically accumulative pesticides such as DDT has been to ban further production, or at least that meant for use in developed countries, and to ban further use once existing stocks were exhausted. As a consequence of such bans, the environmental levels of DDT and many other persistent pesticides have fallen significantly since their concentrations peaked in the 1960s and early 1970s. However, the rapid declines have now stopped, and current levels remain well above zero. As in the case of PCBs, the concentrations of the banned pesticides are now declining only slowly as the environmental burden is cycled among the various media.

Just as troublesome as these industrial chemicals themselves are toxic byproducts produced in small quantities and released into the environment as a consequence of the production, distribution, and use of large quantities of other commercial substances. An example of a toxic byproduct is the "dioxin" that is formed inadvertently in tiny amounts as a byproduct during many industrial processes, including the bleaching of wood pulp and the production of some of the common pesticides and wood preservatives.

The initial strategy used to deal with toxic byproducts was to reduce the releases of these substances into the environment, often by capturing and disposing of a large fraction of their mass before dispersal could occur. An example of this approach is the capture of gaseous pollutants in the smokestacks of electrical power plants; once captured, they are converted to solids and subsequently disposed of in landfills. Approaches to pollution control that focus on capturing rather than reducing the production of toxic byproducts are called **end-of-the-pipe** solutions. One drawback to such strategies is that pollutants are not usually destroyed in the process, but simply made more benign or deposited into a different medium. In the case of the pollutants from power plants, for example, the chemicals are merely put onto land rather than released into the air.

The strategy, called **green chemistry,** that now is supplanting this approach is the reformulation of synthetic routes so that toxic byproducts are not produced in the first place; in other words, the new approach is to "move up the pipe" toward the point of pollutant creation. Examples of the new strategy include the replacement of organic solvents by water as the medium in which the desired products are formed, or perhaps the elimination of a solvent altogether; the substitution of environmentally benign substances to replace heavy-metal catalysts; and the design of products to make them recyclable or safely disposable. For products that themselves pose environmental danger if dispersed, "closed loop recycling" is being put into practice. The overall strategy is to virtually eliminate toxic, persistent substances from the environment by allowing no further releases, and where practical by collecting and destroying existing deposits of these chemicals.

Some environmentalists foresee the eventual adoption of environmental practices in which certain types of substances will not be generated by society at all because their production and use will inevitably present unacceptable hazards in the long run. The aim is that there be "zero discharge" of certain toxic, persistent substances into the environment, guaranteed by the fact that they are never produced.

Many individuals now are trying to play a part in reducing pollution and waste by changing their habits as consumers. Much larger quantities of newspapers, aluminum cans, glass containers, and plastics are being collected separately from garbage and recycled than was the case a decade ago. "Green" products have become commonplace on supermarket shelves in developed countries.

Consumers are turning away from products having excess packaging or which are made from synthetic rather than "natural" materials. In many cases, however, the environmentally optimum choice between two alternative products may not be intuitively obvious. Take for example the two main types of disposable drinking cups: one is constructed of paper derived from trees, and the other is made with polystyrene synthesized from petroleum. Most consumers believe the paper cup to be preferable on environmental grounds. However, a "cradle-to-grave" analysis of the impacts of all the stages of production and disposal shows that the paper cup consumes more chemicals and causes more water pollution than the polystyrene one, mainly because the paper is bleached. Clearly, we need to know the environmental consequences of the may steps involved in producing consumer goods before we can really select which product is superior.

SPHERES OF ENVIRONMENTAL CHEMISTRY

The subject of environmental chemistry includes both the "natural" substances and processes that are of importance in a clean environment, as well as significant pollution problems that have a chemical basis. Although most topics by nature involve more than a single physical medium or a single compound, the material in this book has been subdivided for convenience as follows:

a. atmospheric chemistry and concerns (Chapters 2–4)

b. toxic organic substances (Chapters 5–6)

c. water chemistry and concerns, including heavy metals and the role of soils and sediments (Chapter 7–9)

d. fuels and energy sources, including nuclear (Chapter 10)

Beyond the scope of our discussion of environmental issues is the fact that environmental problems in general have arisen mainly owing to the great increase in world population that has occurred; furthermore, many environmental problems exist that are not considered herein since they are not primarily of chemical origin. Some perspective on the range and importance of environmental issues is contained in the recent "World Scientists' Warning to Humanity" which was issued recently by the Union of Concerned Scientists. This document was signed by more than one thousand prominent scientists, including over one hundred Nobel Prize winners. Their statement is reprinted as Appendix I at the back of this book.

SUGGESTIONS FOR FURTHER READING

1. World Commission on Environment and Development. 1987. *Our Common Future.* New York: Oxford University Press. [This is known informally as "The Brundtland Report."]

2. Gore, A. 1992. *Earth in the Balance: Ecology and the Human Spirit.* New York: Houghton Mifflin.

3. Mungall, C., and D. J. McLaren, eds. 1990. *Planet Under Stress: The Challenge of Global Change.* Toronto: Oxford University Press.

4. Ames, B. N., and L. S. Gold. 1990. Misconceptions on pollution and the causes of cancer. *Angewante Chemie International Edition in English* 29: 1197–1208.

5. Jacobson, J. L., S. W. Jacobson, and H. E. B. Humphrey. 1990. Effects of in utero exposure to polychlorinated biphenyls and related contaminants on cognitive functioning in young children. *Journal of Pediatrics* 116: 38–44.

6. Whitesides, G. 1990. What will chemistry do in the next twenty years? *Angewante Chemie International Edition in English* 29: 1209–1218.

7. Hocking, M. B. 1991. Paper versus polystyrene: A complex choice. *Science* 251: 504–505.

Interchapter:
Toxic Fish Study
in Great Lakes

An Interview with Joseph L. Jacobson

Joseph L. Jacobson is a professor in the Psychology Department at Wayne State University. He was trained as a developmental psychologist at Harvard University, receiving his Ph.D. in 1977. Since 1979 he has been collaborating with his wife, Sandra W. Jacobson, on a prospective, longitudinal study of intellectual and behavioral development in children whose mothers had eaten relatively large quantities of PCB-contaminated Lake Michigan fish.

How manifest (if at all) to the casual observer are the effects of PCBs on the children in your study? Are the deficits sufficient to have a significant effect on the lives of the children?

The effects of PCBs on the children in our study are not manifest to the casual observer. That's why we've described them as subtle differences. We do not know the long-term implications of these deficits since they are relatively subtle. There are three possibilities. The children could outgrow these problems once they get to school and into a more structured academic environment. On the other hand, it is possible that they could continue to have relatively subtle but lasting problems in attention or short-term memory. Or, when that child gets to school, he may have some difficulty in learning reading and arithmetic; if he is then labeled as slow, his academic problems may get exacerbated in a cumulative fashion. It's really not possible to make a prediction based on what we have seen so far as to whether the deficits will go away, continue to be subtle, or get worse over time.

Does your research indicate that children in the general public (those of non-fish eaters) are affected by background levels of PCBs consumed in food by their mothers before birth?

It suggests that is a real possibility. A few of our more highly exposed children had mothers who did not eat fish. PCBs are found in many food sources: fruits, vegetables, beef. There is no way of knowing, other than having a blood test, what level of PCBs you've accumulated over your lifetime. However, the fetus will be exposed to all the PCBs a mother accumulates over her lifetime because they are stored in her body fat. There is no way for the mother to get rid of them except by breastfeeding. When the mother does breastfeed she lowers her own body burden of

PCBs, but of course she transmits them to her infant. So far the big problem with PCBs is with prenatal exposure. Although relatively small quantities go across the placenta, that's where the effect seems to be significant. Postnatal exposure didn't seem to affect the children's development in any significant way that we could determine.

Have any recent advances in instrumentation and analysis allowed you to restudy your blood samples to detect furans or PCBs more quantitatively?

If we had stored blood samples from our original study, we could have reanalyzed our data with the newer instruments. Certainly the analytic chemical analysis of PCBs has improved dramatically since we started. Newer studies are focused on individual PCB molecules and furans and so forth.

Was PCB contamination of fish primarily a problem of the 1970s and 1980s, so that we don't have to worry any more about women of childbearing age having high levels of these chemicals in their bodies? In other words, is the PCB problem now a thing of the past?

The levels of PCBs in Lake Michigan fish have declined markedly since the infants in our study were born in 1980 and 1981. But here you have to keep it in mind that other sources of exposure may be quite substantial, since we found effects in the upper ranges of background level exposures. It probably is still wise for women of child-bearing age to limit their consumption of fatty Great Lakes fish even though the risks have declined.

In light of your study, do you think that enough is currently being done about contamination of our waters and food system?

We are going to be very interested to see how the children in our study are doing now; but it's a big process to analyze 300 children and get all the data cleaned and analyzed in order to learn as much as we can from this study. Modern environmental chemistry is an exciting field with many opportunities, and there are certainly many environmental contaminants that need to be studied and better understood. Investigation and practical measures should continue to be a high-priority concern.

CHAPTER

STRATOSPHERIC CHEMISTRY: THE OZONE LAYER

The ozone layer is a region of the atmosphere that is "Earth's natural sunscreen" since it filters out harmful ultraviolet (UV) rays from sunlight before they can reach us on the surface of our planet and cause damage to humans and other life forms. Any substantial reduction in the amount of overhead ozone, O_3, would threaten life as we know it. Thus the appearance of a large "hole" in the ozone layer over Antarctica represents a major environmental crisis.

The *total* overhead amount of atmospheric ozone at any location is expressed in terms of **Dobson units (DU);** one such unit is equivalent to a 0.001 mm thickness of pure ozone at the density it would possess if it were brought to ground-level (1 atm) pressure. The normal amount of overhead ozone at temperate latitudes is about 350 DU. Because of stratospheric winds, ozone is transported from tropical toward polar regions. Thus the closer to the equator you live, the less the total amount of ozone that protects you from ultraviolet light. Ozone concentrations in the tropics usually average 250 DU, whereas those in subpolar regions are about 450 DU, except of course when holes appear in the ozone layer over such areas. There is also some natural variation of ozone concentration with season.

The Antarctic ozone hole was discovered by Dr. Joe C. Farman and his colleagues in the British Antarctic Survey. They had been recording

13

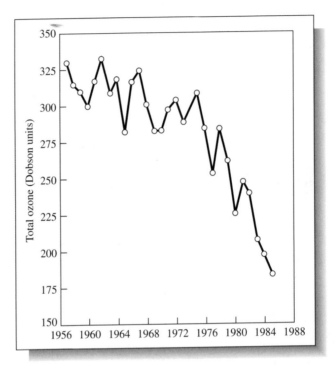

FIGURE 2-1

The yearly variation in total ozone (mean values for October) over Halley Bay Station, Antarctica. (Source: Redrawn from P. Uster in R. R. Jones and T. Wigley, eds. *Ozone Depletion: Health and Environmental Consequences*. Chichester, England: John Wiley and Sons, 1989.)

ozone levels over this region since 1957. Their data indicated that the total amounts of ozone each October had been gradually falling, with precipitous declines beginning in the mid-1970s (Figure 2-1). The period from September to November corresponds to the spring season at the South Pole, and follows a period of very cold 24-hour nights common to polar winters. By the mid-1980s, the springtime loss in ozone at some altitudes over Antarctica was complete, and amounted overall to a loss of more than 50% of the total overhead amount. It is therefore appropriate to speak of a "hole" in the ozone layer which now appears each spring over the Antarctic and which lasts for several months. (See also the contour map on the front cover of this book.) In 1993, the ozone concentration dropped to a record low of 90 DU in early October.

It was not clear for several years after its discovery whether the hole was due to a "natural" phenomenon involving meteorological forces, or was due to a chemical mechanism involving air pollutants. In the latter possibility, the suspect chemical was chlorine, produced mainly from gases that are released into the air in large quantities as a result of their use, for example, in aerosol spray cans and in air conditioners. Scientists had predicted that the chlorine would destroy ozone, but only to a small extent—a few percent—and only after several decades had elapsed. The discovery of the Antarctic ozone hole came as a complete surprise to everyone.

To discover why the hole formed each spring, an American emergency ozone research expedition headed by Dr. Susan Solomon of the National Oceanic and Atmospheric Administration in Boulder, Colorado went to the Antarctic in late winter in August 1986. Using moonlight as a light source, Solomon and her coworkers were able to identify from the specific wavelengths of light absorbed by atmospheric gases that certain molecules were present in the atmosphere far above their heads. As a result of that research and subsequent investigations, it is known that the hole indeed does occur as a result of chlorine pollution. Furthermore, it has been predicted that the hole will continue to reappear each spring for the next several decades, and that a corresponding hole may one day appear above the Arctic region.

As a consequence of these discoveries, governments worldwide moved quickly to legislate a phase-out in production of the chemicals responsible so that the situation does not become much worse with the development of even more severe ozone depletion over populated areas than is already the case, and with the corresponding threat to the health of humans and other organisms that this increase would bring.

In the last few years we have found that the ozone is being depleted not just in Antarctica but worldwide. The extent of this depletion is illustrated in Figure 2-2, which shows the changes that occurred in total ozone concentrations from 1958 to 1991 for the northern middle latitudes. The losses at such middle latitudes—which, as the graph clearly indicates, began in the early 1980s—now amount to about 5% *per decade*. The greatest depletion occurs in the winter-spring season, and amounted to 14% from 1970 to 1993. There is some depletion in the summertime, when most people's exposure to sunlight is at a maximum. For example, the ozone levels above Toronto in July 1993 were 12% lower than the pre-1980 values. The worldwide loss of ozone has become a major environmental concern, since it results in less protection to life at the Earth's surface from the harmful ultraviolet component of sunlight.

In this chapter, we shall investigate the chemical processes involved in the production of the ozone layer, as well as those underlying the phenomenon of ozone depletion.

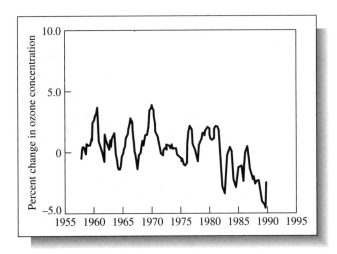

FIGURE 2-2

Changes in ozone concentrations for northern midlatitudes (40°–52° N) relative to the long-term average. (Source: Redrawn from R. Stolarski, *Science* 257 (1992): 727–728.)

REGIONS OF THE ATMOSPHERE

The main components (ignoring the normally ever-present but variable water vapor) of an unpolluted version of the Earth's atmosphere are diatomic nitrogen (N_2, about 78% of the molecules), diatomic oxygen (O_2, about 21%), argon (Ar, about 1%), and carbon dioxide (CO_2, about 0.03%, and climbing as a result of fuel combustion). This mixture of chemicals seems unreactive in the lower atmosphere even at temperatures or sunlight intensities well beyond those naturally encountered at the Earth's surface.

The lack of overt reactivity in the atmosphere is deceptive. In fact, many environmentally important chemical processes occur in air, whether clean or polluted. In the next two chapters, these reactions will be explored in detail. In Chapter 3, reactions that occur in the **troposphere,** the region of the sky that extends from ground level to about 15 kilometers altitude are discussed. In the present chapter we will consider processes in the **stratosphere,** the portion of the atmosphere from 15 to 50 kilometers altitude that lies just above the troposphere. The chemical reactions to be considered are vitally important to the continuing health of the ozone layer, which is found in the bottom half of the stratosphere.

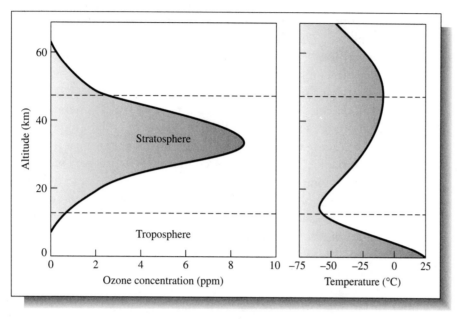

FIGURE 2-3

Variation with altitude of ozone concentration (left) and air temperature (right), for various regions of the lower atmosphere. (Source: Redrawn from "The State of Canada's Environment." Ministry of Supplies and Services, Ottawa, 1991.)

The ozone concentrations and the average temperatures at altitudes up to 50 kilometers in the Earth's atmosphere are shown in Figure 2-3.

The stratosphere is defined as the region that lies between the altitudes where the temperature trends display reversals; the bottom of the stratosphere occurs where the temperature first stops decreasing with height and begins to increase, and the top of the stratosphere is the altitude where the temperature stops increasing with height and begins to decrease.

ENVIRONMENTAL CONCENTRATION UNITS FOR GASES

Concentration units different from the chemists' familiar moles per liter (molarity) are commonly used in environmental science for gases present in air. The most familiar such units are expressed as "parts per _____." Thus, a concentration of 100 molecules of a gas such as carbon dioxide dispersed in one million (10^6) molecules of air would be expressed as "100 parts per million," i.e., "100 ppm." Similarly, "ppb" and "ppt" stand for "parts per billion" (10^9) and "parts per trillion" (10^{12}) respectively. Occasionally the unit pphm—parts per hundred million (10^8)—is used.

It is important to emphasize that for *gases*, these units express the number of *molecules* of a pollutant (i.e., the "solute" in chemists' language) that are present in one million or billion or trillion molecules of air. Since, according to the Ideal Gas Law, the volume of a gas is proportional to the number of molecules it contains, the "parts per" scales also represent the volume a pollutant gas would occupy compared to that of the stated volume of air if the pollutant were to be isolated and compressed until its pressure equaled that of the air. In order to emphasize that the concentration scale is based upon molecules or volumes rather than upon mass, a v (for volume) is sometimes shown as part of the unit, e.g., 100 ppmv or 100 ppm$_v$.

The variation in the concentration of ozone on the ppm scale at different altitudes in the troposphere and stratosphere is illustrated in Figure 2-3a.

THE CHEMISTRY OF THE OZONE LAYER

LIGHT ABSORPTION BY MOLECULES

The chemistry of ozone depletion, and indeed of other processes in the stratosphere, is driven by energy associated with light from the sun. For this reason, we begin our analysis by investigating the relationship between light absorption by molecules and the resulting activation, or energizing, of the molecules that enables them to react chemically.

An object that we perceive as black in color absorbs light at all wavelengths of the visible spectrum—i.e., from about 400 nm (violet light) to about 750 nm (red light). (Wavelength of light are frequently expressed in **nanometers** (nm), where one nanometer is 10^{-9} meter.) Substances differ enormously in their propensity to absorb light of a given wavelength because of differences in the energy levels of their electrons. Diatomic molecular oxygen, O_2, does not absorb visible light significantly, but does absorb some types of **ultraviolet** (UV) light, which is electromagnetic radiation with wavelengths between about 50 and 400 nm. The environmentally relevant portion of the electromagnetic spectrum is illustrated in

Figure 2-4. Notice that the UV region begins at the violet edge of the visible region, hence the name *ultraviolet*. The division of the UV region into components will be discussed later in this chapter. Beyond the ultraviolet range, i.e., of even shorter wavelength, are X-rays. At the other end of the spectrum, beyond the visible region, lies **infrared light,** which will become important to us when we discuss the greenhouse effect in Chapter 4.

An **absorption spectrum** such as that illustrated in Figure 2-5 is a graphical representation that shows the relative fraction of light that can be absorbed by a given type of molecule as a function of wavelength. Here, the light-absorbing behavior of O_2 molecules for the UV region between 125 and 175 nm is shown; some absorption continues beyond

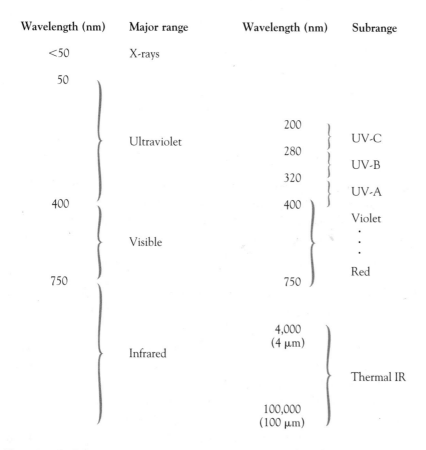

Wavelength (nm)	Major range	Wavelength (nm)	Subrange
<50	X-rays		
50			
	Ultraviolet	200	UV-C
		280	UV-B
		320	UV-A
400		400	
	Visible		Violet
			⋮
			Red
750		750	
	Infrared	4,000 (4 μm)	
			Thermal IR
		100,000 (100 μm)	

FIGURE 2-4

The electromagnetic spectrum. The environmentally interesting ranges are shown.

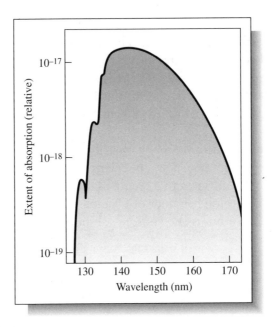

FIGURE 2-5

Absorption spectrum of O_2. (Source: Redrawn from M. J. McEwan and L. F. Phillips. *Chemistry of the Atmosphere*. London: Edward Arnold (Publishers) Ltd., 1975.)

175 nm, but in an ever-decreasing fashion (not shown). Notice that the fraction of light absorbed by O_2 varies dramatically with wavelength; the maximum absorption occurs at about 140 nm. This sort of selective absorption behavior is observed for all atoms and molecules, although the specific regions of strong absorption and of zero absorption vary widely, depending upon the structure of the species and the energy levels of their electrons.

The O_2 gas that lies above the stratosphere is responsible for filtering from sunlight most of the UV light from 120 to 220 nm; the rest is filtered by the O_2 in the stratosphere. Ultraviolet light having wavelengths shorter than 120 nm is filtered in and above the stratosphere by O_2 and other constituents of air such as N_2. Thus, no UV light having wavelengths shorter than 220 nm reaches the Earth's surface, thereby protecting our skin and eyes from damage by this part of sun's output. O_2 also filters some but not all of the UV light in the 220–240 nm range.

Ultraviolet light in the 220–320 nm range is filtered from sunlight mainly by ozone molecules, O_3, that are spread through the middle and lower regions of the stratosphere. The absorption spectrum of ozone in

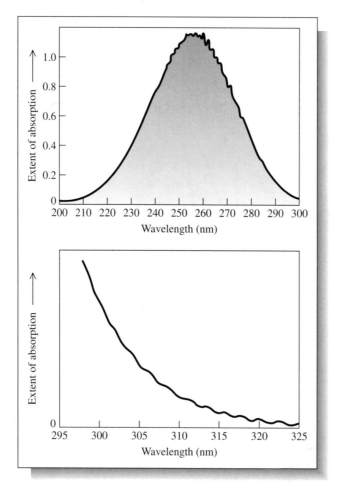

FIGURE 2-6

Absorption spectrum of O₃; *top:* from 200 to 300 nm (Source: Redrawn from M. J. McEwan and L. F. Phillips. *Chemistry of the Atmosphere.* London: Edward Arnold (Publishers) Ltd., 1975.), and *bottom:* from 295 to 325 nm (Source: Redrawn from J. B. Kerr and C. T. McElroy. Evidence for large upward trends of ultraviolet-B radiation linked to ozone depletion. *Science* 262 (1993): 1032–1034. Copyright 1993 by the AAAS.) Note that different scales are used for the extent of absorption in the two cases.

this wavelength region is shown in Figure 2-6. Since its molecular constitution and thus its set of energy levels is different from that of diatomic oxygen, its light absorption characteristics also are quite different. Ozone, aided to some extent by O_2 in the shorter wavelength range, filters out all of the sun's ultraviolet light in the 220–290 nm range, which overlaps the 200–280 nm region known as **UV-C** (see Figure 2-4). However, ozone can only absorb a fraction of the sun's UV light in the 290–320 nm range,

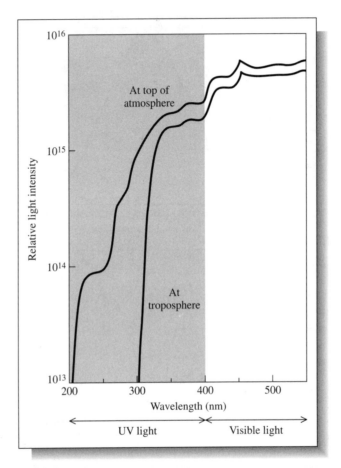

FIGURE 2-7
The intensity of sunlight in the UV and in part of the visible region measured outside the atmosphere and at the Earth's surface. (Source: Reprinted with permission from W. L. Chameides and D. D. Davis, *Chemical and Engineering News* (October 4, 1982): 38–52. Copyright 1988 by American Chemical Society.)

since as you can infer from the bottom part of Figure 2-6, its ability to absorb light of such wavelengths is quite limited. The remaining amount, 10–30% (depending upon latitude), penetrates to the Earth's surface. Thus ozone is not completely effective in shielding us from light in the **UV-B** region, defined as that which lies from 280 to 320 nm. Since the absorption by ozone falls off in an almost exponential manner in this region (see Figure 2-6, *bottom*), the fraction of UV-B that reaches the troposphere increases with wavelength. Because neither ozone nor any other constituent of the clean atmosphere absorbs significantly in the **UV-A** range, i.e., 320–400 nm, most of this, the least biologically harmful type of ultraviolet light, does penetrate to the Earth's surface. The net effect of diatomic oxygen and ozone in screening the troposphere from the UV component of sunlight is illustrated in Figure 2-7. The curve at the left corresponds to the intensity of light received outside the Earth's atmosphere, whereas the curve at the right corresponds to the light that is transmitted to the troposphere (and thus to the surface); the vertical separation between the curves corresponds to the amount of sunlight that is absorbed in the stratosphere and outer regions of the atmosphere.

BIOLOGICAL CONSEQUENCES OF OZONE DEPLETION

The reduction in stratospheric ozone concentration allows more UV-B light to penetrate to the Earth's surface: a 1% decrease in overhead ozone is predicted to result in a 2% increase in UV-B intensity at ground level. This increase in UV-B is the principal environmental concern about ozone depletion, since it leads to detrimental consequences to some life forms, including humans. Exposure to UV-B causes human skin to sunburn and suntan; overexposure can lead to skin cancer. Increasing amounts of UV-B may also adversely affect the human immune system and the growth of some plants and animals.

Most biological effects arise because UV-B can be absorbed by DNA molecules, which then can undergo damaging reactions. By comparing the variation in wavelength of UV-B light of differing intensity arriving at the Earth's surface with the absorption characteristics of DNA as shown in Figure 2-8, it can be concluded that the major detrimental effects of sunlight absorption will occur at about 300 nm. Indeed, in light-skinned people, the skin shows maximum UV absorption from sunlight at about 300 nm.

Almost all skin cancers are due to overexposure to UV-B in sunlight, and so any decrease in ozone is expected to yield eventually an increase in the incidence of skin cancer. Fortunately, most skin cancer is not the often-fatal **malignant melanoma** but rather a slowly spreading type that can be treated. The plot in Figure 2-9, which is based on health data from eight countries of different latitude and that therefore receive different amounts of ground-level UV, shows that the rise in the incidence of non-melanoma skin cancer with exposure to UV is exponential, since the logarithm of the incidence is linearly related to the UV intensity.

The incidence of the malignant form of skin cancer is related to short periods of very high UV exposure, particularly early in life, and

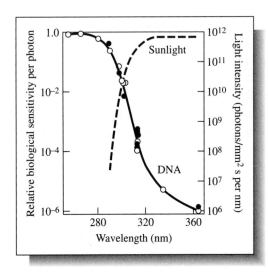

FIGURE 2-8

The absorption spectrum for DNA and the intensity of sunlight at ground level versus wavelength. The degree of absorption of light energy by DNA reflects its biological sensitivity to a given wavelength. (Source: Adapted from R. B. Setlow, *Proceedings of the National Academy of Sciences USA* 71 (1974): 3363–3366.)

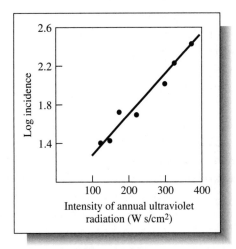

FIGURE 2-9

Incidence (logarithmic scale) for nonmelanoma skin cancer per 100,000 males versus annual UV light intensity. This graph has been constructed from data assembled from various countries. (Source: Redrawn from D. Gordon and H. Silverstone in R. Andrade et al., 1976. *Cancer of the Skin*. Philadelphia: W.B. Saunders, pp. 405–434.)

particularly for fair-skinned, fair-haired, freckled people who burn easily. Recent research indicates that the use of sunscreens that block UV-B but not UV-A may lead to an increase in skin cancer, since sunscreen usage allows people to expose their skin to sunlight for prolonged periods without burning. It is predicted that there will be a 1–2% increase in malignant skin cancer incidence for each 1% decrease in ozone. For people who live at a latitude of about 45° N (i.e., the northern United States and southern Canada) there is expected to be an eventual 11% increase in one type (basal cell carcinoma) of nonmalignant skin cancer, and a 22% increase in the other (squamous cell) type, due to the 6.6% decrease in stratospheric ozone that occurred between 1979 and 1992 over those regions. Any ozone depletion that occurred after 1992 would lead to even greater increases in skin cancer incidence than are stated by these numbers.

Because there is a time lag between exposure to UV and the subsequent manifestation of nonmalignant skin cancers, it is unlikely that effects from ozone destruction are observable as yet; the rise in skin cancer that has occurred in many areas of the world is probably due rather to greater amounts of time spent by people outdoors in the sun over the past few decades. For example, the incidence of skin cancer among residents of Queensland, Australia, most of whom are light-skinned, rose to about 75% of the population as a result of increasing amounts of exposure to sunlight due to lifestyle changes years before ozone depletion began. As a result of this experience, Australia has led the world in public health awareness of the need for protection from ultraviolet exposure.

There is some evidence also that increased UV-B levels give rise to more eye cataracts, particularly among the non-elderly. A 10% increase in UV-B is predicted to result in 6% more cataracts among 50-year-olds. Increased UV-B exposure also leads to a suppression of the human immune system with a resulting increase in the incidence of infectious diseases.

It is speculated that increasing amounts of UV-B interfere with the efficiency of photosynthesis, and thus that plants will respond by producing less leaf, seed, and fruit. All organisms that live in the first five meters or so below the surface in bodies of clear water will also experience increased UV-B exposure and may be at risk. In particular, it is feared that production of the microscopic plants called phytoplankton near the surface of sea water may be at significant risk from increased UV-B; this would affect the marine food chain for which it forms the base. The recent worldwide drop in the population of frogs and other amphibians has now been linked to increasing levels of UV.

PRINCIPLES OF PHOTOCHEMISTRY

As Albert Einstein was the first to realize, light can be considered not only a wave phenomenon but also to have particle-like properties in that it is absorbed (or emitted) by matter only in finite packets now called **photons.** The energy E of each photon is related to the frequency ν and the wavelength λ of the light by the formulas

$$E = h\nu \quad \text{or} \quad E = hc/\lambda \quad \text{since} \quad \lambda\nu = C$$

Here h is Planck's constant ($6.626218 \times 10^{-34} - $ J s) and c is the speed of light (2.997925×10^8 m s^{-1}). From the equation, it follows that *the shorter the wavelength of the light, the greater the energy it transfers to matter when absorbed.* Ultraviolet light is high in energy content, visible light is of intermediate energy, and infrared energy is low in energy. Furthermore, UV-C is higher in energy than UV-B, which in turn is more energetic than is UV-A.

For convenience, the product hc in the equation above can be evaluated on a molar basis to yield a simple formula relating the energy absorbed by 1 *mole* of matter when *each molecule* in it absorbs one photon of a particular wavelength of light. If the wavelength is expressed in nanometers, the value of hc is 119,627 kJ mol^{-1} nm, so the equation becomes

$$E = 119{,}627/\lambda \quad \text{where } E \text{ is in kJ mol}^{-1}$$

The photon energies associated with some wavelengths of critical importance to stratospheric chemistry have been calculated from this formula, and are listed in Table 2-1. These photon energies for UV and visible light are of the same order of magnitude as the enthalpy (heat) changes, $\Delta H°$, of chemical reactions, including those which dissociate atoms from molecules. For example, dissociation of molecular oxygen into its monatomic form requires an enthalpy change of 495 kJ mol^{-1}:

$$O_2 \longrightarrow 2\,O \qquad \Delta H° = 495 \text{ kJ mol}^{-1}$$

diatomic   atomic
oxygen   oxygen

Recall that $\Delta H°$ stands for the enthalpy change determined under standard conditions; to a good approximation, for a dissociation reaction $\Delta H°$ is equal to the energy required to drive the reaction under strato-

TABLE 2-1	PHOTON ENERGIES FOR LIGHT OF DIFFERENT WAVELENGTHS	
λ (nm)	E (kJ mol^{-1})	Comment
220	544	Strong O_2 absorption limit
290	413	Limit of UV-B at surface
320	374	Limit of UV-A region
400	299	Violet limit of visible
750	160	Red limit of visible

spheric pressure and temperature conditions. Since the energy is all to be supplied by one photon per molecule (see below), the corresponding wavelength for the light is

$$\lambda = 119{,}627 \text{ kJ mol}^{-1} \text{ nm}/495 \text{ kJ mol}^{-1} = 241 \text{ nm}$$

Thus any O_2 molecule that absorbs a photon from light of wavelength 241 nm *or shorter* has sufficient excess energy to dissociate.

$$O_2 + \text{UV photon} \longrightarrow 2 \, O$$

If energy in the form of light initiates a reaction, it is called a **photochemical reaction**. The oxygen molecule in the above reaction is variously said to be photochemically dissociated or photochemically decomposed or to have undergone **photolysis**.

Molecules that absorb light generally do not retain the excess energy provided by the photon for very long. Within a tiny fraction of a second, they must either use the energy to react photochemically or else it is dissipated, usually as heat energy that becomes shared among several neighboring molecules as a result of collisions (i.e., molecules must "use it or lose it"). Thus molecules normally cannot accumulate energy from several photons until they receive sufficient energy to react; all the excess energy required to drive a reaction usually must come from a *single* photon. Therefore light of 241 nm or less in wavelength can result in the dissociation of O_2 molecules, but light of higher wavelength does not contain enough energy to promote the reaction at all. The energy from a photon of wavelength greater than 241 nm absorbed by an O_2 molecule is rapidly converted to an increase in the energy of motion of it and of the molecules that surround it.

PROBLEM 2-1

What is the energy, in kilojoules per mole, associated with photons having the following wavelengths?

a. 600 nm   b. 2000 nm

PROBLEM 2-2

The $\Delta H°$ for the decomposition of ozone into O_2 and atomic oxygen is $+105$ kJ mol^{-1}:

$$O_3 \longrightarrow O_2 + O$$
$$\text{ozone}$$

What is the longest wavelength of light that could dissociate ozone in this manner? By reference to Figure 2-4, decide the region of sunlight (UV, visible, or infrared) in which this wavelength falls.

PROBLEM 2-3

Using the enthalpy of formation information given below, calculate the maximum wavelength which can dissociate NO_2 to NO and atomic oxygen. Recalculate the wavelength if the reaction is to result in the complete dissociation into free atoms (i.e., $N + 2 O$).

$\Delta H_f°$ values (kJ mol^{-1}): NO_2: $+33.2$;   NO: $+90.2$;
N: $+472.7$;   O: $+249.2$

THE CREATION AND NONCATALYTIC DESTRUCTION OF OZONE

In this section, the formation of ozone in the stratosphere and its destruction by noncatalytic processes are analyzed. As we shall see, the formation reaction generates sufficient heat to determine the temperature in this region of the atmosphere. Above the stratosphere, the air is very thin and the concentration of molecules is so low that most oxygen exists in atomic form, having been dissociated from O_2 molecules by UV-C photons from sunlight. The eventual collision of oxygen atoms with each other leads to the reformation of O_2 molecules, which subsequently dissociate again when more sunlight is absorbed.

In the stratosphere itself, the intensity of the UV-C light is much less since much of it is filtered by the oxygen that lies above, and since the air is denser, the molecular oxygen concentration is much higher. For this combination of reasons, most stratospheric oxygen exists as O_2 rather than as atomic oxygen. Because the concentration of O_2 molecules is relatively large and the concentration of atomic oxygen is so small, the most likely fate of the stratospheric oxygen *atoms* created by the photochemical decomposition of O_2 is their subsequent collision with undissociated, intact diatomic oxygen *molecules*, thereby resulting in the production of ozone:

$$O \quad + \quad O_2 \quad \longrightarrow \quad O_3 + \text{heat}$$

atomic	diatomic	ozone
oxygen	oxygen	

Indeed, this reaction is the source of all the ozone in the stratosphere. During daylight, ozone is constantly being formed by this process, the rate of which depends upon the amount of UV light and the concentration of oxygen molecules at a given altitude. At the bottom of the stratosphere, the abundance of O_2 is much greater than that at the top because air density increases progressively as one approaches the surface. However, relatively little of the oxygen at this level is dissociated and thus little ozone is formed because almost all the high-energy UV has been filtered from sunlight before it descends to this altitude. For this reason the ozone layer does not extend much below the stratosphere. In contrast, at the top of the stratosphere, the UV-C intensity is greater but the air is thinner and therefore relatively little ozone is produced since the oxygen atoms collide and react with each other rather than with the small number of intact O_2 molecules. Consequently, the density of ozone reaches a maximum where the *product* of UV-C intensity and O_2 concentration is maximum. This maximum *density* of ozone occurs at about 25 km over tropical areas, 21 km over mid-latitudes (as illustrated in the plot in Figure 2-10), and 18 km over subarctic regions. Most of the ozone is located in the region between 15 and 35 km, i.e., the lower and middle stratosphere, known also as the **ozone layer.**

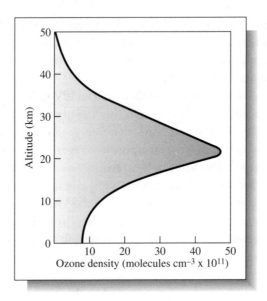

FIGURE 2-10
The density of ozone as a function of altitude.

Note that the zone of maximum ozone *density* lies well below the altitude (35 km) at which the maximum *concentration* of ozone occurs (see Figure 2-3a). At higher altitudes, the air is so thin that, even though the ratio of ozone molecules to air molecules is larger (i.e., the concentration is greater), the absolute number of ozone molecules per liter is considerably reduced compared with their populations in the denser air of lower altitudes. At the lower levels of the atmosphere, there is more ozone per volume of air—greater density—but the ozone molecules are greatly outnumbered by other molecules of air—so there is a lower ozone concentration.

A third molecule, which we will designate as M, such as N_2, is required to carry away the heat energy generated in the collision between atomic oxygen (O) and O_2 that produces ozone. Thus the reaction above is written more realistically as

$$O + O_2 + M \longrightarrow O_3 + M + \text{heat}$$

The release of heat by the reaction results in the temperature of the stratosphere being greater than the air below or above it, as indicated in Figure 2-3b. Indeed, the stratosphere is defined as the region of the atmosphere that lies between these temperature boundaries. Notice from Figure 2-3b that in the stratosphere, the air at a given altitude is cooler than that which lies above it. The general name for this phenomenon is a **temperature inversion**. Because cool air is denser than hot air, it does not rise spontaneously due to the force of gravity; consequently, vertical mixing of air in the stratosphere is a very slow process compared to that in the troposphere. The air in this region therefore is stratified—hence the name stratosphere.

The results for Problem 2-2 show that photons of light in the visible range and even in portions of the infrared range of sunlight possess sufficient energy to split an oxygen atom from a molecule of O_3. However, ozone does not dissociate by absorption of sunlight, even though there is some absorption by O_3, in these wavelength regions. But the UV light with wavelengths shorter than 320 nm that it can and does absorb is effective in provoking a dissociation reaction. Thus, only the absorption of a UV-C or UV-B photon by an ozone molecule in the stratosphere results in the destruction of that molecule:

$$O_3 + UV \text{ photon } (\lambda < 320 \text{ nm}) \longrightarrow O_2 + O^*$$

The oxygen atoms produced in the reaction of ozone with UV light have an electron configuration that differs from that which is that of lowest energy. Atoms or molecules that exist temporarily in such situations are said to be in an **excited state**, and their symbols are marked with a superscript asterisk (*). They possess additional positive energy compared to the lowest energy arrangement of electrons, which is referred to as the **ground state.** Unless an excited atom or molecule quickly reacts with some other atom or molecule, its excess energy usually is lost.

PROBLEM 2-4

By reference to the information in Problem 2-2, calculate the longest wavelength of light that decomposes ozone to O_2 and O^*, the excited state of atomic oxygen that lies 190 kJ mol^{-1} above the ground state.

Most oxygen atoms produced in the stratosphere by photochemical decomposition of ozone or of O_2 subsequently react with intact O_2 molecules to reform ozone. Some oxygen atoms react with intact ozone molecules to destroy them by conversion to O_2:

$$O_3 + O \longrightarrow 2 O_2$$

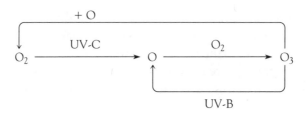

FIGURE 2-11

A schematic summary of the production and noncatalytic destruction reactions for stratospheric ozone.

This reaction is inherently slow, since its activation energy of 18 kJ mol^{-1} is a sizable one, the result being that few collisions occur with sufficient energy to result in reaction.

To summarize, ozone in the stratosphere is constantly being formed, decomposed, and reformed during daylight hours by a series of reactions that proceed simultaneously. Ozone is produced in the stratosphere because there is adequate UV-C from sunlight to dissociate some O_2 molecules and so produce oxygen atoms, most of which collide with other O_2 molecules and form ozone. The ozone gas filters UV-B and UV-C from sunlight, but is destroyed temporarily in this process or by reaction with oxygen atoms. Ozone is not formed below the stratosphere due to a lack of the UV-C required to produce the O atoms necessary to form O_3, because this type of sunlight has been absorbed by O_2 and O_3 in the stratosphere. Above the stratosphere, oxygen atoms predominate and usually collide with other O atoms to reform O_2 molecules. The ozone production and destruction processes are summarized in Figure 2-11.

Even in the ozone layer of the stratosphere, O_3 is not the gas of greatest abundance or even the dominant oxygen-containing species; its concentration never exceeds 10 ppm (see Figure 2-3a). Indeed, if all the ozone were to be collected at ground level at atmospheric pressure, the layer would be only 0.3 mm thick! Thus the term "ozone layer" is something of a misnomer. Nevertheless, this tiny concentration of ozone is sufficient to filter all the remaining UV-C and much of the remaining UV-B from sunlight before it reaches the lower atmosphere. Perhaps the alternative name "ozone screen" is more appropriate than "ozone layer."

CATALYTIC PROCESSES OF OZONE DESTRUCTION

In the early 1960s it was realized that there are mechanisms for ozone destruction in the stratosphere *in addition to* the processes described above.

In particular, there exist a number of atomic and molecular species, designated in general as X, that react efficiently with ozone by abstracting (removing) an oxygen atom from it:

$$X + O_3 \longrightarrow XO + O_2$$

In those regions of the stratosphere where the atomic oxygen concentration is appreciable, the XO molecules react subsequently with oxygen atoms to produce O_2 and to reform X:

$$XO + O \longrightarrow X + O_2$$

The net sum of the two reactions above is

$$O_3 + O \longrightarrow 2\,O_2 \qquad \text{overall reaction}$$

Thus the species X are **catalysts** for ozone destruction in the stratosphere, since they speed up a reaction (here, between O_3 and O) but are eventually reformed intact and are able to begin the cycle again—with, in this case, the destruction of further ozone molecules. As previously discussed, the overall reaction can occur as a simple collision between an ozone molecule and an oxygen atom even in the absence of a catalyst, but almost all such direct collisions are ineffective in producing reaction. The X catalysts greatly increase the efficiency of this reaction. All the environmental concerns about ozone depletion arise from the fact that we are inadvertently increasing the stratospheric concentrations of several X catalysts by release at ground levels of certain gases, especially those containing chlorine. Such an increase in the catalyst concentration will lead to a reduction in the concentration of ozone in the stratosphere.

A factor that minimizes the catalyzed gas-phase destruction of ozone is the requirement for atomic oxygen to complete the cycle by reacting with XO in order to permit the regeneration of the X catalyst in a usable form.

$$XO + O \longrightarrow X + O_2$$

In the region where most ozone is found, namely the lower stratosphere (15–25 km altitude), the concentration of oxygen atoms is very low because little UV-C penetrates to this region and because the O_2 concentration is so high that few oxygen atoms survive for long before being converted to ozone. Thus the gas-phase destruction of ozone by reactions that require atomic oxygen is sluggish in the lower stratosphere. Most ozone destruction by these reactions occurs in the middle and upper stratosphere, where the ozone concentration is lower to start with. Indeed, any decrease in the concentration of ozone at higher altitudes allows more UV penetration to lower altitudes, which produces *more* ozone there; thus there is some "self-healing" of total ozone loss.

Chemically, all the catalysts X are **free radicals,** which are atoms or molecules containing an odd number of electrons. As a consequence of the odd number, one electron is not paired with one of opposite spin character (as occurs for all the electrons in almost all stable molecules). Free

---

LEWIS STRUCTURES OF SIMPLE FREE RADICALS

Most of the free radicals that are important in atmospheric chemistry have their unpaired electron located on a carbon, oxygen, hydrogen or halogen atom. In a formula showing the location and position of bonds, the specific atomic location can be denoted by placing a dot above the relevant atom symbol to represent the unpaired electron, for example, in the notation $\dot{F}$. Characteristically, such an atom forms one less bond than usual—its unpaired electron is not in actual use as a bonding electron. Thus a carbon atom on which an unpaired electron is located forms three rather than four bonds, an oxygen forms one rather than two bonds, and a halogen or hydrogen forms no bonds if it is the radical site. In general, the unpaired electron exists as a nonbonding electron localized on one atom, not as a bonding electron shared between atoms.

For many polyatomic free radicals, the choice of atom to which the unpaired electron is to be assigned in deducing the Lewis structure is obvious from the atom-atom connections. Thus in the hydroperoxy radical HOO, the hydrogen atom cannot be the radical site since it must form a bond to the adjacent oxygen, and neither can the central oxygen since it must form two bonds, one to each neighbor. This leaves the terminal oxygen as the radical site, and we can show the bonding network as H—O—$\dot{O}$. If desired, the unbonded electron pairs can also be shown:

$$H—\overset{..}{\underset{..}{O}}—\overset{.}{\underset{..}{O}}:$$

The procedure is more complicated in molecules that contain multiple bonds. Thus in HOCO it is not initially obvious whether it is the carbon or the terminal oxygen that

radicals are usually very reactive, since there is a driving force for their unpaired electron to pair with one of the opposite spin even if it is located in a different molecule. To indicate that a molecular species is a free radical, we shall henceforth place a superscript dot at the end of its molecular formula, thereby signaling the presence of the unpaired electron: for example, we will use the notation $OH^{\bullet}$ to indicate a free hydroxyl radical. The determination of the appropriate bonding structure for simple free radicals is described in Box 2-1. An analysis of which free radical reactions are feasible in the atmosphere and which are not is given in Box 2-2.

The catalytic destruction of ozone occurs even in a "clean" atmosphere (one unpolluted by artificial contaminants) since small amounts of

carries the unpaired electron. After a little manipulation with various bonding schemes, it becomes clear that the unpaired electron could not be located on an oxygen, since to fulfill its valence requirement of four the carbon would have to form three bonds to the other oxygen! Thus the only reasonable structure is $H—O—\overset{\bullet}{C}=O$.

If only a simple formula rather than a partial or complete Lewis structure is drawn for a radical, the superscript dot is placed *following* the formula and does not indicate which atom carries the unpaired electron. An example is $HCO^{\bullet}$, in which the actual location of the unpaired electron is at carbon, not oxygen.

For a few free radicals involving unusual bonding, such as $NO_2^{\bullet}$, the rules given above generate a Lewis structure that is not the dominant one; further discussion of such systems is beyond the scope of this book.

PROBLEM 2-5

Draw simple Lewis structures, showing the locations of the bonds and of the unpaired electron, for the following free radicals:

a. OH   b. CH_3 c. CF_2Cl

d. H_3COO e. H_3CO f. ClOO

g. ClO   h. HCO   i. NO

the X catalysts have always been present in the stratosphere. The "natural" version of X—that is the species responsible for the greater part of ozone destruction in a nonpolluted stratosphere—is the molecule nitric oxide, NO$^{\bullet}$, which is produced when nitrous oxide, N$_2$O, rises from the troposphere to the stratosphere where it may collide with an excited oxygen atom produced by photochemical decomposition of ozone. Most of

---

BOX 2-2   THE RATES OF FREE RADICAL REACTIONS

It is a characteristic of gas-phase reactions involving simple free radicals as reactants that their activation energy exceeds that imposed by their endothermicity by only a small amount. Thus we can assume, conversely, that all exothermic free radical reactions will have only a small activation energy (Figure 2-12a), and consequently that such reactions will have relatively large rate constants. Therefore, exothermic free radical reactions usually are "fast"(providing, of course, the reactants exist in reasonable concentrations in the atmosphere). An example of an exothermic radical reaction with a small energy barrier is

$$Cl^{\bullet} + O_3 \longrightarrow ClO^{\bullet} + O_2$$

The activation energy here is only 2 kJ mol^{-1}.

In contrast, endothermic reactions in the atmosphere will be much slower since the activation energy barrier must of necessity be much larger (see Figure 2-12b). An example here is the endothermic reaction

$$OH^{\bullet} + HF \longrightarrow H_2O + F^{\bullet}$$

Since for this process $\Delta H^{\circ} = +69$ kJ mol^{-1}, its activation energy is at least equal to this value, and the reaction would be very slow at stratospheric temperatures. Indeed, we can ignore it completely.

The generalization that exothermic radical reactions occur readily is also true when both reactant molecules are free radicals. Thus, for example, the reaction that combines ClO$^{\bullet}$ and NO$_2^{\bullet}$ into chlorine nitrate requires essentially no activation energy, and occurs readily in the atmosphere provided that the concentrations of both radicals are sufficiently large that collisions between the species actually occur:

$$ClO^{\bullet} + NO_2^{\bullet} \longrightarrow ClONO_2$$

these collisions will yield $N_2 + O_2$, but a few of the collisions will result in the following reaction:

$$N_2O + O^* \longrightarrow 2 NO^{\bullet}$$

nitrous   nitric
oxide   oxide

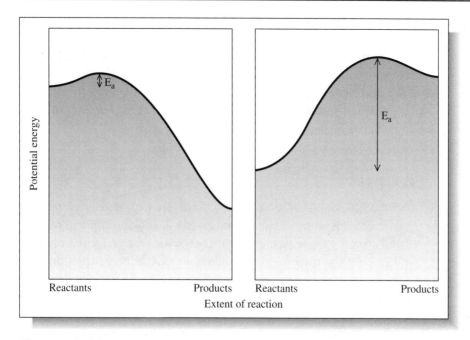

FIGURE 2-12
Potential energy profiles for typical free-radical reactions, showing (a) exothermic, and (b) endothermic patterns.

PROBLEM 2-6

Draw an energy profile diagram, i.e., one similar to Figure 2-12b, for the abstraction from water of a hydrogen atom by ground-state atomic oxygen, given that the reaction is endothermic by about 69 kJ mol^{-1}. On the same diagram, show the energy profile for the reaction of O* (with H_2O to give the same products, given that O*) lies above ground state atomic oxygen (O) by 190 kJ mol^{-1}. From these curves, predict why abstraction by O* is fast but that by O is extremely slow.

The NO˙ molecules which are the products of this reaction catalytically destroy ozone, i.e., they act as X in the mechanism described above:

$$NO˙ + O_3 \longrightarrow NO_2˙ + O_2$$

$$NO_2˙ + O \longrightarrow N˙O_2$$

overall

$$O_3 + O \longrightarrow 2 O_2$$

We can ignore the possibility that NO˙ produced in the troposphere will migrate to the stratosphere, since as explained in Chapter 3, the gas is efficiently oxidized to nitric acid, which is then readily washed out of the tropospheric air since it is so soluble in rain. The tropospheric concentration of N_2O, and hence the amount of it that rises to the stratosphere, currently is rising by about 0.2% per year, for reasons that are explained in Chapter 4.

PROBLEM 2-7

Not all XO molecules such as $NO_2˙$ survive long enough to react with oxygen atoms; some are photochemically decomposed to X and atomic oxygen, which then reacts with O_2 to reform ozone. Write out the three steps (including one for ozone destruction) for this process and add them together to deduce the net reaction. Does this sequence destroy ozone overall?

Although NO˙ is the most important catalyst X in the middle stratosphere, the hydroxyl free radical OH˙ plays the dominant role in ozone destruction at higher altitudes:

$$OH˙ + O_3 \longrightarrow HOO˙ + O_2$$

$$HOO˙ + O \longrightarrow OH˙ + O_2$$

overall

$$O_3 + O \longrightarrow 2 O_2$$

The hydroxyl radical originates from the reaction of excited oxygen atoms O* with water or methane molecules:

$$O* + CH_4 \longrightarrow OH˙ + CH_3˙$$

methane   hydroxyl
radical

PROBLEM 2-8

Write a balanced equation for the reaction by which O* produces OH˙ from water vapor.

While the rate of production of ozone from oxygen depends only upon the concentrations of O_2 and O_3 and of UV light at a given altitude, what determines the rate of ozone destruction is somewhat more complex. Ozone decomposition by UV-B or by X catalysts depends upon ozone's concentration times either the sunlight intensity or the catalyst concentration; in general, the concentration of ozone will rise until the net rate of destruction just meets the rate of production, and then will remain constant at this level as long as the intensity of sunlight remains the same. When the net rate of production of a chemical equals its net rate of consumption, it is said to be in a **steady state** since its concentration does not vary with time. If, however, the rate of destruction is temporarily increased by the presence of more catalyst, the steady-state concentration of ozone must then decrease until the rates of formation and destruction are again equal.

As noted earlier, measurements have indicated that there has been a worldwide decrease of several percent in the steady-state ozone concentration in the stratosphere during the 1970s and 1980s; scientists believe that this depletion, and the phenomenon of ozone holes in polar regions, are due mainly to increasing concentrations of the X catalysts.

Experimentally, the amount of UV-B reaching ground level increases by a factor of three to six in the Antarctic during the early part of the spring because of the appearance of the ozone hole. Abnormally high UV levels have also been detected in southern Argentina when ozone-depleted stratospheric air from the Antarctic traveled over the area. Calculations suggest that substantial increases in ground level UV-B intensity have also occurred in the late winter months as far south as 30° N in the Northern Hemisphere due to ozone depletion. The situation over this region is complicated by the fact that UV-B is absorbed by the ground-level ozone produced by pollution reactions (as explained in Chapter 3), thereby masking any changes in UV-B that are due to small amounts of stratospheric ozone depletion. The first measurements of UV-B trends in nonpolar regions were reported in late 1993 by James Kerr and Thomas McElroy of Canada's Atmospheric Environment Service. They found that levels of UV-B over Toronto increased *annually* by about 5% in winter and about 2% in summer from 1989 to 1993.

ATOMIC CHLORINE AND BROMINE AS X CATALYSTS

The decomposition of chlorine-containing gases in the stratosphere generates an ever-increasing supply of atomic chlorine to this region. As the stratospheric chlorine concentration increases, so does the potential for ozone destruction, since $Cl^•$ is an efficient catalyst of the X type.

However, synthetic gases are not the only suppliers of chlorine to the ozone layer. There always has been chlorine in the stratosphere as a result of the slow upward migration of the methyl chloride gas, CH_3Cl, (more formally called *chloromethane*) produced at the Earth's surface (mainly in the oceans as a result of the interaction of chloride ion with decaying vegetation); this gas is only partially destroyed in the troposphere. When intact molecules of methyl chloride reach the stratosphere, they are photochemically decomposed by UV-C or attacked by $OH^•$ radicals. In either case, atomic chlorine, $Cl^•$, is eventually produced.

$$CH_3Cl \xrightarrow{\text{UV-C}} Cl^• + CH_3^•$$

chloromethane   atomic
                                chlorine

or $$OH^• + CH_3Cl \longrightarrow \ldots \longrightarrow Cl^• + \text{other products}$$

Chlorine atoms are efficient X catalysts for ozone destruction:

$$Cl^• + O_3 \longrightarrow ClO^• + O_2$$
$$ClO^• + O \longrightarrow Cl^• + O_2$$

overall $$O_3 + O \longrightarrow 2\,O_2$$

Each chlorine atom can catalytically destroy many tens of thousands of ozone molecules in this manner. At any given time, however, the great majority of stratospheric chlorine normally exists not as $Cl^•$ or as chlorine monoxide, $ClO^•$, but as a nonradical form that is *inactive* as a catalyst for ozone destruction. The two main inactive molecules containing chlorine in the stratosphere are hydrogen chloride gas, HCl, and chlorine nitrate gas, $ClONO_2$.

The chlorine nitrate is formed by the combination of chlorine monoxide and nitrogen dioxide; after a few days or hours, a given $ClONO_2$ molecule is photochemically decomposed back to its components, and thus the catalytically active $ClO^•$ is reformed.

$$\text{ClO}^{\bullet} + \text{NO}_2^{\bullet} \underset{\text{sunlight}}{\overset{}{\rightleftarrows}} \text{ClONO}_2$$

chlorine   nitrogen   chlorine
monoxide   dioxide   nitrate

The other catalytically inactive form of chlorine, HCl, is formed when atomic chlorine reacts with stratospheric methane:

$$\text{Cl}^{\bullet} + \text{CH}_4 \longrightarrow \text{HCl} + \text{CH}_3^{\bullet}$$

hydrogen
chloride

This reaction is only slightly endothermic, so it proceeds at a slow but significant rate (see Box 2-2). Eventually, each HCl molecule is reconverted to the active form, i.e., chlorine atoms, by reaction with the hydroxyl radical:

$$\text{OH}^{\bullet} + \text{HCl} \longrightarrow \text{H}_2\text{O} + \text{Cl}^{\bullet}$$

When the first predictions concerning stratospheric ozone depletion were made in the 1970s, it was not realized that about 99% of stratospheric chlorine usually is tied up in the inactive forms. When the existence of inactive chlorine was discovered in the early 1980s, the predicted amounts of stratospheric ozone loss were lowered appreciably. As we shall see, however, inactive chlorine can become temporarily activated and destroy ozone, a discovery which was not made until the late 1980s.

Although there has always been some chlorine in the stratosphere due to the natural release of CH_3Cl from the surface, in recent decades it has been completely overshadowed by much larger amounts of chlorine produced from synthetic chlorine-containing gaseous compounds that are released into air during their production or use. Most of these substances are chlorofluorocarbons, abbreviated CFCs; their nature, production, usage, and replacements will be discussed in detail later in this chapter.

As with methyl chloride, large quantities of methyl bromide, CH_3Br, are also produced naturally and that some of it eventually reaches the stratosphere, where it is decomposed photochemically to yield atomic bromine. Methyl bromide is used commercially as a soil fumigant (see Chapter 6), and on that account its release into the troposphere is increasing.

Like chlorine, bromine atoms destroy ozone:

$$Br^{\cdot} + O_3 \longrightarrow BrO^{\cdot} + O_2$$

atomic
bromine

Almost all the bromine in the stratosphere remains in active form, since the inactive form, hydrogen bromide, HBr, is efficiently decomposed photochemically by sunlight back to atomic bromine. In addition, the formation of HBr from attack of atomic bromine on methane is a slower reaction than is the analogous process involving atomic chlorine:

$$Br^{\cdot} + CH_4 \longrightarrow HBr + CH_3^{\cdot}$$

hydrogen
bromide

A lower percentage of stratospheric bromine exists in inactive form than chlorine because of the slower speed of this reaction; the bromine reaction is quite endothermic and therefore is very slow. On an atom-for-atom basis, then, stratospheric bromine is more efficient at destroying ozone than is chlorine, but there is much less of it in the stratosphere.

When molecules such as HC1 and HBr eventually diffuse from the stratosphere back into the upper troposphere, they dissolve in water droplets and are subsequently carried to lower altitudes and are transported to the ground by rain. Thus although the lifetime of chlorine and bromine in the stratosphere is long, it is not infinite and the catalysts are eventually removed.

THE OZONE HOLES

As discussed previously, scientists discovered in 1985 that stratospheric ozone over Antarctica is reduced by about 50% for several months each year due mainly to the action of chlorine. An episode of this sort, during which there is said to be a hole in the ozone layer, can occur from September to early November, corresponding to spring at the South Pole. The holes have been appearing since about 1979, as shown in Figure 2-1, which illustrates the variation in October ozone concentrations in the Antarctic as a function of year. Extensive research in the late 1980s has led to an understanding of the chemistry of this phenomenon.

The ozone hole occurs as a result of special winter weather conditions in the lower stratosphere, where ozone concentrations usually are highest, that temporarily convert *all* the chlorine that is stored in the catalytically inactive forms HCl and $ClONO_2$ into the active forms $Cl^{\cdot}$ and $ClO^{\cdot}$. Consequently, the high concentration of active chlorine causes a large, though temporary, depletion of ozone.

The conversation of inactive to active chlorine occurs at the surface of crystals formed by a solution of water and nitric acid, HNO_3, the latter being formed by combination of $OH^{\cdot}$ with $NO_2^{\cdot}$ gases. Condensation of gases into liquid droplets or solid crystals doesn't normally occur in the stratosphere since the concentration of water in that region is exceedingly small. However, the temperature in the lower stratosphere drops so low ($-80°$ C) over the South Pole in the sunless winter months that some condensation does occur. The usual stratospheric warming mechanism—the release of heat by the $O_2 + O$ reaction—is absent because of the lack of production of atomic oxygen from O_2 when there is total darkness. Because the polar stratosphere becomes so cold during the total darkness at midwinter, the air pressure drops since according to the Ideal Gas Law it is proportional to the Kelvin temperature. This pressure phenomenon, in combination with the Earth's rotation, produces a vortex, a whirling mass of air in which wind speeds can exceed 300 km per hour. Since matter cannot penetrate the vortex, the air inside it is isolated and remains very cold for many months. At the South Pole, the vortex is sustained well into the springtime (October). The vortex around the North Pole usually breaks down in February or early March, before much sunlight has returned to the area.

The crystals produced by condensation of the gases within the vortex form **polar stratospheric clouds**, or **PSCs**. As the temperature drops, the first crystals to form are those of the trihydrate of nitric acid, $HNO_3 \cdot 3H_2O$. When the air temperature drops a few degrees below $-80°$ C, a second type of crystal, higher in the proportion of water to nitric acid and larger in size, also forms.

Gas-phase HCl molecules attach themselves to the surface of the ice particles and become a part of the solid. When gaseous molecules of chlorine nitrate, $ClONO_2$, collide with such HCl molecules, molecular chlorine gas is formed:

$$HCl(s) + ClONO_2(g) \xrightarrow{\text{surface of crystals}} Cl_2(g) + HNO_3(s)$$

| hydrogen chloride | chlorine nitrate | diatomic chlorine | nitric acid |

This process is illustrated schematically in Figure 2-13. During the dark winter months, molecular chlorine accumulates and becomes the predominant chlorine-containing gas. Once a little sunlight reappears in the very early Antarctic spring, the Cl_2 is decomposed by the UV component of the light into atomic chlorine:

$$Cl_2 + UV\ light \longrightarrow 2\ Cl^{\bullet}$$

diatomic   atomic
chlorine   chlorine

As well, reaction of water in the larger type of crystals with other $ClONO_2$ molecules gives HOCl, which in sunlight breaks down into $Cl^{\bullet}$ atoms and $OH^{\bullet}$ free radicals.

$$H_2O(s) + ClONO_2(g) \xrightarrow{\substack{\text{surface of} \\ \text{large crystals}}} HOCl(g) + HNO_3(s)$$

chlorine   hypochlorous
nitrate   acid

$$HOCl + UV\ light \longrightarrow OH^{\bullet} + Cl^{\bullet}$$

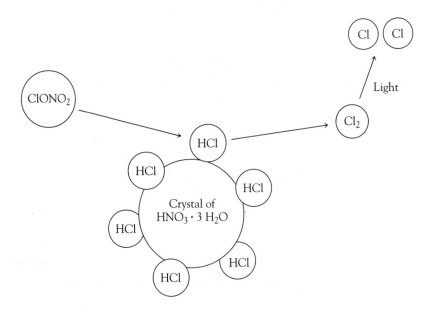

FIGURE 2-13
A scheme illustrating the production of molecular chlorine from inactive forms in the spring stratosphere over the Antarctic.

PROBLEM 2-9

Some scientists have suggested that the reaction of HCl gas with $ClONO_2$ does not occur in a single step on the crystal surface. Rather, chlorine nitrate first is converted to HOCl by the above reaction, and then gaseous HOCl reacts with HCl present on the crystal surface to yield Cl_2 and water. Write the latter reaction and add it to the former. Is your net reaction equivalent to the reaction shown above between HCl and $ClONO_2$?

Since stratospheric temperatures above the Antarctic remain below $-80°$ C even in the early spring, the crystals persist for months. Any of the $Cl^{\bullet}$ that is converted back to HCl by the reaction with methane is subsequently reconverted to Cl_2 on the crystals and then to $Cl^{\bullet}$ by sunlight. Inactivation of chlorine monoxide by conversation to chlorine nitrate does not occur since all the $NO_2^{\bullet}$ necessary for this reaction is temporarily bound as nitric acid in the crystals. Many of the large crystals move downward under the influence of gravity into the upper troposphere, thereby removing $NO_2^{\bullet}$ from the lower stratosphere over the South Pole, and further preventing the deactivation of chlorine.

Only when the PSCs and the vortex have vanished does chlorine return predominantly to the inactive forms. The liberation of nitric acid from the crystals into the gas phase results in its conversion to $NO_2^{\bullet}$ by the action of sunlight:

$$HNO_3 + UV \longrightarrow NO_2^{\bullet} + OH^{\bullet}$$

In addition, air containing $NO_2^{\bullet}$ mixes with polar air once the vortex breaks down in late spring. The nitrogen dioxide quickly combines with chlorine monoxide to form the catalytically inactive chlorine nitrate. Consequently, the catalytic destruction cycles cease operation, and the ozone concentration builds back up to its normal level a few weeks after the PSCs have disappeared and the vortex has ceased. Thus the ozone hole closes for another year. (Before this happens, however, some of the ozone-poor air mass can can move away from the Antarctic and mix with surrounding air, temporarily lowering the stratospheric ozone concentrations in adjoining geographic regions, such as Australia, New Zealand, and the southern portions of South America.)

The chemical mechanism by which atomic chlorine catalytically destroys ozone in the lower stratosphere over the South Pole starts with the usual reaction of chlorine with ozone:

Step 1:
$$Cl^{\cdot} + O_3 \longrightarrow ClO^{\cdot} + O_2$$

In Figure 2-14, the experimental $ClO^{\cdot}$ and O_3 concentrations are plotted as a function of latitude for part of the Southern Hemisphere during the spring of 1987. As anticipated, the two species display opposing trends, i.e., they anticorrelate very closely. At sufficient distances away from the South Pole (90° S), the concentration of ozone is relatively high and that of $ClO^{\cdot}$ is low, since chlorine is mainly tied up in inactive forms. However, as one travels nearer to the Pole, the concentration of $ClO^{\cdot}$ suddenly becomes high and simultaneously that of O_3 falls off sharply: most of the chlorine has been activated and most of the ozone has consequently been destroyed. The latitude at which the concentrations both change sharply marks the beginning of the ozone hole, which continues through to the South Pole. See also the contour maps for Antarctic O_3 and $ClO^{\cdot}$ concentrations on the back cover of this book.

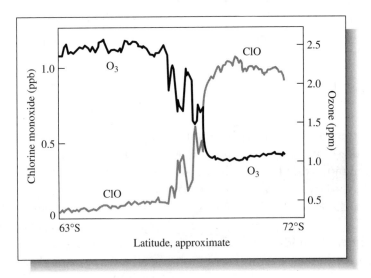

FIGURE 2-14

Ozone and chlorine monoxide concentrations versus latitude near the South Pole on September 16, 1987. (Source: Reprinted with permission from P. S. Zurer, *Chemical and Engineering News* (May 30, 1988): 16. Copyright 1988 by American Chemical Society.)

In the lower stratosphere—the region where the PSCs form and chlorine is activated—the concentration of free oxygen *atoms* is small; few atoms are produced there on account of the scarcity of the UV-C light that is required to dissociate O_2; furthermore, any atomic oxygens produced in this way immediately collide with the abundant O_2 molecules to form O_3. Thus, ozone destruction mechanisms based upon the $O_3 + O \longrightarrow 2 O_2$ reaction, even when catalyzed, are not important here. Instead of reacting with atomic oxygen as in the ozone-destruction sequence higher in the stratosphere, the ClO$^{\bullet}$ molecules combine to form dichloroperoxide, ClOOCl (or Cl_2O_2):

Step 2: $2\ ClO^{\bullet} \longrightarrow Cl-O-O-Cl$
                                             dichloroperoxide

The rate of this reaction becomes important to ozone loss under these conditions because the chlorine monoxide concentration rises steeply due to the activation of the chlorine. Once the intensity of sunlight has risen to an appreciable amount in the Antarctic spring, the dichloroperoxide molecule ClOOCl absorbs UV and splits off one chlorine atom. The resulting ClOO$^{\bullet}$ radical is unstable, and so it subsequently decomposes (with a half-life of less than a day) to release the other chlorine atom:

Step 3: $ClOOCl + UV\ light \longrightarrow ClOO^{\bullet} + Cl^{\bullet}$

Step 4: $ClOO^{\bullet} \longrightarrow O_2 + Cl^{\bullet}$

Adding steps 2, 3, and 4 we see that the net result is the conversion of two ClO$^{\bullet}$ molecules to atomic chlorine:

$$2\ ClO^{\bullet} \longrightarrow \ldots \xrightarrow{UV} \ldots \longrightarrow 2Cl^{\bullet} + O_2$$

Thus by these steps, ClO$^{\bullet}$ returns to the ozone-destroying form Cl$^{\bullet}$ even without the intervention of atomic oxygen. If we add the above reaction to two times step 1, we obtain the overall reaction

$$2\ O_3 \longrightarrow 3\ O_2$$

Thus a complete catalytic ozone destruction cycle exists in the lower stratosphere under these special weather conditions, i.e., when a vortex is present. About three-quarters of the ozone destruction in the Antarctic ozone hole occurs by the mechanism set forth as steps 1 through 4 above, in which chlorine is the catalyst. This ozone destruction cycle, which contributes so greatly to the creation of the ozone hole, has as its overall reaction the catalyzed combination of two O_3 molecules, not the reaction of O_3 with atomic oxygen. The slow step in the mechanism is number 2, which is the combination of 2 $ClO^{\bullet}$ molecules; since this step is second-order in $ClO^{\bullet}$ concentration (i.e., its rate is proportional to the square of the $ClO^{\bullet}$ concentration), it proceeds at a substantial rate, and hence the destruction of ozone is significant, only when $\left[ClO^{\bullet}\right]$ is high. The "sudden" appearance of the ozone hole is consistent with the quadratic rather than linear dependence of ozone destruction upon chlorine concentration by the Cl_2O_2 mechanism. Let us hope that there are not many more environmental problems whose effects will display such nonlinear behavior and which would similarly surprise us!

A minor route for ozone destruction that involves bromine also operates in the ozone hole. In the first steps of this mechanism, two ozone molecules are destroyed, one by a chlorine atom and one by a bromine atom. The $ClO^{\bullet}$ and $BrO^{\bullet}$ molecules produced in these processes then collide with each other and rearrange their atoms to yield O_2 and atomic chlorine and bromine. Again, the net reaction converts two molecules of ozone into three of O_2 without the need for atomic oxygen.

PROBLEM 2-10

Write out the mechanism by which one bromine atom and one chlorine atom catalytically destroy ozone, as in the process described above. Add up the steps to determine the overall reaction.

PROBLEM 2-11

Suppose that the concentration of chlorine continues to rise in the stratosphere, but that the relative increase in bromine does not increase proportionately. Will the dominant mechanism involving dichloroperoxide or the "chlorine plus bromine" mechanism of Problem 2-10 become relatively more important or less important as the destroyer of ozone in the Antarctic spring?

PROBLEM 2-12

> Why is the mechanism involving dichloroperoxide of negligible importance in the destruction of ozone, compared to that which proceeds by ClO˙ + O, in the upper levels of the stratosphere?

Overall, an ozone destruction rate in the lower stratosphere above Antarctica of about 2% *per day* occurs each September due to the combined effects of the major and the minor catalytic reaction sequences. As a result, by early October almost all the ozone is wiped out between altitudes of 15 and 20 km, just the region in which its concentration normally is highest over the Pole. This result is illustrated in Figure 2-15, which shows the measured ozone partial pressure of ozone as a function of

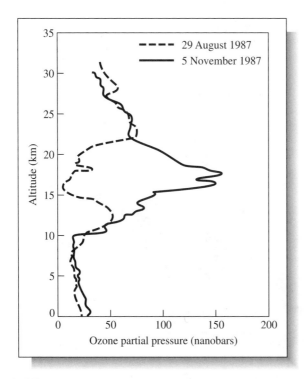

FIGURE 2-15

The vertical distribution of ozone over McMurdo, Antarctica in late winter (August 29) and midspring (November 5) of 1987. (Source: Redrawn from B. J. Johnson, T. Deschler, and R. A. Thompson, *Geophysical Research Letters* 19 (1992): 1105–1108. Copyright by the American Geophysical Union.)

altitude over the Antarctic before (solid line) and during (dashed line) the appearance of the hole in 1987. Because (as explained later) the stratospheric concentration of chlorine will continue to increase until at least the early 2000s, the size of the ozone hole may continue to increase.

Episodes of partial ozone depletion over portions of the Arctic region have been observed recently during the spring. The phenomenon is less severe than in Antarctica, no *complete* hole having yet been observed: the reasons for this are that the stratospheric temperature over the Arctic does not fall as low nor for as long, and air circulation to surrounding areas is not as limited. The polar stratospheric clouds of ice crystals form less frequently and do not last as long; usually only $HNO_3 \cdot 3H_2O$ crystals, which are not large enough to fall out of the stratosphere and thereby **denitrify** it, are formed. The vortex containing the cold air mass above the Arctic usually breaks up by late winter, and therefore $NO_2^\bullet$-containing air mixes with vortex air before much sunlight returns to the polar

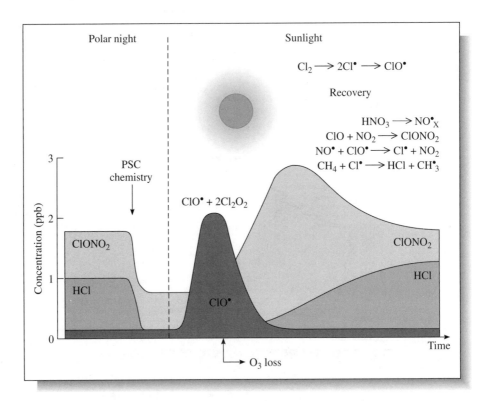

FIGURE 2-16

The evolution of stratospheric chlorine chemistry with time above the Arctic in winter and spring. (Source: Redrawn from C. R. Webster et al., *Science* 261 (1993): 1130.)

region in the spring. Since the stratospheric air temperature usually rises above $-80°$ C by early March, the nitric acid in the crystals is converted back to gaseous nitrogen dioxide before the intense spring sunlight can drive the Cl_2O_2 mechanism. Due to increased $NO_2^{\bullet}$ from both these sources, the activated chlorine is mostly transformed back to $ClONO_2$ before it can destroy much ozone. Thus the total extent of ozone destruction over the Arctic area is much less than that over the Antarctic; typically only about 5% of the overhead ozone is lost, compared to about 50% in Antarctica in the spring. However, since as explained in Chapter 4, increasing the atmospheric concentration of CO_2 further cools the stratosphere, it is expected that in the future lower temperatures may occur and persist in the Arctic stratosphere and give rise to significant ozone destruction.

The annual variation with time of the chlorine chemistry in the stratosphere above the North Pole is illustrated in Figure 2-16. Notice

Ozone destruction step

$$O_3 + Cl^{\bullet} \longrightarrow O_2 + ClO^{\bullet}$$

Atomic chlorine reconstitution

Midstratosphere

$$Cl^{\bullet} + O \longrightarrow Cl^{\bullet} + O_2$$

Ozone hole/Low stratosphere

$$2 ClO^{\bullet} \longrightarrow ClOOCl$$
$$ClOOCl + UV \longrightarrow ClOO^{\bullet} +$$
$$ClOO^{\bullet} \longrightarrow Cl^{\bullet} + O_2$$

Inactivation of chlorine

$$Cl^{\bullet} + CH_4 \longrightarrow HCl + CH_3^{\bullet}$$
$$ClO^{\bullet} + NO_2^{\bullet} \longrightarrow ClONO_2$$

Activation of chlorine on particle surfaces

$$HCl(s) + ClONO_2(g) \longrightarrow Cl_2(g) + HNO_3(s)$$
$$Cl_2 + UV \longrightarrow 2 Cl^{\bullet}$$

$$H_2O(s) + ClONO_2(g) \longrightarrow HOCl(g) + HNO_3(s)$$
$$HOCl + UV \longrightarrow OH^{\bullet} + Cl^{\bullet}$$

FIGURE 2-17
A summary of ozone destruction reaction cycles that involve chlorine.

that although HCl is converted completely in the PSCs, the $ClONO_2$, which is present in excess, is not completely eliminated since the larger, H_2O-dominated crystals on which it can form HOCl are not present. Once the PSCs disappear as air temperatures rise, chlorine nitrate initially dominates since it forms rapidly from $ClO^{\bullet}$ and nitrogen dioxide. The reaction of atomic chlorine with methane is a slower process, and thus the HCl concentration is slower to rise.

The various reactions that lead to catalytic ozone destruction by atomic chlorine by various mechanisms are summarized in Figure 2-17.

THE ROLE OF CHEMICALS IN OZONE DESTRUCTION

CHLOROFLUOROCARBONS (CFCS)

As mentioned previously, the recent increase in stratospheric chlorine is due primarily to the use and release into the atmosphere of **chlorofluorocarbons,** compounds containing chlorine, fluorine, and carbon (only) and which are commonly called **CFCs.** The stratospheric chlorine concentration currently is about 3.5 ppb, which is twice as great as it was in the 1970s and six times the background level of 0.6 ppm, paralleling the growth in the use of CFCs. In the 1980s, about 1 million tonnes (i.e., metric tons, 1000 kg each) of CFCs were released annually to the atmosphere. These compounds are nontoxic, nonflammable, nonreactive, and have useful condensation properties (suiting them for use as coolants, for example); on account of these favorable characteristics they have found a multitude of uses in modern society. There are several CFCs of commercial importance; for convenience we shall refer to them by their code numbers 11, 12, and so on, the significance of which is discussed in Box 2-3. (Note that "Freon" is a commercial trade name used by Du Pont, a large corporation producing industrial chemicals, to refer to CFCs—it is a term often used informally in discussions of environmental air chemistry.)

The material called CFC-12, which is pure CF_2Cl_2, boils at $-30°$ C and is therefore a gas at room temperature, although it is readily liquefied under pressure. Beginning in 1930, it was used as the circulating fluid in refrigerators, replacing the toxic gases ammonia and sulfur dioxide. This allowed the widespread domestic use of such appliances to replace iceboxes (old-fashioned insulated cabinets cooled by blocks of ice). Until recently it was also used extensively in automobile air conditioners, from which much was released (about 0.5 kilograms per year per vehicle) to the atmosphere during use and servicing. Today special equipment is often used to capture the CFCs when air conditioners in cars are serviced.

After World War II, it was discovered that by vaporizing CFC-12 from the liquid state, it could be used to create bubbles in rigid plastic foams. The tiny bubbles of embedded CF_2Cl_2 make these products good thermal insulators because the gas is a poor conductor of heat. In such rigid foams, the CFC-12 is trapped for decades before its release to the atmosphere occurs. However, in the formation of foam sheets such as those used in white trays of the sort used to package fresh meat products and formerly in hamburger "clamshell" cartons used by fast-food restaurants, the CFC-12 is immediately released.

Commercially, CFC-12 is produced by reacting carbon tetrachloride with gaseous hydrogen fluoride, HF:

$$CCl_4 \ + \ 2HF \ \longrightarrow \ CF_2Cl_2 + \ 2\,HCl$$
$$\underset{\text{tetrachloride}}{\underset{\text{carbon}}{}} \qquad \underset{\text{fluoride}}{\underset{\text{hydrogen}}{}}$$

In fact, this reaction produces a mixture of CFC-12 and the compound $CFCl_3$, called CFC-11. The latter is a liquid which boils near room temperature; the mixture of CFCs is separated into pure components by fractional distillation. CFC-11 was used to blow holes in soft foam products such as pillows, carpet padding, cushions, car seats, and padding. It was also employed to make the rigid urethane foam products that are used to insulate refrigerators, freezers, and some buildings. The utilization of insulating foam products has risen in recent decades in view of continuing emphasis on energy conservation.

Formerly, both CFC-11 and CFC-12 were employed extensively as the propellants in aerosol spray cans. Because of concern for the ozone layer, this usage was essentially eliminated in the late 1970s (except for a few crucial medical applications) by the United States, Canada, Norway, and Sweden. Most other industrialized countries now have also instituted similar measures. However, utilization of CFCs in spray cans continued elsewhere in the world. In the late 1980s, this usage amounted to about one-fifth of worldwide CFC consumption, and was the largest single source of CFC released to the atmosphere. Gases such as butane, often combined with a flame suppressant, have replaced CFCs in many aerosol spray cans made in developed countries.

The other CFC of major, although lesser, environmental concern is $CF_2Cl—CFCl_2$, called CFC-113. It was used extensively to clean the grease, glue, and solder residues from electronic circuit boards after their fabrication, to the extent of about 2 kilograms per square meter! The evaporation of the solvent and its emission to the atmosphere which was allowed in the past has now been reduced substantially by recovery and recycling procedures. Many major manufacturers have changed their fabrication processes so that no cleaning liquid of any type now has to be used.

The three CFCs listed above are described as **hard** because no tropospheric **sink**, that is, no natural removal process such as dissolving in rain, exists for any of them. All three types rise after a few years to the stratosphere, where over decades they eventually migrate to its upper part and there undergo photochemical decomposition by UV-C from sunlight, thereby releasing chlorine atoms. For example,

BOX 2-3   FORMULAS AND CODES FOR CARBON COMPOUNDS (INCLUDING CFCs)

Since carbon is a Group IV element and a nonmetal, its atoms each form four covalent bonds in most stable molecules. The simplest molecule of carbon is methane, CH_4, which is emitted into air preponderantly as a result of biological decay processes. Stable compounds also are formed when any or all of the four hydrogen atoms are substituted by halogen atoms. Examples are CH_2Cl_2, CF_2Cl_2, and CHF_2Cl.

Carbon atoms can combine with each other to form extensive chains. The simplest chain has two carbon atoms held together by a single bond. Each carbon in such a unit forms three other bonds in order to bring the total bonds for each carbon to four. If all the six atoms joined to the carbons are hydrogen, the gaseous compound C_2H_6, ethane, results.

The formula for such compounds can be shown one carbon at a time in order that the nature of the atoms bonded to each carbon is displayed. Thus ethane can also be written as CH_3CH_3, as $CH_3—CH_3$ or as $H_3C—CH_3$ to show the carbon-carbon bond. Molecules containing only carbon and hydrogen, and in which all the linkages are single bonds, are called alkanes. All compounds with one or more of the hydrogens in an alkane substituted by halogens are stable species. Examples include CH_3CFCl_2 and $CH_2F—CF_3$.

The chemical formulas for individuals CFCs, such as CFC-11, can be deduced from their code numbers by adding 90. The resulting digits correspond, respectively, to the number of carbon, hydrogen, and fluorine

$$CF_2Cl_2 + UV\text{-}C \longrightarrow CF_2Cl^{\bullet} + Cl^{\bullet}$$

eventually another $Cl^{\bullet}$ is released

CFCs do not absorb sunlight with wavelengths greater than 290 nm, and generally require that of 220 nm or less for photolysis. The CFCs must rise to the midstratosphere before decomposing since UV-C does not penetrate much to lower altitudes. Because vertical motion in the stratosphere is slow, their atmospheric lifetimes are long: 60 years on average for CFC-11 molecules, 105 years for CFC-12. CFC-11 is photochemically decomposed at lower altitudes than is CFC-12, and is therefore more able to destroy ozone at low stratospheric altitudes where the concentration of O_3 is greatest. Currently, CFC-11 and CFC-12 account for almost all the ozone depletion from CFCs.

---

atoms present in one molecule. For example, adding 90 to 11 gives 101, so it follows that CFC-11 contains one carbon, zero hydrogens, and one fluorine. Since that the total number of noncarbon atoms adds up to $2n + 2$, where n is the number of carbons, the number of chlorines in these substituted alkanes can be deduced by difference, that is, by substracting from $2n + 2$ the number of hydrogen plus fluorine atoms. Thus, for CFC-11, $2n + 2 = 2(1) + 2 = 4$, so the number of chlorine atoms is $4 - (0 + 1) = 3$, and its formula is $CFCl_3$.

PROBLEM 2-13

Deduce the formulas for the compounds with the following code numbers:

a. 12   b. 113   c. 123   d. 134

PROBLEM 2-14

Deduce the code numbers for each of the following compounds :

a. CH_3CCl_3 b. CCl_4 c. CH_3CFCl_2

PROBLEM 2-15

Reactions of the type

$$OH^{\cdot} + CF_2Cl_2 \longrightarrow HOF + CFCl_2^{\cdot}$$

are conceivable tropospheric sinks for CFCs. Can you deduce why they don't occur, given that C—F bonds are much stronger than O—F?

Another carbon-chlorine compound that lacks a tropospheric sink is carbon tetrachloride, CCl_4, which also is photochemically decomposed in the stratosphere, and which accounts for about 8% of ozone depletion by chlorine. Commercially, it is used as a solvent and as an intermediate to the manufacture of CFC-11 and CFC-12, during the production of which some is lost to the atmosphere. It was also formerly used as a dry cleaning solvent, but because of its toxicity, this application has been phased out in North America and Western Europe, though apparently not as yet in Eastern Europe.

A rather different sort of compound, CH_3—CCl_3, methyl chloroform or 1,1,1-trichloroethane, is produced in large quantities and is used in metal cleaning in such a way that much of it is released into the atmosphere. Although much of it is removed from the troposphere by reaction with the hydroxyl radical, enough survives to migrate to the stratosphere and to contribute significantly to the depletion of ozone.

CFC REPLACEMENTS

The hard CFCs (i.e., containing no hydrogen) and CCl_4 have no tropospheric sinks. They are not soluble in water and thus they are not rained out from air. They are not attacked by the hydroxyl radical or any other atmospheric gases. And they are not photochemically decomposed by either visible or UV-A light.

The compounds being proposed as their replacements all contain hydrogen atoms bonded to carbon, and consequently a majority (though not 100%) of the molecules will be removed from the troposphere by means of the sequence of reactions which begins with hydrogen abstraction by $OH^{\cdot}$:

$$OH^{\bullet} + H\!-\!\overset{\mid}{\underset{\mid}{C}}\!- \quad\longrightarrow\quad H_2O + C\text{-based free radical}$$

$$\downarrow$$

$$CO_2 \text{ and HCl eventually}$$

The temporary replacements for CFCs that are being employed in the 1990s and the early years of the 21st century contain *hydrogen, chlorine, fluorine,* and *carbon* and are called **HCFCs** (hydrofluorochlorocarbons); they are also sometimes called **soft** CFCs, since tropospheric sinks exist by which they may be degraded.

One HCFC currently in use is CHF_2Cl, the gas called HCFC-22 (or just CFC-22) according to the same numbering scheme used for hard CFCs. It is already employed in most domestic air conditioners and in some refrigerators and freezers. It has found some use in replacing CFC-11 in blowing foams such as those used in food containers. Since it contains a hydrogen atom and thus is mainly removed from air before it can rise to the stratosphere, its long-term ozone-reducing potential is only 5% of that of CFC-11. This advantage is offset, however, by the property of HCFC-22 that it decomposes to release chlorine more quickly than does CFC-11, so its short-term potential for ozone destruction is greater than that implied by this percentage. In particular, HCFC-22 results in 15% as much ozone destruction as does CFC-11 in the first 15 years after release. But because most HCFC-22 is destroyed within a few decades after its release, it is responsible for almost no longer-term ozone destruction, thus yielding an overall 5% value. However most concerns about stratospheric ozone destruction center on the next few decades, before significant reduction of stratospheric chlorine will occur from the phase-out of hard CFCs.

Another promising candidate to replace CFCs is $CHCl_2\!-\!CF_3$, called HCFC-123. It probably will replace CFC-11 in polyurethane foam production and in some refrigeration applications. Since most of it is destroyed in the troposphere by reaction with the hydroxyl radical, its potential for ozone destruction is small. Insulation with rigid foam in refrigerators and freezers may ultimately be replaced by vacuum insulation (as in vacuum-insulated bottles), although it is more expensive and less durable.

Several HCFCs with a methyl group attached to a halogenated carbon, such as $CH_3\!-\!CFCl_2$ and $CH_3\!-\!CF_2Cl$, have desirable properties as replacements for hard CFCs, but their hydrogen content is so great that they are flammable. There is a fine balance between achieving sufficient hydrogen content to ensure efficient hydroxyl radical attack, and at the same time precluding the presence of so much hydrogen that the chemical is flammable.

Reliance exclusively on HCFCs as CFC replacements would eventually lead to a renewed buildup of stratospheric chlorine, as the volume of HCFC consumption would presumably rise with increasing world population and affluence. Products entirely free of chlorine, and that therefore pose no hazard to stratospheric ozone, are proposed as the ultimate replacements for CFCs and HCFCs. Fully fluorinated compounds are unsuitable since they have no tropospheric or stratospheric sinks, and as described in Chapter 4, they would contribute to global warming for very long periods of time. Most attention has been paid to **hydrofluorocarbons, HFCs.** The compound $CH_2F—CF_3$, called HFC-134a, has an atmospheric lifetime of decades before finally succumbing to $OH^•$ attack. It is the favored candidate to replace CFC-12 in refrigerators and in some types of air conditioners; indeed most automobile manufacturers now use it in the air conditioners of new cars, and some manufacturers of refrigeration equipment use it in commercial installations. The HFCs eventually react to form hydrogen fluoride; this molecule contains the very strong H—F bond and for that reason does not break up, for example under $OH^•$ attack, to give fluorine atoms which would destroy ozone (see Problem 2-16).

BROMINE-CONTAINING COMPOUNDS

Mention should be made of "halon" chemicals, bromine-containing substances such as CF_3Br and CF_2BrCl. Because they have no tropospheric sinks, they eventually rise to the stratosphere. There they are photochemically decomposed, with the release of atomic bromine (and chlorine), which, as we have already discussed, is an efficient X catalyst for ozone destruction. Bromine from these sources currently accounts for about 5% of ozone depletion. Since they are nontoxic and leave no residues upon evaporation, halons are very useful for fighting fires, particularly in inhabited, enclosed spaces such as military aircraft, and those housing electronic equipment, such as computer centers. The substitution of other chemicals for halons in the testing of the extinguishers would drastically reduce halon emissions to the atmosphere, since only a minority of the releases are from the fighting of actual fires.

Some molecules of another commercial bromine-containing compound, methyl bromide, CH_3Br, eventually make their way to the stratosphere. There each one is photodissociated into atomic bromine and subsequently catalytically destroys ozone. There have been initiatives by the United States Environmental Protection Agency to add methyl bromide to the list of ozone-depleting chemicals and to ban its use by the year 2000. This initiative has been strongly opposed by farmers, since this chemical is widely used to sterilize soil before planting and to fumigate some crops after their harvest; it is also used to control termites. Currently practicable alternatives to the use of methyl bromide in these applications are apparently less effective and more costly. In addition, it was discovered in 1994 that a sizeable fraction of atmospheric CH_3Br is due to release during fires involving biomass.

PROBLEM 2-16

Fluorine atoms can be liberated in the stratosphere as a result of the decomposition of CFCs and other compounds.

a. Write the set of reactions by which the fluorine atoms could operate as X catalysts in the destruction of ozone.

b. Deduce why such destruction of ozone by fluorine is not significant, given that the H—F bond is much stronger than is the H—OH bond.

INTERNATIONAL AGREEMENTS ON CFC USAGE

In contrast to almost all other environmental problems, such as global warming (Chapter 4), international agreement on remedies to stratospheric ozone depletion has been obtained and implemented in a fairly short period of time. As discussed above, the use of CFCs in most aerosol products was banned in the late 1970s in North America and some Scandinavian countries. This decision was taken on the basis of *predictions*, first made by Sherwood Rowland and Mario Molina, chemists at the University of California, Irvine concerning the effect of chlorine on the thickness of the ozone layer; at the time there was no experimental indication of any depletion. Both Rowland and Molina have continued to carry out research on ozone depletion. Indeed, Molina's research group discovered the ClOOCl mechanism that operates in the polar ozone hole. Rowland is very active in promoting the international control of stratospheric ozone depletion.

The growing awareness of the seriousness of chlorine buildup in the atmosphere has led to two recent international agreements to reduce and eventually phase out CFC production in the world. The breakthrough came at a conference in Montreal, Canada in 1987 which gave rise to the "Montreal Protocol"; this agreement was further strengthened by decisions at follow-up conferences in London in 1990 and Copenhagen in 1992. There is now international agreement that all CFC and halon production will end in developed countries by the year 1995. (Developing countries have been allowed several additional years to accomplish this.) Similarly, production of carbon tetrachloride and methyl chloroform, CH_3CCl_3, will be phased out. It is also anticipated that HCFC usage will be phased out by 2040 at the latest. The dramatic pattern of CFC growth and reduction in worldwide production of CFCs that occurred in the 1980s is illustrated in Figure 2-18 (*top*). The use of CFCs as aerosol pro-

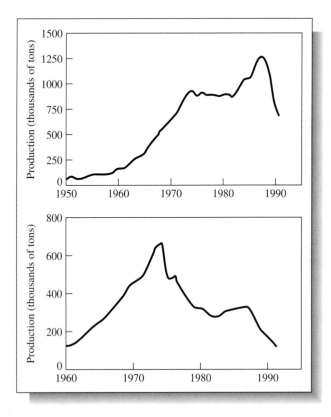

FIGURE 2-18

The production of CFCs. *Top:* total production. *Bottom:* Production for aerosol propellant use. (Source: Redrawn from L. R. Brown, H. Kane, and E. Ayres. 1993. *Vital Signs 1993.* New York: W. W. Norton.)

pellants peaked in the early 1970s (Figure 2-18, *bottom*). The large increase in CFC production in the early and middle 1980s was due to their use in blowing foam.

If these agreements are carried out, it is predicted that the stratospheric total chlorine level (3.4 ppb in 1991) should peak at 4–5 ppb and stay almost constant at this value from 1995 to 2005, and then should slowly decline thereafter. The slowness in the decline is due to (a) the long time it takes the CFC molecules to rise to the upper stratosphere and to absorb a photon and dissociate to atomic chlorine, and (b) the slowness of the removal of chlorine from the stratosphere. The Antarctic ozone hole probably will continue to appear until 2075, that is, until the chlorine concentration is reduced to 2 ppb. Without the international agreements, catastrophic increases in chlorine, to many times the present level, would have occurred, particularly since CFC usage in Third World countries would have increased dramatically. A further doubling of stratospheric chlorine levels would probably have led to the formation of an ozone hole each spring over the Arctic region.

In order to prevent their release into the atmosphere, some hard CFCs in air conditioners, refrigerators, and perhaps even in hard foam are being recovered, cleaned of contaminants and reused. Unfortunately, most of the CFCs produced in the past have already been released to the atmosphere. As of 1991, for example, 59% of all the CFC-11 ever produced was present in the troposphere, 11% of it still resided in foams and 1% in refrigeration equipment; furthermore, 8% was present in the stratosphere but only 19% had already been removed from that region.

Recently, several scientists have suggested a temporary cure for the Antarctic ozone hole that could be implemented until chlorine levels fall sufficiently. Their idea is to inject massive amounts of ethane, C_2H_6, and/or propane, C_3H_8, into the stratosphere. Either gas should combine rapidly with atomic chlorine to convert it back to inactive hydrogen chloride:

$$Cl^{\bullet} + C_2H_6 \longrightarrow HCl + C_2H_5^{\bullet}$$
$$\text{ethane}$$

However, other scientists have postulated that the carbon-centered free radicals produced as byproducts in these reactions will engage in reaction mechanisms that will further deplete ozone.

In the early 1990s, the theory that CFCs cause most of the stratospheric ozone depletion was ridiculed by a number of American political commentators, none of them a current expert in the field of atmospheric science. Their argument was based mainly upon the fact that natural sources, especially seawater and volcanoes, spew much more chlorine into the atmosphere than do CFCs, and their conclusion was that ozone depletion must therefore be due to natural, rather than artificial, causes.

A fact that invalidates the above argument is that the natural sources emit almost all their chlorine into the *troposphere* rather than into the stratosphere. The sodium chloride emitted into ground-level air over oceans, and the HC1 emitted into the high troposphere and the very low stratosphere by volcanoes, are both water soluble, and hence are rained out before they can rise to levels of the stratosphere where they could destroy ozone. Thus the total *stratospheric* chlorine introduced by natural processes is a small fraction of that which originates from CFCs.

PROBLEM 2-17

Given that their C—H bond energies are lower than that in CH_4, can you rationalize why ethane, C_2H_6, or propane, C_3H_8, is a better choice than methane to inactivate atomic chlorine in the stratosphere?

PROBLEM 2-18

No controls on the release of CH_3Cl, CH_2Cl_2, or $CHCl_3$ have been proposed. What does that imply about their atmospheric lifetimes, compared to those for CFCs, CCl_4, and methyl chloroform?

FURTHER CHEMICAL PROCESSES OF THE LOWER STRATOSPHERE

As previously discussed, the concentration of oxygen atoms in the lower stratosphere (15–25 km) is small, and consequently ozone destruction mechanisms based upon the $O_3 + O \longrightarrow 2 O_2$ reaction, even when catalyzed, are not significant. Instead, most ozone loss in this region occurs via the overall reaction

$$2 O_3 \longrightarrow 3 O_2$$

We have already seen that this reaction readily occurs under the special conditions in the Antarctic ozone hole. Elsewhere in the lower stratosphere, the same overall process is catalyzed, though at a much lower rate than in the hole, by the $OH^{\bullet}$ and $HOO^{\bullet}$ radicals, both of which react with ozone in a two-step sequence:

$$OH^{\bullet} + O_3 \longrightarrow HOO^{\bullet} + O_2$$
$$HOO^{\bullet} + O_3 \longrightarrow OH^{\bullet} + 2\,O_2$$

Notice that the second step of this mechanism is not identical to that which operates in the middle and upper stratosphere, but nevertheless does regenerate the hydroxyl radical.

The radical $HOO^{\bullet}$ can react reversibly with $NO_2^{\bullet}$ to produce a molecule of $HOONO_2$:

$$HOO^{\bullet} + NO_2^{\bullet} \rightleftharpoons HOONO_2$$

Thus, addition of nitrogen oxides to the *lower* stratosphere could lead to an *increase* in the steady-state ozone concentration, since the rate of destruction by $HOO^{\bullet}$ in the step described above (i.e., $HOO^{\bullet} + O_3 \longrightarrow OH^{\bullet} + 2O_2$) is thereby slowed. In contrast, the addition of nitrogen oxides to the middle and upper stratosphere *reduces* the ozone concentration since $NO^{\bullet}$ reacts there as X with atomic oxygen to complete the catalytic two-step destruction mechanism. The relevance of these findings to the issue of whether aircraft should be allowed to fly in the stratosphere is discussed in Box 2-4.

Clearly the stratosphere is a complicated system, and predictions concerning ozone depletion can be made only by sophisticated analyses which consider the effects of many reactions occurring simultaneously. It is not easy even to predict whether a given condition, such as air flight in this region, will result in an increase or a decrease in the amount of overhead ozone! Only by extensive research over the last two decades have scientists been able to make semiquantitative predictions regarding the causes of ozone loss. For example, we now know that at any given time much of the nitrogen oxides in the stratosphere is tied up in inactive form as nitric acid, HNO_3, as a result of the reaction of nitrogen dioxide with hydroxyl free radical:

$$OH^{\bullet} + NO_2^{\bullet} \longrightarrow HNO_3$$

BOX
2-4

CAN SUPERSONIC AIRCRAFT DEPLETE STRATOSPHERIC OZONE?

The surprising discovery that adding nitrogen oxides to the lower stratosphere could potentially increase its O_3 concentration is in sharp contrast to the predictions made in the early 1970s when the United States was considering construction of a massive fleet of supersonic transport airplanes (SSTs). One argument against the project was that emission of nitric oxide from the engine exhaust of the SSTs which, like the Concorde, fly in the lower stratosphere, would endanger the ozone layer.

Another pollutant released from SSTs is water vapor, which produces increased levels of hydroxyl radical and thereby destroys more ozone. In addition, the sulfur dioxide released from SST exhaust would form small sulfuric acid droplets that also could increase ozone depletion by a mechanism discussed later in this section.

Currently, the aviation industry again is considering the construction of substantial numbers of supersonic aircraft. To increase speed and improve fuel economy, they would fly not in the lower stratosphere but in its intermediate region. Since emissions there of $NO^{\bullet}$ and $NO_2^{\bullet}$ would decrease ozone, some method will have to be found to substantially reduce the exhaust levels of these gases before construction of such a fleet is seriously considered. Alternatively, allowing more aircraft to fly in the lower stratosphere will so substantially increase the concentrations there of water and $NO_2^{\bullet}$ from their exhausts that PSCs above the Arctic will form at significantly higher temperatures than at present, and will persist longer into the spring months when ozone depletion would occur. Thus an Arctic ozone hole could conceivably result from a substantial increase in pollutants from supersonic aircraft.

The nitric acid eventually undergoes photochemical decomposition in daylight hours to reverse this reaction and to produce species that are catalytically active in ozone destruction.

Some scientists have expressed concern that chlorine-activating reactions could occur not only on ice crystals but also on the surfaces of other particles present in the lower stratosphere, thereby accelerating global ozone destruction. In particular, the reactions may take place on cold liquid droplets consisting mainly of sulfuric acid that occur in the lower stratosphere over all latitudes. The dominant source of the H_2SO_4 at these altitudes is direct injection of sulfur dioxide into the stratosphere from volcanoes, followed by oxidation. Indeed, measurable ozone depletion was noted for several years after a massive eruption of the Mexican volcano El Chichón in 1982. Further research is currently underway to establish whether some or all of the widespread losses in stratospheric ozone that have occurred in recent years are due to reactions on the liquid droplets. One estimate, based upon calculations (rather than on direct experimental investigation), is that almost half the current loss of ozone over northern mid-latitudes is due to reactions on sulfates, and slightly less could be due to springtime depletions originating in the Arctic.

An indirect mechanism may contribute to ozone depletion, one that involves the sulfuric acid droplets and that arises from an unusually high rate of denitrification of stratospheric air. Ozone itself converts some nitrogen dioxide, $NO_2^{\bullet}$, to nitrogen trioxide, $NO_3^{\bullet}$, which then combines with other $NO_2^{\bullet}$ molecules to form dinitrogen pentoxide, N_2O_5:

$$NO_2^{\bullet} + O_3 \longrightarrow \underset{\substack{\text{nitrogen} \\ \text{trioxide}}}{NO_3^{\bullet}} + O_2$$

$$NO_2^{\bullet} + NO_3^{\bullet} \longrightarrow \underset{\substack{\text{dinitrogen} \\ \text{pentoxide}}}{N_2O_5}$$

This process normally is reversible, but in the presence of stratospheric liquid droplets, a conversion to nitric acid occurs:

$$N_2O_5 + H_2O \xrightarrow{\substack{H_2SO_4 \\ \text{droplets}}} 2\,HNO_3$$

By this mechanism, much of the $NO_2^{\bullet}$ that normally would be available to tie up chlorine monoxide as the nitrate, $ClONO_2$, becomes unavailable for this purpose, and hence a greater proportion of the chlorine atoms would occur in the catalytically active form.

Overall, current ozone losses *globally* amount to more than 2% per decade. This is much greater than the 5% *total* loss by 2050 that had been predicted by scientists in the early 1980s on the basis of processes involving solely gas-phase reactions. The greater degree of ozone loss may be due to the culmination of all the processes involving stratospheric crystals and droplets. Indeed, it is found that almost all the ozone loss occurs in the

BOX 2-5
RECENT RESEARCH ON OZONE DESTRUCTION

Because of its great environmental importance, much current research is under way to further unravel the mysteries of stratospheric ozone destruction. Two of the more important results of this research are discussed below.

By late 1993, J. A. Pyle and his associates at the University of Cambridge in England had concluded that ozone destruction in the late spring over northern latitudes may well be initiated in the lower stratosphere by the photochemical decomposition of $ClONO_2$. The catalytic destruction cycle, which requires no atomic oxygen, is as follows:

$$ClONO_2 + sunlight \longrightarrow Cl^{\cdot} + NO_3^{\cdot}$$
$$Cl^{\cdot} + O_3 \longrightarrow ClO^{\cdot} + O_2$$
$$NO_3^{\cdot} + sunlight \longrightarrow NO^{\cdot} + O_2$$
$$NO^{\cdot} + O_3 \longrightarrow NO_2^{\cdot} + O_2$$
$$ClO^{\cdot} + NO_2^{\cdot} \longrightarrow ClONO_2$$

overall $2\,O_3 \longrightarrow 3\,O_2$

This mechanism is especially important in the spring when deactivation of chlorine occurs; initially the active chlorine is converted almost exclusively to $ClONO_2$ rather than predominantly to HCl since the former is much faster to reform than is the latter. Under these conditions, the concentration of $ClONO_2$ may be great enough to lead to significant ozone destruction by this mechanism.

Research by Mario Molina, now at the Massachusetts Institute of Technology, and his associates has confirmed that activation of chlorine can occur on the surface of sulfuric acid droplets as well as on ice crystals. The temperatures of the liquid droplets must be low enough that significant uptake by them of gaseous HCl occurs, or no net reaction takes place. They also find that reaction on the PSC ice crystals occurs in a thin aqueous layer present at the surface of the crystal. In the layer, the dissolved hydrogen chloride gas forms ions:

$$HCl(g) \xrightarrow{\text{aqueous layer}} H^+(aq) + Cl^-(aq)$$

Meanwhile, gaseous $ClONO_2$ reacts at the surface with water molecules to produce HOCl:

$$ClONO_2(g) + H_2O(aq) \longrightarrow$$
$$HOCl(aq) + H_2O(aq)$$

Reaction of the two forms of dissolved chlorine produces molecular chlorine, which escapes to the gas phase:

$$Cl^-(aq) + HOCl(aq) \longrightarrow$$
$$Cl_2(g) + OH^-(aq)$$

lower stratosphere, just the altitudes where the crystals and droplets predominate and where processes involving them are most important. Some of the most recent research involving ozone destruction in the lower stratosphere is summarized in Box 2-5.

SYSTEMATICS OF STRATOSPHERIC CHEMISTRY: A REVIEW

At first sight, the many chemical reactions discussed in this chapter may seem unconnected and difficult to remember. However, there are regularities in this group of equations that can facilitate learning.

In particular, many of the species in the stratosphere have only one *loosely bound oxygen atom*; it is readily detached from the rest of the molecule (denoted Y) in several characteristic ways. In Table 2-2 we list the molecules Y—O that contain loose oxygen. In every case, dissociation of this oxygen atom requires much less energy than that required to break any of the remaining bonds, so the resulting Y units remain intact. Notice that the Y species, other than O_2, are the free radicals we previously called X ozone destruction catalysts. In terms of electronic structure, all "loose" oxygens are joined by a single bond to another electronegative atom that possesses one or more nonbonding electron pairs. The interaction between the nonbonded electron pairs on this atom with those on the oxygen weakens the single bond.

TABLE **2-2**	MOLECULES CONTAINING LOOSE OXYGEN ATOMS	
Molecule Y—O	Structure of Y—O	Comment
O_3	O_2—O	The most loose oxygen
$HOO^•$	HO—O	
$ClO^•$	Cl—O	
$BrO^•$	Br—O	
$NO_2^•$	ON—O	The least loose oxygen

The characteristic reactions involving loose oxygen are collected below:

1. Reaction with Atomic Oxygen

Here the oxygen atom detaches the loose oxygen atom by combining with it:

$$Y—O + O \longrightarrow Y + O_2$$

These reactions are all exothermic since the $O{=}O$ bond is much stronger than the Y—O bond.

2. Photochemical Decomposition

The YO species absorbs UV-B or longer wavelength light from sunlight, and subsequently releases the loose oxygen atom:

$$Y—O + \text{ sunlight } \longrightarrow Y + O$$

3. Reaction with NO˙

Nitric oxide abstracts the loose oxygen atom:

$$Y—O + NO˙ \longrightarrow Y + NO_2˙$$

This reaction is exothermic since the ON—O bond strength is the greatest of those involving a loose oxygen. (Recall the general principle that exothermic free radical reactions are relatively fast.)

4. Abstraction of Oxygen from Ozone

Abstraction of the loose oxygen atom from ozone (only) to form the Y—O species is characteristic of OH˙, Cl˙, Br˙, and NO˙. Thus all these radicals act as catalytic ozone destroyers X:

$$O_2—O + X˙ \longrightarrow O_2 + XO˙$$

The reaction involving ozone is exothermic since it contains the weakest of the bonds involving a loose oxygen. The other YO species do not undergo this reaction to an important extent either because it is endothermic and therefore its rate is negligibly slow or because the X species react more quickly with other chemicals.

5. Combination of Two YO Molecules

If the concentration of YO species becomes high, they may react by the collision of two of them (identical or different species). If at least one is O_3 or $HOO^{\cdot}$, so that an unstable chain of three or more oxygen atoms is created, then the loose oxygens combine to form one or more molecules of O_2:

$$2\,O_2-O \longrightarrow 3\,O_2$$

$$2\,HO-O^{\cdot} \longrightarrow HOOH + O_2$$

$$HO-O^{\cdot} + O-O_2 \longrightarrow OH^{\cdot} + 2\,O_2$$

$$HO-O^{\cdot} + {}^{\cdot}O-Cl \longrightarrow HOCl + O_2$$

When neither is ozone or $HOO^{\cdot}$, the two YO molecules combine to form a larger molecule, which subsequently often decomposes photochemically:

$$2\,NO_2^{\cdot} \longrightarrow N_2O_4$$

$$2\,ClO^{\cdot} \longrightarrow ClOOCl \xrightarrow{UV} \ldots \longrightarrow 2\,Cl^{\cdot} + O_2$$

$$ClO^{\cdot} + NO_2^{\cdot} \underset{\text{sunlight}}{\rightleftarrows} ClONO_2$$

$$ClO^{\cdot} + BrO^{\cdot} \longrightarrow \ldots \longrightarrow Cl^{\cdot} + Br^{\cdot} + O_2$$

The YOOY molecules have little thermal stability, and even at moderate temperatures may dissociate back to their YO components before light absorption and photolysis has time to occur.

In addition to the reactions involving loose oxygens, another characteristic stratospheric process is hydrogen atom abstraction. The species $OH^{\cdot}$, O^*, $Cl^{\cdot}$, and $Br^{\cdot}$ readily abstract a hydrogen atom from stable molecules such as methane, provided that the reaction is exothermic.

PROBLEM 2-19

Which of the following species do(es) not contain a "loose oxygen"?

a. $HOO^{\cdot}$ b. $OH^{\cdot}$ c. $NO^{\cdot}$ d. O_2 e. $ClO^{\cdot}$

PROBLEM 2-20

From which Y—O species

a. does $NO^•$ abstract an oxygen atom?

b. does atomic oxygen abstract an oxygen atom?

c. does sunlight detach an oxygen atom?

d. do the Y—O—O—Y species (with identical Y groups) form in the stratosphere?

e. is O_2 produced when two identical YO species react?

PROBLEM 2-21

Using the principles above, predict what would be the likely fate of $BrO^•$ molecules in a region of the atmosphere which was particularly

a. high in atomic oxygen,

b. high in $ClO^•$,

c. high in $BrO^•$ itself,

d. high in sunlight intensity.

PROBLEM 2-22

Using the principles above, deduce what reaction(s) could be sources of atmospheric

a. $ClONO_2$,   b. ClOOCl,   c. $Cl^•$ atoms.

PROBLEM 2-23

Draw the Lewis structure for the free radical $FO^•$. On the basis of this structure, could you predict whether it contains a "loose oxygen"? If it does, why isn't it an active participant in stratospheric chemistry? (Hint: See your answer to Problem 2-16.)

PROBLEM 2-24

What is the expected product when $ClO^•$ reacts with $NO^•$? What are the possible fates of the product(s) of this reaction? Devise a mechanism incorporating (a) this reaction, (b) the reaction of Cl$^{\cdot}$ with ozone, and (c) the photochemical decomposition of $NO_2^{\cdot}$ to NO$^{\cdot}$ and atomic oxygen. Is the net result of this cycle, which operates in the lower stratosphere, the destruction of ozone?

REVIEW QUESTIONS

1. Which three gases constitute most of Earth's atmosphere?

2. What range of altitudes constitutes the troposphere? the stratosphere?

3. What is the current estimate of the loss in stratospheric ozone per decade?

4. What is the wavelength range for visible light? Does ultraviolet light have shorter or longer wavelengths than visible light?

5. Which atmospheric gas is primarily responsible for filtering sunlight in the 120–220 nm region? Which if any gas absorbs most of the sun's rays in the 220–320 region? Which absorbs primarily in the 320–400 nm region?

6. What is the name given to the finite "packets" of light absorbed by matter?

7. What are the equations relating photon energy E to light's frequency ν and wavelength λ?

8. What is meant by the expression "photochemically dissociated" as applied to stratospheric O_2?

9. Write the equation for the chemical reaction by which ozone is formed in the stratosphere. What are the sources for the different forms of oxygen used here as reactants?

10. Explain why the density of ozone peaks at about 25 km altitude.

11. Write the two reactions that, aside from the catalyzed reactions, contribute most significantly to ozone destruction in the stratosphere.

12. What is meant by the phrase "excited state" as applied to an atom or molecule? Symbolically, how is an excited state signified?

13. Explain why the phrase "ozone layer" is a misnomer.

14. What is a Dobson Unit? How is it used?

15. Define the term "free radical," and give two examples relevant to stratospheric chemistry.

16. What are the two steps, and the overall reaction, by which species X such as ClO$^{\cdot}$ catalytically destroy ozone in the middle and upper stratosphere?

17. What is meant by the term "steady state" as applied to the concentration of ozone in the stratosphere?

18. Explain why, atom for atom, stratospheric bromine destroys more ozone than does chlorine.

19. Describe the process by which chlorine becomes activated in the Antarctic ozone-hole phenomenon.

20. What is the principal four-step mechanism by which chlorine destroys ozone in the spring over Antarctica?

21. Explain why ozone holes have not yet been observed over the Arctic.

22. What are two effects to human health that scientists believe will result from ozone depletion?

23. Define the term CFCs and write the formulas for the two most common CFCs. Give one use for each CFC you mention.

24. Define what is meant by a tropospheric "sink." What general term denotes CFCs lacking such a sink?

25. Explain what HCFCs are, and state what sort of reaction provides a tropospheric sink for them. Is their destruction in the troposphere 100% complete? Why are HCFCs not considered to be a suitable long-term replacements for hard CFCs?

26. What types of chemicals are proposed as long-term replacements for CFCs?

27. Chemically, what are "halons"? What is their main use?

28. What gases are being phased out according to the Montreal Protocol agreements?

29. Describe why the injection of ethane into the Antarctic stratosphere might help heal the ozone hole.

30. Explain why ozone desctruction via the reaction of O_3 with atomic oxygen does not occur to a significant effect in the lower stratosphere.

31. Write the mechanism and overall reaction by which ozone destruction occurs in the lower stratosphere by a catalytic process that does not involve chlorine or nitric oxide.

32. Describe how the natural release of SO_2 by volcanoes could trigger chlorine-catalyzed ozone destruction in the stratosphere. In your answer, be sure to include the reaction by which HCl is converted to $Cl^{\cdot}$. Why would the conversion on aerosol particles of N_2O_5 to nitric acid increase the rate of chlorine activation?

SUGGESTIONS FOR

FURTHER READING

1. Molina, M. J., and F. S. Rowland. 1974. Stratospheric sink for chlorofluoromethanes: Chlorine atom-catalyzed destruction of ozone. *Nature* 249: 810–812.

2. Solomon, S. 1990. Progress toward a quantitative understanding of Antarctic ozone depletion. *Nature* 347: 347–354.

3. McFarland, M. 1989. Chlorofluorocarbons and ozone. *Environmental Science and Technology* 23: 1203–1207.

4. Taylor, J.-S. 1990. DNA, sunlight, and skin cancer. *Journal of Chemical Education* 67: 835–841.

5. Stolarski, R. S. 1988. The Antarctic ozone hole. *Scientific American* January: 30–36.

6. Toon, O. B., and R. P. Turco. 1991. Polar stratospheric clouds and ozone depletion. *Scientific American* June: 68–74.

7. Rowland, F. S. 1991. Stratospheric ozone depletion. *Annual Reviews of Physical Chemistry* 42: 731–768.

8. Brune, W. H. et al. 1991. The potential for ozone depletion in the Arctic polar stratosphere. *Science* 252: 1260–1266.

9. Turco, R. P., O. B. Toon, and P. Hamill. 1989. Heterogeneous physicochemistry of the polar ozone hole. *Journal of Geophysical Research* 94: D16493–D16510.

10. Gribbin, J. 1988. *The Hole in the Sky*. New York: Bantam Books.

11. Kerr, J. B., and C. T. McElroy. 1993. Evidence for large upward trends of ultraviolet-B radiation linked to ozone depletion. *Science* 262: 1032–1034.

12. Khalil, M. A. K., and R. A. Rasmussen. 1993. The environmental history and probable future of fluorocarbon 11. *Journal of Geophysical Research* 98: 23091–23106.

13. Madronich, S., and F. R. de Gruijl. 1993. Skin cancer and UV radiation. *Nature* 366: 23.

INTERCHAPTER:
CHEMISTRY OF THE OZONE HOLE

AN INTERVIEW WITH SUSAN SOLOMON

Susan Solomon received her Ph.D. in chemistry from the University of California at Berkeley in 1981. That same year she joined the National Oceanic and Atmospheric Administration as a research scientist. In 1986–1987, Dr. Solomon led the ground-based expeditions to McMurdo Station, Antarctica, to probe the cause of the ozone hole. She is a member of the National Academy of Sciences and is widely considered one of the world's leading experts on ozone depletion.

How has the study of ozone chemistry changed since the discovery of the ozone "hole" over Antarctica? What questions do you predict that scientists will be asking in the next ten years?

The discovery of the ozone hole precipitated a revolution in scientific thinking that may happen once a century in a scientific discipline. In particular, it demonstrated that a global environmental change was not only a possibility for the distant future, but a reality of the twentieth century.

We were expecting to see a few percent change in upper-atmosphere ozone levels sometime around the year 2050, and instead the British Antarctic Survey showed that ozone levels had already fallen to about 50%, at least in one part of the world. On a more technical level, it showed that the "gas-phase think" that we stratospheric chemists had been using for decades needed major revision to include the key role of reactions on the surfaces of stratospheric particles.

In the next ten years, ozone scientists will be asking two important questions: What is the cause of the observed ozone losses outside of Antarctica (especially at mid-latitudes), and how much worse will the ozone depletion get as we go through the period around the turn of the twenty-first century when the chlorine abundance in the stratosphere will reach the largest values ever observed?

Is there a real possibility that an ozone hole will form over the Arctic in coming years?

Most researchers agree that there is a possibility that an Arctic ozone hole may form sometime in the next several decades, although this hasn't happened yet. There has already been significant ozone depletion in the Arctic, on the order of 10 to 20%, but that's much less than the 40 to 60% observed in Antarctica. The differences between the two hemispheres basically hinge on the fact that the Arctic stratosphere is generally warmer than the Antarctic, especially in the spring. This means that the

temperatures are critical because both sunlight and the clouds that form at low temperatures are needed to deplete ozone rapidly. Temperatures in the stratosphere vary from year to year, though, just as they do on the ground. It is plausible to suppose that an unusually cold stratospheric spring could lead to an Arctic ozone hole. It is also expected that the combined effects of other anthropogenic changes would lead to a generally colder stratosphere in the future, which could induce the formation of an Arctic ozone hole. A key change is the increasing levels of carbon dioxide and other greenhouse gases. Although these warm the surface climate, they are expected to cool the stratosphere.

Do scientists yet understand why some ozone depletion occurs in nonpolar areas? If so, what is the reason?

There are probably two different factors influencing ozone depletion in nonpolar regions. One is simply the dilution of ozone-depleted air as it makes its way from the poles down to lower latitudes. This must be happening to some extent. How large an extent depends upon how isolated the air in the polar regions is: what we don't yet know for sure is whether or not the polar air can be mixed with air in lower latitudes once per winter, twice per winter, or three times per winter, but we know that it happens at least once, and that in itself will deplete ozone on a hemispheric scale to some extent if there is polar ozone loss. In addition, there is strong evidence that some of the same reactions that occur on the frozen surfaces in polar regions can also occur on the liquid surfaces that are found in the mid-latitude stratosphere. These are also believed to contribute to mid-latitude ozone loss. We need to better understand how global air masses interact in order to predict what will happen in the future.

Why is it so important to evaluate and address this problem now even though the effects are relatively minor to date?

It's important to evaluate this problem because the lifetimes of the gases in question are very long. CFCs stay in the atmosphere from 50 to 500 years, depending upon which CFC you are considering. Therefore, if we wait until we have such extensive ozone depletion that severe environmental problems are clear, we will then have to try to live with those problems for decades or even centuries before they will begin to get better.

What will be the likely future of the ozone layer given current international agreements? What is the role of HCFCs and methyl bromide?

Given current international agreements, the chlorine loading of the stratosphere will probably get about 10% larger than it is now within the next decade and then slowly begin to decrease. It will take about 40 to 60 years before the Antarctic ozone hole is likely to begin to go away. Nearly all of the most ozone-damaging compounds will be phased out by 1996. While HCFCs do contain some chlorine, they are much less damaging to the ozone layer than the CFCs that they are replacing. Most people believe that they will be used only temporarily until other technologies are developed that will likely be CFC-free, or that they will come to be used only in very small amounts in specialty applications over the longer term. Methyl bromide is a chemical used mainly for agricultural purposes, but it also has some natural sources. It is believed to be as damaging to the ozone layer as some of the compounds that have already been phased out. However, methyl bromide lives only for two years or less in the atmosphere, so that any impacts on the ozone layer can be reversed relatively rapidly once they are established by stopping the emissions at that point. The decision regarding whether to use HCFCs and methyl bromide and in what amounts and for how long is a policy, and not a science, question.

Do commercial aircraft flying in the low stratosphere deplete the ozone layer?

Emissions of nitrogen oxides by commercial aircraft in the lower stratosphere can cause some ozone depletion, but probably much less than is caused by CFCs. The current fleet of stratospheric aircraft is very small, but it is possible that more planes will be used in future. More research is needed to fully understand the impacts not just of nitrogen oxides released by such planes, but also the soot, sulfur gases, water vapor, and so on.

Could ozone depletion be successfully combated by sending rockets carrying ozone or other chemicals into the stratosphere?

It takes a lot of energy to break the O_2 bond and make ozone. The largest nuclear power plant in the United States could not make enough ozone in year to fill in that year's ozone hole, even if we were willing to spend the money to try to do so. Furthermore, ozone is a very reactive gas and decomposes on surfaces, so that you can't put it in a container in the first place, let alone transport it wherever you want. So making the ozone to fill the ozone hole is not a very practical solution. Other intervention approaches also have to deal with the problem of volume. The ozone hole is big: about twice the size of the continental United States and the thickness of Mount Everest. That's a big volume of gas to process or add to. Nobody has yet come up with any solid mitigation strategies for ozone depletion.

Do you think that all the important processes that may deplete stratospheric ozone are already known, or is more research needed?

There are many things we still don't know about ozone depletion. We don't fully understand mid-latitude ozone depletion. We don't know what the effects of changes in ozone will be for climate, both in the stratosphere and on the ground. We don't know how much ozone depletion might result if a massive volcanic eruption on the scale of the Tamboura event were to inject particles into the chlorine-perturbed stratosphere of today. We don't know if or when the ozone layer will begin to rehabilitate. Even though controls have been put in place, it seems prudent to verify that those controls will have the intended effect. We don't know which substitute molecules are best to use. We don't know much about the impacts of increased UV on the ecosystem.

GROUND-LEVEL AIR CHEMISTRY AND AIR POLLUTION

The best known examples of air pollution are the smogs that occur in many cities throughout the world. For example, the reactants that produce one type of smog are mainly emissions from cars; indeed the operation of motor vehicles produces more air pollution than any other single human activity. The most obvious manifestation of such smog is a brownish-gray haze, which is due to the presence in air of small water droplets containing products of chemical reactions that occur among pollutants in air. This haze, familiar to most of us who live in urban areas, also now extends periodically to once-pristine areas such as the Grand Canyon in Arizona, Yosemite National Park in California, and even Acadia National Park in Maine. Smog often has an unpleasant odor due to some of its gaseous components. More seriously, the intermediates and final products of the reactions in smog can affect human health, and can cause damage to plants, animals, and some materials. Although the quality of air is improving with time in most developed countries, it is worsening in the larger cities of developing countries.

One of the most important features of the Earth's atmosphere is that it is an oxidizing environment, a phenomenon due to the large concentration of diatomic oxygen, O_2, that it contains. Almost all the gases that are released into the air, whether they are "natural" substances or "pollutants," are eventually completely oxidized in air and the end products subsequently deposited on the Earth's surface. Thus the oxidation reactions are vital to the cleansing of the air.

In this chapter, the chemistry of tropospheric air is examined. The detrimental effects on animals, plants, and materials of polluted air, including that which we encounter indoors, and methods by which air pollution can be combated, are described. Then the detailed reactions that occur in clean air and the processes encountered in the polluted air of many modern cities are analyzed. Although the mechanisms of the reactions in polluted air appear complex, they follow the same set of principles described for clean air. It is only by understanding the science underlying such complicated environmental problems that we can hope to solve them. Before analyzing any of these chemical processes in depth, however, we shall consider, as a practical preliminary step, how concentrations of air pollutants can be converted from one set of units to another.

THE INTERCONVERSION OF GAS CONCENTRATIONS

There is as yet no consensus regarding the appropriate units by which to express concentrations of substances in air. In Chapter 2, ratios involving numbers of molecules—the "parts per" system—were emphasized as a measure; there, units of ppm, ppb, and so on, were employed. Other measures, differing distinctly from the "parts per" system as well as from one another in basic meaning, are often also encountered:

Molecules of a gas per cubic centimeter of air

Micrograms of a substance (whether gaseous or not) per cubic meter of air

Moles of a gas per liter of air

Given the lack of a consensus on a single appropriate scale, it is important to be able to convert gas concentrations from one set of units to another.

Since the number of moles of a substance is proportional to the number of molecules of it (Avogadro's Number, 6.02×10^{23}, is the proportionality constant) and since the partial pressure of a gas is proportional to the number of moles of it, a concentration for example of 2 ppm for any pollutant gas present in air means

2 molecules of the pollutant in 1 million molecules of air;

2 moles of the pollutant in 1 million (10^6) moles of air (i.e., the pollutant is $2/10^6$ of the mole total, or 2×10^{-6});

2×10^{-6} atmospheres partial pressure of pollutant, given a total air pressure of 1 atmosphere;

2 liters of pollutant in 1 million liters of air (when the partial pressures and temperatures of pollutant and air have been adjusted to be equal)

Let us convert a concentration of 2 ppm to its value in molecules (of pollutant) per cubic centimeter (cm^3) of air, for conditions of 1 atmosphere total air pressure and 25°C. Since the value of the numerator, 2 molecules, in the new concentration scale is the same as that in the original, all we need to do is to establish the volume, in cubic centimeters, that 1 million molecules of air occupy. This volume is easy to evaluate using the Ideal Gas Law ($PV = nRT$), since we know that

$$
\begin{aligned}
P &= 1.0 \text{ atmosphere} \\
T &= 25 + 273 = 298 \text{ K} \\
n &= 10^6 \text{ molecules}/6.02 \times 10^{23} \text{ molecules mol}^{-1} \\
&= 1.66 \times 10^{-18} \text{ mol}
\end{aligned}
$$

and the gas constant $R = 0.082$ L atm mol^{-1} K^{-1}.

$$
\begin{aligned}
\text{Now} \quad PV &= nRT \\
\text{so} \quad V &= nRT/P \\
&= 1.66 \times 10^{-18} \text{ mol} \times 0.082 \text{ L atm mol}^{-1} \text{ K}^{-1} \times 298\text{K}/1 \text{ atm} \\
&= 4.06 \times 10^{-17} \text{ L}
\end{aligned}
$$

Since 1 L = 1000 cm^3, then V = 4.06×10^{-14} cm^3. Since 2 molecules of pollutant occupy 4.06×10^{-14} cm^3, it follows that the concentration in the new units is 2.0 molecules/4.06×10^{-14} cm^3, or 4.9×10^{13} molecules per cm^3.

In general, the most straightforward strategy to use in order to change the value of a concentration a/b from one scale to its value p/q on another is to independently convert the numerator a of the initial unit to the numerator p of the final unit (both of which involve only the pollutant) and then convert the denominator b to its new value q (both of which involve the total air sample).

To convert a value in ppm or molecules per cm^3 to moles per liter, we must change the molecules of pollutant to the number of moles of pollutant; for a pollutant concentration, again of 2 ppm, we can write

$$moles\ of\ pollutant = 2\ molecules \times 1\ mole/6.02 \times 10^{23}\ molecules$$
$$= 3.3 \times 10^{-24}\ moles$$

Thus the molarity is 3.3×10^{-24} moles/4.06×10^{-17} L, or 8.2×10^{-8} M.

An alternative way to approach the conversions accomplished above is to use the definition that 2 ppm means 2 liters of pollutant per 1 million liters of air, and to find the number of moles and molecules of pollutant contained in a volume of 2 liters at the stated pressure and temperature.

A unit often used to express concentrations in polluted air is **micrograms per m^3.** If the pollutant is a pure substance, we can interconvert such values into the molarity and the "parts per" scales, provided that the pollutant's molar mass is known.

Consider as an example the conversion of 320 micrograms/m^3 to the ppb scale, given that the pollutant is SO_2, the total air pressure is 1.0 atm, and the temperature is 27°C. Initially the concentration is

$$\frac{320\ micrograms\ of\ SO_2}{1\ m^3\ of\ air}$$

First we convert the numerator from grams of SO_2 to moles of SO_2 since we can then easily obtain the number of molecules of SO_2 (note: 1 mole SO_2 has a mass of 64.1 g):

$$320 \times 10^{-6}\ g\ SO_2 \times \frac{1\ mole\ SO_2}{64.1\ g\ SO_2} \times \frac{6.02 \times 10^{23}\ molecules\ of\ SO_2}{1\ mole\ SO_2}$$

$$= 3.01 \times 10^{18}\ molecules\ of\ SO_2$$

Then using the Ideal Gas Law, we can change the volume of air to moles of air and then molecules of air, using the equivalencies 1 L = 1 dm^3 = $(0.1\ m)^3$:

$$n = PV/RT = \frac{1.0\ atm \times 1.0\ m^3 \times \dfrac{1\ L}{(0.1\ m)^3}}{0.082\ L\ atm\ mol^{-1}\ K^{-1} \times 300\ K} = 40.7\ moles$$

Now 40.7 moles $\times$ 6.02 $\times$ 10^{23} molecules mol^{-1} = 2.45 $\times$ 10^{25} molecules, or 2.45 $\times$ 10^{16} billion molecules of air. Thus the SO_2 concentration is

$$\frac{3.01 \times 10^{18} \text{ molecules of } SO_2}{2.45 \times 10^{16} \text{ billion molecules of air}} = 123 \text{ ppb}$$

Note that the conversion of moles to molecules was not strictly necessary, as Avogadro's Number cancels from numerator and denominator. As stated previously, ppb refers to the ratio of the number of moles as well as to the ratio of the number of molecules.

It is vital in all interconversions to distinguish between quantities associated with the pollutant and those of air.

PROBLEM 3-1

Convert a concentration of 32 ppb for any pollutant to its value on

a. the ppm scale,

b. the molecules per cm^3 scale

c. the molarity scale.

Assume 25°C and a total pressure of 1.0 atm.

PROBLEM 3-2

Convert a concentration of 6.0 $\times$ 10^{14} molecules per cm^3 to the ppm scale and to the moles per liter (molarity) scale. Assume 25°C and 1.0 atm. total air pressure.

PROBLEM 3-3

Convert a concentration of 40 ppb of ozone, O_3, into

a. the number of molecules per cm^3, and

b. the number of micrograms per m^3.

Assume the air mass temperature is 27°C and its total pressure is 0.95 atmospheres.

URBAN OZONE: THE PHOTOCHEMICAL SMOG PROCESS

THE ORIGIN AND OCCURRENCE OF SMOG

Many urban areas of North America and elsewhere undergo episodes of air pollution during which relatively high levels of ground-level ozone, O_3—an undesirable constituent of air at low altitudes—are produced as a result of the light-induced reaction of pollutants. This phenomenon is called **photochemical smog,** and is sometimes characterized as "an ozone layer in the wrong place" to contrast it with the stratospheric ozone depletion problems discussed in Chapter 2. (The word "smog" itself is a combination of "*smoke*" and "*fog*.") The process of smog formation actually involves hundreds of different reactions, involving dozens of chemicals, occurring simultaneously; indeed, urban atmospheres have been referred to as "giant chemical reactors." The most important reactions that occur in such air masses will be discussed in detail later in this chapter.

The photochemical smog phenomenon was first observed in Los Angeles in the 1940s, and has generally been associated with that city ever since; however, pollution controls have partially alleviated the smog problem in L.A. in recent decades. Quantitatively, most countries individually, as well as the World Health Organization (WHO), have established goals for maximum allowable ozone concentrations in air of about 100 ppb (averaged over a one-hour period): for example the standard in the United States is 120 ppb and that in Canada is 82 ppb. The ozone level in very clean air amounts to only a few percent of these values. By way of contrast, the peak levels of ozone in Los Angeles air used to reach 680 ppb, but peak levels have now declined to 300 ppb.

The chief original reactants in an episode of photochemical smog are the nitric oxide, NO·, and unburnt hydrocarbons (i.e., compounds containing only carbon and hydrogen) that are emitted into the air as pollutants from internal combustion engines. The concentrations of these chemicals are orders of magnitude greater than found in clean air. Recently it has been realized that gaseous hydrocarbons are also present in urban air as a result of the evaporation of solvents, liquid fuels, and other organic compounds; collectively the substances including hydrocarbons and their derivatives that readily vaporize are called **volatile organic compounds,** or VOCs. Another vital ingredient in photochemical smog is sunshine, which serves to increase the concentration of free radicals that participate in the chemical processes of smog formation.

The final products of smog are ozone, nitric acid, and partially oxidized and in some cases nitrated organic compounds.

$$\text{VOCs} + \text{NO}^{\bullet} + \text{sunlight} \longrightarrow \longrightarrow \text{mixture of } O_3, HNO_3, \text{organics}$$

Substances such as $NO^{\bullet}$, hydrocarbons and other VOCs that are initially emitted into air are called **primary pollutants;** those into which they are transformed, such as O_3 and HNO_3, are called **secondary pollutants.** A summary of the emissions of the primary pollutants SO_2, $NO^{\bullet}$, and VOCs from various sources in the United States and Canada is given in Figure 3-1.

The most reactive VOCs in urban air are hydrocarbons that contain a $C{=}C$ bond, since they can add free radicals. Other hydrocarbons are also present and can react, but the rate of their reaction is slow; however

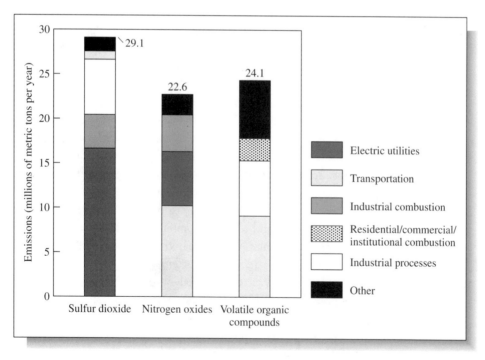

FIGURE 3-1

Combined emissions of air pollutants in the United States and Canada from various sectors. (Source: Redrawn from M. Placet and D. G. Streets in Chapter 1 of "The Causes and Effects of Acidic Deposition. Volume II: Emissions and Control." *The National Acid Precipitation Assessment Program.* Washington, D.C.: U.S. Government Printing Office, 20402, 1987.)

their reaction can become important in late stages of photochemical smog episodes.

Nitrogen oxide pollutant gases are produced whenever a fuel is burnt in air with a hot flame. At such high temperatures, some of the nitrogen and oxygen gases in air combine to form nitric oxide, $NO^{\bullet}$:

$$N_2 + O_2 \xrightarrow{\text{flame}} \underset{\text{nitric oxide}}{2\,NO^{\bullet}}$$

The greater the flame temperature, the more $NO^{\bullet}$ that is produced. The nitric oxide is gradually oxidized to nitrogen dioxide, $NO_2^{\bullet}$, over a period of minutes to hours, depending upon the concentration of the pollutant gases. Collectively, $NO^{\bullet}$ and $NO_2^{\bullet}$ in air is referred to as $\mathbf{NO_x}$. The small levels of NO_x in clean air result in part from the operation of the above reaction in the very energetic environment of lightning flashes and in part from the release of NO_x and NH_3 from biological sources, as discussed in Chapter 4. Because the reaction between N_2 and O_2 has a high activation energy, it is negligibly slow except at the very high temperatures such as occur in the modern combustion engines in vehicles—particularly when traveling at high speeds—and in power plants.

In order that a city be subject to photochemical smog, several conditions must be fulfilled. First there must be substantial vehicular traffic in order to emit sufficient $NO^{\bullet}$, hydrocarbons, and other VOCs into the air. Second, there must be warmth and ample sunlight in order for the crucial reactions, some of them photochemical, to proceed at a rapid rate. Finally, there must be relatively little movement of the air mass so that the reactants are not diluted. For reasons of geography (e.g., the presence of mountains) and dense population, cities such as Los Angeles, Denver, Mexico City, Tokyo, Athens, and Rome all fit the bill splendidly and are subject to frequent smog episodes.

The air in Mexico City is so polluted by ozone and smog, and by lead from compounds added to gasoline, that it is estimated to be responsible for thousands of premature deaths annually; indeed, in the center of the city residents can purchase pure oxygen from booths to help them breathe more easily! In 1990, Mexico City exceeded the WHO air guideline on 310 days; in 1992, ozone levels reached as high as 400 ppb. Athens and Rome, as well as Mexico City, now attempt to limit traffic during smog episodes.

Due to long-range transport of primary and secondary pollutants in air currents, many areas which themselves generate few emissions are subject to regular episodes of high ground-level ozone and other smog oxidants. Indeed, some rural areas which lie in the path of such polluted air masses experience higher levels of ozone than do nearby urban areas, because in the cities some of the ozone is eliminated by reaction with nitric oxide released by cars into the air. An example is the farmland in southwest Ontario, which often receives ozone-laden air from industrial regions in the United States that lie across Lake Erie from it; as a result of the damage caused by this pollution, crops such as white beans can no longer be grown successfully in this area. Elevated levels of ozone also affects materials: it hardens rubber, reducing the useful lifespan of consumer products such as automobile tires, and bleaches color from some materials such as fabrics.

REDUCING GROUND-LEVEL OZONE AND PHOTOCHEMICAL SMOG

In order to improve the air quality in those urban environments that are subject to photochemical smog, the amount of reactants, principally NO_x and hydrocarbons containing $C\!=\!C$ bonds plus other VOCs, emitted into the air must be reduced. For economic and technical reasons, the most common strategy has been to reduce hydrocarbon emissions. However, except in downtown Los Angeles, the percentage reduction in ozone and other oxidants achieved thereby usually has been much less than was the percentage reduction in hydrocarbons. This happens because usually there is initially an overabundance of hydrocarbons relative to the amount of nitrogen oxides, and the hydrocarbon reduction simply reduces the excess without slowing down the reactions significantly. In other words, it is usually the nitrogen oxides, rather than the $C\!=\!C$ containing hydrocarbons, that are the rate-limiting species; this is especially true for rural areas which lie downwind of polluted urban centers. Due to the large number of reactions that occur in polluted air, the functional dependence of smog production upon reactant concentration is complicated indeed, and the net consequence of making moderate decreases in primary pollutants is difficult to deduce without computer simulation. Nevertheless, the simple control strategies that have been put in place in the United States have resulted in some reduction in ozone levels in the past few decades, notwithstanding the huge increase in total miles driven—up to 100% in the last 25 years—that has occurred. The progress that has been made in reducing both ozone levels in the Los Angeles area, as well as the sources of the pollutants that produce smog, are summarized in the diagrams shown in Figure 3-2.

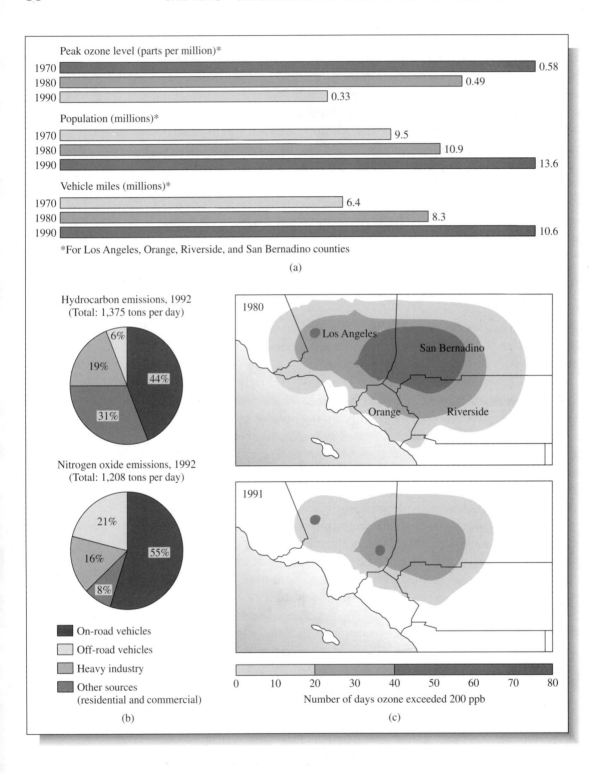

Peak ozone level (parts per million)*

1970 — 0.58
1980 — 0.49
1990 — 0.33

Population (millions)*

1970 — 9.5
1980 — 10.9
1990 — 13.6

Vehicle miles (millions)*

1970 — 6.4
1980 — 8.3
1990 — 10.6

*For Los Angeles, Orange, Riverside, and San Bernadino counties

(a)

Hydrocarbon emissions, 1992
(Total: 1,375 tons per day)

6%
19%
44%
31%

Nitrogen oxide emissions, 1992
(Total: 1,208 tons per day)

21%
16%
8%
55%

■ On-road vehicles
□ Off-road vehicles
▨ Heavy industry
▨ Other sources
(residential and commercial)

(b)

1980

Los Angeles
San Bernadino
Orange
Riverside

1991

0   10   20   30   40   50   60   70   80
Number of days ozone exceeded 200 ppb

(c)

Furthermore, some urban areas such as Atlanta, Georgia, and others located in the southern United States incorporate or border upon heavily wooded areas whose trees emit enough reactive hydrocarbons to sustain smog and ozone production even when the concentration of **anthropogenic** hydrocarbons, that is those which result from human activities, is low. Deciduous trees and shrubs emit mainly the gas isoprene, whereas conifers emit pinene and limonene; all three hydrocarbons contain $C{=}C$ bonds. In urban atmospheres, the concentration of these compounds normally is much less than that of the anthropogenic hydrocarbons and it is not until the latter are reduced substantially that the influence of these substances becomes noticeable. In areas affected by the presence of vegetation, then, only the reduction of emissions of nitrogen oxides will reduce photochemical smog production substantially.

Although hydrocarbons with $C{=}C$ bonds are the most reactive type in photochemical smog processes, others play a significant role after the first few hours of an episode have passed and the concentration of free radicals has risen. For this reasons, control of emissions of *all* VOCs is required in areas with serious photochemical smog problems. Gasolines, which are a complex mixture of hydrocarbons, are now formulated in order to reduce their evaporation, since gasoline vapor has been found to contribute significantly to atmospheric concentrations of hydrocarbons. New regulations proposed for California (with Los Angeles especially in mind) would limit the use of hydrocarbon-containing products such as barbeque starter fluid, household aerosol sprays, and oil-based paints that consist partially of a hydrocarbon solvent that evaporates into the air as the paint dries. The air quality in this region has improved because of current emission controls, but the increase in vehicles miles driven and the hydrocarbon emissions from non-transportation sources such as solvents have thus far prevented a more complete solution.

The creation of nitric oxide in a combustion system can be lessened by lowering the temperature of the flame. However, in recent years, a more complete control of NO_x emissions from gasoline-powered cars and trucks has been attempted using catalytic converters placed just ahead of

FIGURE 3-2

Air pollution in the Los Angeles area. (Source: Redrawn from J. M. Lents and W. J. Kelly, "Cleaning the Air in Los Angeles." *Scientific American* (October 1993): 32–39.) (a) Decadal trends in peak ozone, population, and vehicle miles. (b) Sources of emissions (1992) of hydrocarbons and nitrogen oxide (NO_x). (c) Number of days ozone levels exceed 200 ppb.

the mufflers in the exhaust system. The original **two-way converters** dealt only with the carbon-based gases, including carbon monoxide, CO, completing their combustion to the end product, carbon dioxide. However by use of a surface impregnated with a rhodium catalyst, the modern **three-way converter** first changes nitrogen oxides back to elemental nitrogen and oxygen, using unburnt hydrocarbons and CO and H_2 as reducing agents.

$$2\,NO^{\bullet} \longrightarrow N_2 + O_2$$

Then, with a palladium and/or platinum catalyst, the carbon-containing gases are oxidized almost completely to CO_2 and water.

$$2\,CO + O_2 \longrightarrow 2\,CO_2$$
$$\text{carbon} \atop \text{monoxide}$$

An oxygen sensor in the exhaust system is monitored by a computer chip that controls the air/fuel ratio of the engine to ensure a high level of conversion of the pollutants. (The whole process is illustrated in Figure 3-3.) Some progress has been reported recently in the use of less valuable metals, such as copper and chromium, instead of the expensive platinum-group metals as catalysts; although the latter are recycled, a portion of them is inevitably lost.

Once an engine has warmed up, properly working three-way catalysts eliminate 80–90% of the hydrocarbons, CO, and NO_x from the

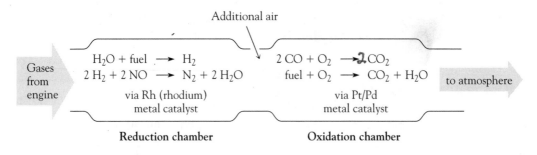

FIGURE 3-3

Schematic representation of the three-way catalytic converter. Reduction of NO_x to N_2 occurs in the first chamber, and oxidation of carbon-containing substances to CO_2 occurs in the second.

engine before the exhaust gases are released into the atmosphere. Before the engine has warmed up, however, and also when there are episodes of sudden acceleration or deceleration, the converters cannot operate effectively and there are burst of emissions from the tailpipe. In addition, older cars (with no converters or just two-way converters) still on the road continue to pollute the atmosphere with nitrogen oxides even during normal operation. Recently it has been established that vehicles that have been misfueled with leaded gasoline even once—the result is damage to the catalytic converter—or whose converters have been otherwise tampered with, produce most of the emissions: 50% of the hydrocarbons and carbon monoxide are released from 10% of the cars. For this reason, some governments have recently instituted mandatory inspections of exhaust systems. The maximum amounts of emissions that can legally be released from light-duty motor vehicles such as cars have gradually been decreased in order to improve air quality; the emission standards applicable to North America are listed in Table 3-1.

In North America, approximately equal amounts of NO_x are emitted currently from vehicles and from electric power plants, and taken together they constitute almost all the anthropogenic sources of these gases. To reduce their NO_x production, some power plants in the United States use special burners designed to lower the temperature of the flame. Other plants have been fitted with large-scale versions of catalytic converters in order to change NO_x back to N_2 before the release of stack gases into air. The reduction of NO_x to N_2 in some of these catalytic systems is accomplished by adding ammonia, NH_3, to the gas stream, since

TABLE 3-1　EMISSION STANDARDS, IN GRAMS PER MILE, FOR CARS AND OTHER LIGHT-DUTY MOTOR VEHICLES[a]

Constituency	Hydrocarbons 1980	Hydrocarbons 1993	CO 1980	CO 1993	NO_x 1980	NO_x 1993
United States	0.41	0.41 (0.125)	7.0	3.4 (1.7)	2.0	1.0 (0.2)
California	0.39	0.25	9.0	3.4	1.0	0.4
Canada		0.41 (0.25)		3.4 (3.4)		1.0 (0.4)
Mexico	4.8	0.40	53	3.4	—	1.0

[a]Values in parentheses are standards proposed for 1996.

this highly reduced compound of nitrogen combines with the partially oxidized compound $NO^\bullet$ to produce N_2 gas in the presence of oxygen:

$$4\,NH_3 + 4\,NO^\bullet + O_2 \longrightarrow 4\,N_2 + 6\,H_2O$$

ammonia   nitric
oxide

Tight control is needed to regulate the addition of ammonia in order to prevent its inadvertent oxidation to NO_x. In a related technology, reduced nitrogen in the form of the compound urea, $CO(NH_2)_2$, is injected directly into the combustion flame to combine there, rather than later in the presence of a catalyst, with $NO^\bullet$ to produce N_2.

$$2\,CO(NH_2)_2 + 4\,NO^\bullet + O_2 \longrightarrow 4\,N_2 + 2\,CO_2 + 4\,H_2O$$

urea   nitric
oxide

The photochemical production of ozone also occurs during dry seasons in rural tropical areas where the burning of biomass for the clearing of forests or brush is widespread. Although most of the carbon is transformed immediately to CO_2, some methane and other hydrocarbons are released, as is some NO_x. Ozone is produced when these hydrocarbons react with the nitrogen oxides under the influence of sunlight.

ACID RAIN

What goes up must come down, though in the case of acid rain, what comes down is worse than what went up!

One of the most serious environmental problems facing many regions of the world today is **acid rain.** This generic term covers a variety of phenomena, including acid fog and acid snow, all of which correspond to atmospheric precipitation of substantial acid. As will be discussed in a subsequent section of this chapter, acid rain has a variety of ecologically damaging consequences and the presence of acidity in air probably also has direct effects on human health.

The phenomenon of acid rain was discovered in Great Britain in the late 1800s, but then it was essentially forgotten until the 1960s. It refers to precipitation that is significantly *more* acidic than "natural" (i.e., unpolluted) rain, which itself is mildly acidic due to the presence in it of dissolved atmospheric carbon dioxide, which forms carbonic acid:

$$CO_2(g) + H_2O(aq) \rightleftharpoons H_2CO_3(aq)$$
$$\text{carbonic acid}$$

The H_2CO_3 then partially ionizes to release a proton, with a resultant reduction in the pH of the system:

$$H_2CO_3(aq) \rightleftharpoons H^+ + HCO_3^-$$
$$\text{carbonic acid} \qquad\qquad \text{bicarbonate ion}$$

Due to this source of acidity, the pH of unpolluted, "natural" rain is about 5.6. Only rain that is appreciably more acidic than this, that is, with a pH of less than 5, is considered to be truly "acid" rain.

The two predominant acids in acid rain are sulfuric acid, H_2SO_4, and nitric acid, HNO_3. Generally speaking, acid rain is precipitated far downwind from the source of the primary pollutants, namely sulfur dioxide, SO_2, and nitrogen oxides, NO_x. These acids are created during the transport of the air mass that contains the primary pollutants. Thus acid rain is a pollution problem that does not respect state or national boundaries because of the Long Range Transport of Atmospheric Pollutants (LRTAP). For example, most acid rain that falls in Norway originates as sulfur and nitrogen oxides emitted in other countries in Europe. Similarly, concerns have been expressed in the United States that substantial air pollution will result in the Southwest from the new so-called Carbon I and II coal-fired power plants located in Mexico just south of San Antonio, Texas; the power plants emit sulfur dioxide into the air because their stack gases are not cleansed before release.

In the next section, the sources of SO_2 air pollution and its potential abatement are discussed. Later in the chapter, the reaction conditions and mechanism by which atmospheric sulfur dioxide is converted into sulfuric acid are analyzed in detail.

THE SOURCES AND ABATEMENT OF SULFUR DIOXIDE POLLUTION

On a global scale, most SO_2 is produced by volcanoes and by the oxidation of sulfur gases produced by the decomposition of plants. Because this "natural" sulfur dioxide is mainly emitted high into the atmosphere or far from populated centers, the background concentration of the gas in clean air is quite small. However, a sizable amount of sulfur dioxide is presently emitted into ground-level air, particularly over land masses in the Northern Hemisphere. The main anthropogenic source of SO_2 is the combustion, especially in electric power generating plants, of coal, a solid which, depending upon the source from which it is mined, contains 1 to 5% sulfur. About half of the sulfur is trapped as "inclusions" in the mineral content of the coal; if the coal is pulverized before combustion, this type of sulfur can be mechanically removed. The other half of the sulfur, however, is bonded in the complex carbon structure of the solid and cannot be removed without expensive processing.

Sulfur occurs to the extent of a few percent in crude oil, but it is reduced to the level of only a few hundredths of one ppm in products such as gasoline. Sulfur dioxide is emitted into air directly as SO_2 or indirectly as H_2S by the petroleum industry when oil is refined and natural gas is cleaned before delivery. Indeed, the predominant component in natural gas wells is sometimes H_2S rather than CH_4! The substantial amounts of hydrogen sulfide obtained from its removal from oil and natural gas are often converted to solid, elemental sulfur, an environmentally benign substance, using the gas-phase process called the **Claus reaction:**

$$2\ H_2S + SO_2 \longrightarrow 3\ S + 2\ H_2O$$

$$\text{hydrogen} \qquad\qquad\qquad \text{sulfur}$$
$$\text{sulfide}$$

One-third of the molar amount of hydrogen sulfide extracted from the fossil fuel is first combusted to sulfur dioxide to provide the other reactant for this process. It is very important to remove hydrogen sulfide from gases before their dispersal in air because it is a highly poisonous substance, more so than is sulfur dioxide. Hydrogen sulfide is also a common pollutant in the emissions from pulp and paper mills.

Several other gases containing sulfur in a highly reduced state are emitted as air pollutants in petrochemical processes; these include CH_3SH, $(CH_3)_2S$, and CH_3SSCH_3. The term **total reduced sulfur** is used to refer to the total concentration of sulfur from H_2S and these three compounds.

The maximum amount of sulfur that can be contained in gasoline may well be further restricted in the future, at least in California, since its presence decreases the effectiveness of catalytic converters; apparently molecules containing reduced sulfur can occupy some of the catalytic sites and thus "clog" them chemically, restricting their ability to convert NO_x, CO, or hydrocarbons to harmless products.

Large point sources (individual sites that emit large amounts of a pollutant) of SO_2 are also associated with the nonferrous smelting (i.e., conversion of ores to free metals) industry. Many valuable and useful metals, such as copper and nickel, occur in nature as sulfide ores. In the first stage of their conversion to the free metals, they are usually "roasted" in air to remove the sulfur, which is converted to SO_2 and then often released into the air. For example,

$$2\,NiS + 3\,O_2 \longrightarrow 2\,NiO + 2\,SO_2$$

nickel	nickel	sulfur
sulfide	oxide	dioxide

Ores such as copper sulfide can be smelted in a process that uses pure oxygen forced into the smelting chamber, and the very concentrated sulfur dioxide obtained from the process can be readily extracted, liquefied and sold as a byproduct. The SO_2 concentration in the waste gases from conventional roasting processes (such as that used for nickel) is high; consequently it is feasible to pass the gas over an oxidation catalyst that converts much of the SO_2 to sulfur trioxide, onto which water can be sprayed to produce commercial concentrated sulfuric acid.

$$2\,SO_2(g) + O_2(g) \longrightarrow 2\,SO_3(g)$$

$$SO_3(g) + H_2O(aq) \longrightarrow H_2SO_4\,(aq)$$

sulfur trioxide　　　　　　sulfuric acid

The latter reaction (which represents only initial reactants and end product) is in fact accomplished in two steps (not shown) in order to ensure that none of the substances escape into the environment: first the trioxide is combined with sulfuric acid, and then water is added to the resulting solution.

When the emitted sulfur dioxide is dilute, as in the case of power plant emissions, its extraction by oxidation is not feasible. Instead, the SO_2 gas is removed by an acid-base reaction between it and calcium carbonate in the form of wet, crushed limestone. The emitted gases are either passed through a slurry of the wet solid, or are bombarded by jets of the slurry. About 90% of the gas can be removed by such a **scrubber** process, more formally known as **flue-gas desulfurization.** The product is a slurry of calcium sulfite and sulfate which then is usually buried in a landfill. The reactions are as follows:

$$CaCO_3 + SO_2 \longrightarrow CaSO_3 + CO_2$$

calcium   sulfur   calcium
carbonate dioxide   sulfite

$$CaSO_3 + \tfrac{1}{2} O_2 \longrightarrow CaSO_4$$

calcium   calcium
sulfite   sulfate

In some applications, the calcium sulfate is recycled back to calcium carbonate and reused rather than being buried. Alternatively, in "fluidized bed combustion" systems, the limestone is mixed with the pulverized coal before combustion occurs.

The alternative to sulfur dioxide extraction—to simply allow the pollutant gas to be emitted into the air—can cause devastation from SO_2 to the plant life in the surrounding area unless extremely high smoke stacks are used. The tallest such stacks in the world are located at Sudbury, Ontario, and reach 400 meters high. However, using tall stacks simply solves a local SO_2 problem at the expense of creating a problem downwind. (For example, emissions from mainland North America can be detected in Greenland.)

The 1991 Air Quality Accord between the United States and Canada requires both countries to reduce substantially their sulfur dioxide emissions. Such emissions in the United States are now restricted in accordance with the Clean Air Act; by the year 2000 there should be a substantial reduction in SO_2 emissions compared to values from the 1970s and 1980s. (An 88% reduction of the emissions of toxic substances from air, compared to 1990 levels, is also mandated by the Clean Air Act.) There is an SO_2 tonnage limit for each power plant that emits this gas, and a cap on overall national emissions as well. In Canada, the federal government is still trying to persuade the provincial governments to pass laws that would bring about comparable reductions.

Sulfur dioxide emissions from power plants can also be reduced by substituting oil, natural gas, or low-sulfur coal, but these fuels are usually more expensive than high-sulfur coal.

THE ECOLOGICAL EFFECTS OF ACID RAIN AND OF PHOTOCHEMICAL SMOG

The primary air pollutant $NO^{\bullet}$ is not especially soluble in water, and the acid that the sulfur dioxide produces upon dissolving in water is a weak one; thus the primary pollutants $NO^{\bullet}$ and SO_2 themselves do not make rainwater particularly acidic. However, some of these primary pollutants are converted over a period of hours or days into the secondary pollutants sulfuric acid, H_2SO_4, and nitric acid, HNO_3, both of which are very soluble in water and are strong acids. Indeed, virtually all the acidity in acid rain is due to the presence of these two acids. In Eastern North America, sulfuric acid predominates because much electrical power is generated from power plants that use high-sulfur coal. In Western North America, nitric acid attributable to vehicle emissions is predominant, since the coal mined and burnt there is low in sulfur.

PROBLEM 3-4

> The acid dissociation constant for the second stage of ionization of H_2SO_4 is 1.2×10^{-2} mol L^{-1}. Calculate the percentage of HSO_4^- that is ionized in acid rain that has a pH of 4.0. Repeat the calculation for a pH of 3.0. Is the trend shown by these calculations consistent with qualitative predictions made according to Le Chatelier's Principle (which states that the position of equilibrium shifts so as to minimize the effect of any stress)?

Figure 3-4 shows contour maps of the average pH in precipitation in different regions of Northern America and western Europe. In general, pH values have been falling (the rain has become more acidic) with passing years. The lowest pH recorded, 2.4, occurred for a rainfall in April 1974 in Scotland. In North America, the greatest acidity occurs in the eastern United States and in southern Ontario, since both regions lie in the path of air originally polluted by emissions from power plants in the Ohio valley. Currently the average pH of rainfall in the eastern United States lies between 3.9 and 4.5. These value are not so acidic as to cause such pronounced effects such as burning when the rain comes into contact with human skin, but as we shall see later the acidity does have important environmental effects.

In addition to the delivery of acids to ground level during precipitation, a comparable amount is deposited on the Earth's surface by means of **dry deposition,** the process in which nonaqueous chemicals are deposited onto solid and liquid surfaces at ground level when air containing them passes over the surfaces and deposits them as pollutants. Much of the original SO_2 gas is never oxidized in the air, but rather is

FIGURE 3-4

pH contours for rainfall (in 1985) for (a) central North America and (b) western Europe. Regions of very acidic rainfall are stippled. (Source: Redrawn from E. G. Nisbet, "Leaving Eden: To Protect and Manage the Earth." Cambridge, England, Cambridge University Press, 1991.)

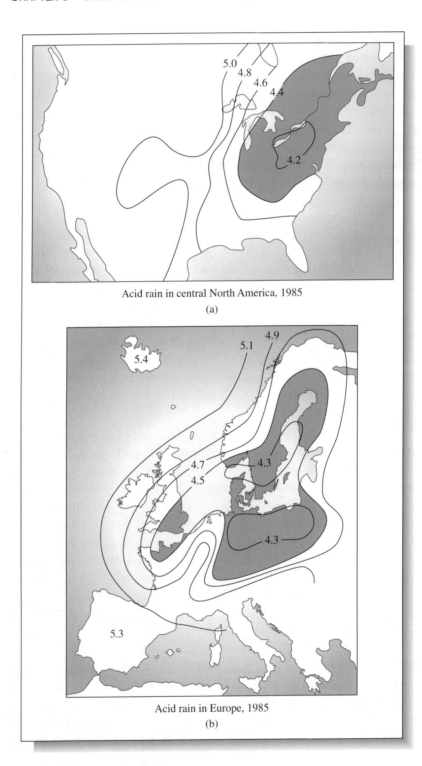

Acid rain in central North America, 1985

(a)

Acid rain in Europe, 1985

(b)

removed by dry deposition from air before reaction can occur: oxidation and conversion to sulfuric acid occurs after deposition. (**Wet deposition** processes encompass the transfer of pollutants to the Earth's surface by rain, snow, or fog, that is, by aqueous solutions.)

The extent to which acid precipitation affects biological life in a given area depends strongly upon the composition of the soil and bedrock in that area. Strongly affected areas are those having granite or quartz bedrock, since the associated soil has little capacity to neutralize the acid. Thus, the largest areas susceptible to acid rain are the pre-Cambrian Shield regions of Canada and Scandinavia. In contrast, if the bedrock is limestone or chalk, the acid can be efficiently neutralized ("buffered") since these rocks are composed of calcium carbonate, $CaCO_3$, which acts as a base and reacts with acid:

$$CaCO_3(s) + H^+(aq) \longrightarrow Ca^{2+}(aq) + HCO_3^-(aq)$$

$$\text{calcium carbonate} \qquad\qquad\qquad \text{bicarbonate ion}$$

$$HCO_3^-(aq) + H^+(aq) \longrightarrow H_2CO_3(aq) \longrightarrow CO_2(g) + H_2O(aq)$$

$$\text{bicarbonate ion} \qquad\qquad \text{carbonic acid}$$

The reactions here proceed almost to completion due to the excess of H^+. Thus the rock dissolves, producing carbon dioxide and calcium ion to replace the hydrogen ion. These same reactions are responsible for the deterioration of limestone and marble statues; fine detail, such as ears, noses and other facial features, are gradually lost as a result of reaction with acid and with sulfur dioxide itself.

Because of acid rainwater falling and draining into them, tens of thousands of lakes in the Shield regions of both Canada and Sweden have become strongly acidified, as have lesser numbers in the United States, Great Britain, and Finland. Lakes in Ontario are particularly hard hit, since they lie directly in the path of polluted air and because the soil there contains little limestone. In a few cases, attempts have been made to neutralize the acidity by adding limestone or calcium hydroxide to them; this process must be repeated every few years to sustain an acceptable pH.

Acidified lakes characteristically have elevated concentrations of dissolved aluminum, Al^{3+}, which is leached from rocks by the hydrogen ions (H^+); under neutral pH conditions, the aluminum is immobilized in the rock by its insolubility. (The chemistry underlying these processes is further discussed in Chapter 7.) Scientists now believe that both the acidity itself and the high concentrations of aluminum are responsible for the devastating decreases in fish populations that have been observed in many acidified water systems. Different types of fish and aquatic plants vary in their tolerances for aluminum and acid, so the biological composition of a lake varies as it gradually becomes increasingly acidic. Generally speaking, fish reproduction is severely diminished at low levels of acidity, which however can be tolerated by adult fish. Very young fish, hatched in early spring, also are subject to the shock of very acidic water that occurs when the acidic winter snow all melts in a short time and enters the water systems. Few species survive and reproduce when the pH drops much below 5; healthy lakes have a pH of about 7 or a little higher. As a result, many lakes and rivers in affected areas are now devoid of their valuable fish; for example, some rivers in Nova Scotia are too acidic for Atlantic salmon. The water in acidified lakes often is crystal clear due to the death of most of the flora and fauna.

In recent years it has also become clear that air pollution can have a severe effect on trees. The cause-and-effect relationship has been very difficult for scientists to untangle, however. Acidification of the soil can leach nutrients from it and, as occurs in lakes, solubilize aluminum. This element may interfere with the uptake of nutrients by trees and other plants. Apparently both the acidity of the rain falling on affected forests, and the ozone and other oxidants in the air to which they are exposed, pose a significant stress to the trees. This stress alone will not kill them, but when combined with drought, temperature extremes, disease, or insect attack, the trees become much more vulnerable. Typically, the die-off of branches starts from the tops of the trees. The phenomenon of forest decline was first observed on a large scale in western Germany.

Forests at high altitudes are most affected by acid precipitation, possibly because they are exposed to the base of low-level clouds, where the acidity is most concentrated. Fogs and mists are more acidic still than precipitation, since there is much less total water to dilute the acid. For example, white birch trees along the shores of Lake Superior are experiencing dieback in regions where acid fog occurs frequently. Deciduous trees affected by acid rain gradually die from their tops downward; the outermost leaves dry and fall early in the year and are not replenished the following spring. The trees become weakened as a result of these changes, and become more susceptible to other stressors.

Ground-level ozone itself has an effect on some agricultural crops due to its chemical attack on plants. Apparently the ozone reacts with the ethene (ethylene) gas that the plants emit, generating free radicals that then damage plant tissue. As in the case of trees, air pollution acts as a stressor to plants. The collective damage to North American crops, for example alfalfa in the United States and white beans in Canada, is estimated at a billion dollars a year.

PARTICULATES IN AIR POLLUTION

The black smoke released into the air by a diesel truck is often the most obvious form of pollution that we routinely encounter. The smoke is composed of particulate matter. **Particulates** are the tiny solid or liquid particles suspended in air and which are usually individually invisible to the naked eye. Collectively, however, small particles often form a haze that restricts visibility. An **aerosol** is a collection of particulates, whether solid particles or liquid droplets, dispersed in air. A true aerosol (as opposed, say, to the output of a hair-spray dispenser) has very small particles: their diameters are less than 0.0001 m, i.e., 100 μm. There are many common names for atmospheric particles: "dust" and "soot" refer to solids, whereas "mist" and "fog" refer to liquids, the latter denoting a high concentration of water droplets.

Substances that dissolve within the body of a particle are said to be **absorbed** by it; those which simply stick to the surface of the particle are said to be **adsorbed.** An important example of the latter effect is represented by the adsorption of large organic molecules onto carbon ("soot") particles, as discussed later in Chapter 6. Indeed, much of the mass in the air of suspended particles of all sizes consists of carbon.

One of main sources of carbon-based atmospheric particulates is the exhaust from diesel engines. Only recently have emissions from diesel trucks begun to be controlled. Due to increasingly tight regulations, starting in 1994 many new diesel trucks and buses in North America will require aftertreatment of engine exhaust. Medium-sized trucks probably will use soot oxidation catalysts, whereas buses may well employ soot filters.

Intuitively, one might think that all particles should settle out and be deposited onto the Earth's surface rapidly under the influence of gravity, but this is not true for the smaller ones. The rate, in distance per second, at which particles settle increases with the square of their diameter (a particle half the diameter of another, in other words, falls four times more slowly). The small ones fall so slowly they are suspended almost indefinitely in air (unless they stick to some object they encounter). More detail concerning the rates at which particles fall through the atmosphere is discussed in Box 3-1.

Particles whose diameters are less than 2.5 μm are called **fine particulates,** and usually remain airborne for days or weeks; those of larger

---

BOX 3-1   THE FALLING RATE OF PARTICLES

Quantitatively, for spherical particles whose diameter d exceeds about 10^{-6} m (i.e., 1 μm), the settling velocity v is given by Stoke's Law:

$$v = g\, d^2\, (\rho_1 - \rho_2)/18\eta$$

Here g is the acceleration due to gravity, η is the viscosity of air, ρ_1 is the density of the particle and ρ_2 is the density of air. For example, let us use this formula to calculate the settling velocity of a particle whose diameter is 2 μm and whose density equals that of water, given that $\eta = 170 \times 10^{-6}$ g/cm sec and $g = 980$ cm/sec^2. We will assume that the density of air is negligible compared to that of water.

Since $\rho_1 >> \rho_2$, then $(\rho_1 - \rho_2)$ is approximately equal to ρ_1. Since water has a density of about 1 g/cm^3, we will take this to be the value of ρ_1. Thus, since

Thus this particle descends about 0.01 centimeters a second, which is about 10 meters a day. From Stoke's Law it can be concluded that the residence time in air of a particle increases rapidly as the radius decreases. As a rule, the residence times of particles in tropospheric air can be estimated by assuming that they fall about 10 meters a day before reaching the ground.

PROBLEM 3-5

How much faster would a particle of diameter 0.01 cm fall compared to one having a diameter of 0.1 μm? What would be the ratio of atmospheric residence times of the two particles, assuming both began their descent from the same altitude?

$$v = g\, d^2\, (\rho_1 - \rho_2)/18\eta$$

$$v = \frac{980\ \text{cm sec}^{-2} \times (2 \times 10^{-6}\ \text{m} \times 100\ \text{cm m}^{-1})^2 \times 1.0\ \text{g cm}^{-3}}{18 \times 170 \times 10^{-6}\ \text{g cm}^{-1}\ \text{sec}^{-1}}$$

$$v = 1.3 \times 10^{-2}\ \text{cm sec}^{-1}$$

diameter than 2.5 μm are called **coarse particulates** and settle out fairly rapidly. (About 100 million particles of diameter 2.5 μm would be required to cover a dime.) In addition to this **sedimentation** process, particles also can be removed from air by absorption into falling raindrops.

The larger solid particles, such as those that constitute "dust," originate mainly from nonchemical processes. Important examples include natural sources such as volcanic eruptions and human activities such as stone crushing in quarries; land cultivation, which results in particles of topsoil being picked up by the wind constitutes, another source of dust. Large particles are of less concern to human health than are small ones for a combination of several reasons:

1. Since they settle out quickly, human exposure to them via inhalation is reduced.

2. When inhaled, coarse particles are efficiently filtered by the nose and throat and generally do not travel as far as the lungs. In contrast, inhaled fine particles usually travel to the lungs and can be adsorbed on cell surfaces there; such particles are said to be **respirable.**

3. The surface area per unit mass of large particles is smaller than that of small ones, and thus, gram for gram, their ability to transport adsorbed gas molecules to any parts of the respiratory system and there to catalyze chemical and biochemical reactions is correspondingly smaller.

4. Devices such as electrostatic precipitators and baghouse filters (a finely woven fabric bag through which air is forced) that are used to remove particulates from air are efficient only for coarse particles. Thus although a device may remove 95% of the total particulate mass, the reduction of surface area and of respirable particles is a much lower fraction; see Problem 3-7.

The concentration of particulates in air was formerly reported as the **total suspended particulates,** abbreviated TSP. Since the matter involved usually is not homogeneous, no molar mass for it is appropriate and thus concentrations are given as the mass of particles per volume of air, usually as μg/m^3. A common air quality standard for TSP is 75 μg/m^3. Because only the fine particles are respirable and thus are of most importance to human health, a more appropriate index is the concentration of particulates with less than some particular diameter. A concentration in air of *particulate matter* (PM) with diameters less than 10 μm would be described by the notation "PM$_{10}$." In 1984, the United States

Environmental Protection Agency changed their measure of environmentally relevant particulate matter from TSP to PM_{10}; the latter amounts to about half the TSP mass in most cities. The current air standard for PM_{10} in the United States is 150 $\mu g/m^3$.

PROBLEM 3-6

Let k be a given measure of length; then suppose a cubic particle of dimension $3k \times 3k \times 3k$ is split up into 27 of those with size $k \times k \times k$. Calculate the relative increase in surface area when this occurs. From your answer, deduce whether the total surface area of a given mass of atmospheric particle is larger or smaller when it occurs as a large number of small particles rather than a small number of large ones.

PROBLEM 3-7

An air-filtering device is tested and is found to remove all particles larger than 1 μm in diameter, but almost none of the smaller ones. Calculate the percent of the surface area removed by the device for a sample of particulates, 95% of the mass of which is particles of diameter 10 μm and 5% of which is particles of diameter 0.1 μm. Assume all particles are spherical and of equal density.

Mineral pollutants represent a part of the particulate content of air, too. Since many of the large particles in atmospheric dust, particularly in the air of rural areas, originate as soil or rock, their elemental composition is generally similar to that of the Earth's crust, and contain large concentrations of Al, Ca, Si, and O in the form of aluminum silicates (some of which incorporate the calcium ion). Near and above oceans, the concentration of solid NaCl in medium-sized particles is very high, since sea spray leaves sodium chloride particles airborne when the water evaporates. Pollen released from plants also consists of coarse particles in the 10–100 μm range.

Coarse particulates begin life as even coarser matter, originating chiefly from the disintegration of larger systems. Fine particulates, on the other hand, are produced in large part from the coagulation of even smaller particles which were in turn formed from gaseous molecules. For example, sulfuric acid travels in air not as a gas but as an aerosol of fine droplets, since H_2SO_4 has such a great affinity for water molecules. Many of the detrimental health effects associated with **acid air** may result from breathing this highly acidic aerosol.

Both sulfuric and nitric acids in air often eventually encounter ammonia gas that is released as a result of biological decay processes occurring at ground level. The acids undergo an acid-base reaction with the ammonia, which transforms them into ammonium sulfate and ammonium nitrate salts. For example,

$$H_2SO_4(aq) + 2\,NH_3(g) \longrightarrow (NH_4)_2SO_4(aq)$$

sulfuric acid   ammonia   ammonium sulfate

Although these salts initially are formed in aqueous particles, evaporation of the water results in the formation of solid particles. The predominant ions in fine particles are sulfate (including hydrogen sulfate, HSO_4^-) and ammonium, NH_4^+, although on the west coast of North America nitrate instead often is the predominant anion since more pollution results initially from nitrogen oxides than from sulfur dioxide. Because HNO_3 has a much higher vapor pressure than does H_2SO_4—it tends to remain in the vapor state—there is less condensation of nitric acid onto preexisting particles than occurs with H_2SO_4. Aerosols dominated by SO_4^{2-} are often called **sulfate aerosols.**

Particles whose diameter is about that of the wavelength of visible light, that is, 0.4–0.8 μm, can interfere with the transmission of light in air, reducing visual clarity, long-distance visibility, and the amount of light reaching the ground. For instance, a high concentration in air of particles of diameters between 0.1 μm and 1 μm produces a haze. Indeed, one conventional technique of measuring the extent of particulate pollution in an air mass is to determine its haziness. The widespread haze in the Arctic atmosphere in winter is due to sulfate aerosols that originate from the burning of coal, especially in Russia and Europe, and the enhanced haziness in summertime over much of North America is due by and large to sulfate aerosols arising from industrialized areas in the United States and Canada. Fine particles also are largely responsible for the haze associated in Los Angeles and other locations subject to episodes of photochemical smog. The smog aerosols contain nitric acid that has been neutralized to salts such as ammonium nitrate; since this salt has a low vapor pressure, it is much more likely to condense on the surface of particles than is the more volatile HNO_3 itself. Also present in these aerosols are intermediate carbon-containing products from photo-chemical smog reactions.

HEALTH EFFECTS OF OUTDOOR AIR POLLUTION

As would be expected, the major effects to human health from air pollution occur in the lungs. For example, asthmatics suffer worse episodes of their disease when the sulfur dioxide or the ozone concentration rises in the air that they breathe. And as discussed below, the most serious health problems are those that arise from the combination of large concentrations of soot-based particulate matter and sulfur dioxide or its oxidation products.

In the middle decades of the twentieth century, several Western industrialized cities experienced wintertime episodes of smog from soot and sulfur pollution that were so serious that the death rate increased noticeably. For example, in London in December 1952 about 4,000 people died within a few days as a result of the high concentrations of these pollutants that had built up in a stagnant, foggy air mass trapped by a temperature inversion (see Chapter 2) close to the ground. Those at most risk were elderly persons already suffering from bronchial problems, and young children. A ban of household coal-burning, from which most of the pollutants originated, has now largely eliminated such problems. Scientists are still unsure whether the main sulfur-containing agent that caused such serious problems in London was the SO_2, the sulfuric acid droplets, or the sulfate particulates.

Today, due to pollution controls, soot and sulfur smogs are no longer a major problem in Western countries. For example, deaths from bronchitis have fallen by over half in the United Kingdom, the result of changes in smoking habits and in air quality. However, the quality of winter air in some areas of what was the Eastern bloc, such as southern Poland, northern Czechoslovakia, and eastern Germany, is still very poor on account of the burning of large amounts of high-sulfur (up to 15% S) "brown" coal for both industrial and home heating purposes. For example, although the acceptable limit for the concentration of SO_2 in air is 80 $\mu g/m^3$ in many countries, including the United States, the level of this gas in Prague has surpassed 3,000 $\mu g/m^3$ on occasion. Indeed, four out of five children admitted to the hospital in some areas of Czechoslovakia in the early 1990s were there for treatment of respiratory problems.

Although smog episodes from sulfur-based chemicals have been eliminated in the West, many residents in these countries still are chronically exposed to measurable levels of sulfuric acid and sulfates due to the long-range transport of these substances from industrialized regions that still do emit SO_2 into the air. Recent research has shown a positive correlation between atmospheric concentrations of oxidized sulfur and ozone and hospital admissions for respiratory problems in southern Ontario. There is some evidence that the acidity of the pollution is the main active agent in causing lung dysfunction, including wheezing and bronchitis in children. Asthmatic individuals appear to be adversely affected by acidic sulfate aerosols, even at very low concentrations.

Recently it has been speculated that pollution due to SO_2 and sulfates causes a decrease in resistance to colon and breast cancer in people living in northern latitudes. The suggested mechanism of this action is a reduction in amount of available UV-B that is necessary to form vitamin D, which is a protective agent for both types of cancers. Since sulfur dioxide absorbs UV-B and sulfate particles scatter it, significant concentrations of either substance in air will reduce the amount of UV-B reaching ground level.

Photochemical smog, which arises from nitrogen oxides, is now more important than is sulfur-based smog (which we discussed above) in many cities, particularly those of high population and vehicle density. It consists of gases such as ozone, and an aqueous phase containing water-soluble organic and inorganic compounds.

Ozone itself is a harmful air pollutant. In contrast to sulfur-based chemicals, its effect on the robust and healthy is as serious as on those with preexisting respiratory problems. Experiments with human volunteers have shown that ozone produces transient irritation in the respiratory system, giving rise to coughing, nose and throat irritation, shortness of breath, and chest pains upon deep breathing. Thus even healthy, young people often experience such symptoms while exercising outdoors by cycling or jogging during smog episodes. Indeed, there is evidence that the daily race times of cross-country runners increase with increasing ozone concentration in the air that they inhale. One recent study indicated that a few percent of the day-to-day fluctuations in mortality rate in Los Angeles is explained by variations in the concentrations of air pollutants. It is not yet clear what, if any, long-term lung dysfunction results from exposure to ozone, and indeed this is a controversial subject among scientists. One anticipated effect is a decreased resistance to disease from infection because of the destruction of lung tissue. Many scientists believe that chronic exposure to high levels of urban ozone leads to the premature aging of lung tissue. At the molecular level, ozone readily attacks substances containing components with $C{=}C$ bonds, such as occur in biological tissues of the lung. It may be the case, too, that the fine particulates produced in smog have a deleterious health effect on humans.

Finally, we note that although serious episodes of soot-based smog have been largely eliminated in Western industrialized countries, the air pollution parameter that correlates most strongly with increases in the rate of disease or mortality in most such regions is the concentration of respirable particulates. It appears that particulate-based air pollution has a greater effect on human health than that produced directly by pollutant gases. Plots of daily mortality in cities versus their PM_{10} values for the same or the previous day are consistently linear though with some scatter; an increase by 100 $\mu g/m^3$ in the PM_{10} index of a city is estimated to increase its mortality rate by 6–17%. Other surveys indicate that 4–9% of the current mortality rate in the United States is associated with exposure to sulfate aerosol. In a recent survey of six American cities, a 26% higher rate of premature death was found in that city with the greatest particulate pollution, as compared with the cleanest. Even air that meets the current American standards for respirable particulates—a PM_{10} of less than 150 $\mu g/m^3$ averaged over 24 hours—seems to affect health adversely. Research is still under way to determine whether a specific component of the particulate matter causes these effects or whether it is simply the concentration of all respirable particles that is to blame.

HEALTH EFFECTS OF INDOOR AIR POLLUTION

Although pollutant concentrations vary significantly from building to building, the levels of common air pollutants often are greater indoors than outdoors; see Figure 3-5 for the ranges normally encountered for several common substances. Since most people spend more time indoors than outdoors, exposure to indoor air pollutants is an important environmental problem. Indeed, the inadequate ventilation practices of developing countries that burn coal, wood, crop residues, and other unprocessed biomass fuels create smoke and carbon monoxide pollution that leads to respiratory problems and ill health among huge numbers of people in these countries, particularly among women and young children.

FORMALDEHYDE

The most important and the most controversial indoor organic air pollutant gas is formaldehyde, $H_2C{=}O$. It is a widespread trace constituent of the atmosphere since it occurs as a stable intermediate in the oxidation of methane and of other VOCs. While its concentration in outdoor air is normally too small to be important—about 0.01 ppm in urban areas, except during episodes of photochemical smog—the level of formaldehyde gas *indoors* is often orders of magnitude greater, averaging about 0.1 ppm, and in certain cases exceeding 1 ppm.

The chief sources of indoor exposure to this gas are emissions from cigarette smoke and from synthetic materials that contain formaldehyde **resins** (a type of plastic) used in UFFI (urea formaldehyde foam insulation) and as an adhesive in plywood and particleboard (chipboard); formaldehyde itself is used in the dyeing of carpets and fabrics. Many useful resins (which are, chemically speaking, rigid polymeric materials) are prepared by combining formaldehyde with another organic substance; in the first few months and years after their creation, however, such materials release small amounts of free formaldehyde gas. Consequently, new prefabricated homes (mobile trailer homes are a good example) that contain chipboard generally have much higher levels of

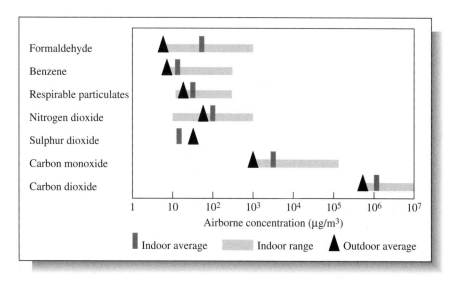

FIGURE 3-5

Indoor and outdoor concentrations for common pollutants (horizontal scale indicates orders of magnitude). (Source: Redrawn from Chapter 13 of "The State of Canada's Environment." Ministry of Supplies and Services, Ottawa, 1991.)

formaldehyde in their air than do older, conventional homes. Many manufacturers of pressed-wood products have now modified their production processes in order to reduce the rate at which formaldehyde is released. Formaldehyde-containing resins are used also to treat many consumer paper products in order to increase their strength when wet.

Formaldehyde has a pungent odor, with a detection threshold in humans of about 0.1 ppm; its odor can often be detected in stores that sell carpets and synthetic fabrics. At levels slightly higher than 0.1 ppm, many people report problems of irritation to their eyes, especially if they wear contact lenses, and to their noses, throats, and skin. The formaldehyde in cigarette smoke can cause eye irritation. At much higher levels of formaldehyde, that are not likely to be encountered except in occupational settings, breathing becomes difficult. Formaldehyde in air may cause children to develop more respiratory infections and allergies and asthma, although evidence for these effects is very controversial.

Formaldehyde is a carcinogen (a cancer-causing agent) in test animals and may also be carcinogenic in humans; it was classified as a "probable human carcinogen" by the United States Environmental Protection Agency in 1987. The expected cancer sites are in the respiratory system, including the nose. Studies of human populations exposed to formaldehyde have led to no clear-cut conclusions concerning an increase in cancer frequency due to nonoccupational exposure to formaldehyde. From animal studies, an upper limit to the possible effect in humans can be estimated: it corresponds to an increase in the cancer rate of one or two cases per ten thousand people in ten years of living in a high-formaldehyde house or trailer. However, the lower limit to the effect could well be a zero increase in cancer rate. In summary, no scientific consensus has yet been reached on the dangers to human health of low-level exposure to formaldehyde.

NITROGEN DIOXIDE AND CARBON MONOXIDE

Both nitrogen dioxide, $NO_2^\bullet$, and carbon monoxide, CO, are gases released as a result of combustion processes, including those that take place in homes and offices when fossil fuels are burned.

Indoor concentrations of $NO_2^\bullet$ often exceed outdoor values in homes that contain stoves, space heaters, and water heaters that are fueled by gas. The flame temperature in these appliances is sufficiently high that some atmospheric nitrogen and oxygen combine to form $NO^\bullet$, which eventually is oxidized to nitrogen dioxide. Thus in one recent study it was established that $NO_2^{\bullet}$ levels in homes that use gas for cooking or that have a kerosene stove average 24 ppb, compared to 9 ppb for homes that have neither. Peak concentrations near gas cooking stoves can exceed 300 ppb.

Nitrogen dioxide is soluble in biological tissue and is an oxidant, so its effects on health, if any, are expected to occur in the respiratory system. There have been many epidemiologic (statistical) studies of the effects on respiratory illness in children owing to exposure to low levels of $NO_2^{\bullet}$ emitted by gas appliances, but the results are not mutually consistent and are inadequate for establishing a cause-effect relationship. One recent study, conducted by researchers at Harvard University, found that a 15 ppb increase in the mean $NO_2^{\bullet}$ concentration in a home leads to about a 40% increase in lower respiratory symptoms among children aged 7 to 11 years. Nitrogen dioxide is the only oxide of nitrogen that is detrimental to health at concentrations likely to be encountered in residences.

Carbon monoxide is a colorless, odorless gas whose concentration indoors can be greatly increased by the incomplete combustion of carbon-containing fuels such as wood, gasoline, kerosene, or gas. High indoor concentrations usually are the result of a malfunctioning combustion appliance, such as a kerosene heater.

Average indoor and outdoor CO concentrations usually amount to a few parts per million, though elevated values in the 10–20 ppm range are common in parking garages due to the carbon monoxide emitted by motor vehicles. People such as traffic police who work outdoors in areas of high vehicular traffic can be exposed to elevated CO levels for long periods. The introduction of **oxygenated** substances, which are hydrocarbons in which some of the atoms have been substituted with oxygen, into gasoline was expected to reduce CO emissions from vehicles. Early results from Denver, the first city to use oxygenates (the study confined itself only to winter emissions) indicate that no reductions have as yet been observed.

The major danger from carbon monoxide arises from its ability, when inhaled, to complex strongly with the hemoglobin in blood and thus to impair its ability to transport oxygen to cells. On average, non-smokers have about 1% of their hemoglobin tied up as the complex with CO (called carboxyhemoglobin); the value for smokers is double this value or more because of the carbon monoxide that they inhale during smoking and that arises from the incomplete combustion of the cigarettes. Studies have shown that increased mortality from heart disease can result even if only several percent of hemoglobin is chronically tied up as the CO complex. Exposure to very high concentrations of CO results in headache, fatigue, unconsciousness, and eventually death (if such exposure is sustained for long periods). In developing countries, carbon monoxide poisoning is a serious hazard when biomass fuels are used to heat poorly ventilated rooms in which people sleep.

ENVIRONMENTAL TOBACCO SMOKE

It is well established that smoking tobacco is the leading cause of lung cancer, and is one of the main contributors to heart disease. Nonsmokers are often exposed to cigarette smoke, although in lower concentrations than smokers since it is diluted by air. This **environmental tobacco smoke,** ETS, has been the subject of many investigations in order to ascertain whether or not it is harmful to people who are exposed to it.

ETS consists of both gases and particles. The concentration of some toxic products of partial combustion is actually *higher* in sidestream smoke than in mainstream since combustion occurs at a lower temperature in the smoldering cigarette than in one through which air is being inhaled. Of course, since the sidestream smoke is usually diluted by air before being inhaled, the concentrations reaching the lungs of nonsmokers are much lower than those reaching the lungs of smokers themselves.

The chemical constitution of tobacco smoke is complex: it contains thousands of components, several dozen of which are carcinogens. The gases include carbon monoxide, nitrogen dioxide, formaldehyde, the polycyclic aromatic hydrocarbons to be discussed in Chapter 6, other VOCs, and radioactive elements such as polonium. The particulate phase, called the **tar,** contains nicotine and the less volatile hydrocarbons, and much of it is respirable. Many people experience irritation of their eyes and airways from exposure to ETS; the gaseous components of ETS, especially formaldehyde, hydrogen cyanide, acetone, toluene, and ammonia, cause most of the odor and irritation. Exposure to ETS aggravates the symptoms of many who suffer from asthma or angina pectoris (chest pains brought on by exertion). Some recent studies have established correlations between the rate of acute respiratory illness and the level of indoor $PM_{2.5}$ (which would include the total amount of respirable particulates from all sources, including tobacco smoke). "Passive smoking"—which involves inhalation of sidestream as well as already exhaled smoke—is believed by some scientists to cause bronchitis, pneumonia and other infections in up to 300,000 infants in the United States each year.

In 1993, the United States Environmental Protection Agency classified ETS as a known human carcinogen, and estimated that it causes about 3,000 lung cancer deaths annually. ETS is also considered to be responsible for killing 40,000 Americans annually from heart disease. A recent British study estimated that ETS kills 140,000 Europeans annually through cancer and heart disease.

ASBESTOS

The term **asbestos** refers to a family of six naturally occurring silicate minerals that are fibrous in character. Chemically, they are composed of long double-stranded networks of silicon atoms connected through intervening oxygen atoms; the net negative charge of this silicate structure is neutralized by the presence of cations such as magnesium. Thus the most commonly used form of asbestos, chrysotile, has the formula $Mg_3Si_2O_5(OH)_4$. It is a white solid whose individual fibers are curly. Chrysotile is mined mainly in Quebec, and is the principal type of asbestos used in North America. It has been employed in huge quantities because of its resistance to heat, its strength, and its relatively low cost. Common applications of asbestos include its use as insulation and spray-on fireproofing material in public buildings, in automobile brake-pad lining, as an additive to strengthen cement used for roofing and pipes, and as a woven fiber in fireproof cloth.

The use of asbestos has been sharply reduced because it is now recognized from studies on the health of asbestos miners and other workers with asbestos to be a human carcinogen. It causes mesothelioma, an incurable cancer of the lung, abdomen, and heart. Airborne asbestos fibers and cigarette smoke act **synergistically** (their combined effect is greater than the sum of their individual effects) in causing lung cancer. There is much controversy concerning whether chrysotile should be banned outright from further use and whether or not existing asbestos in buildings should be removed. Many experts feel that asbestos should be left in place unless it becomes damaged enough that there is a chance that its fibers will become airborne. Indeed, the removal of asbestos insulation can increase dramatically the levels of airborne asbestos in a building unless extraordinary precautions are taken. One scientist stated: "Removing asbestos is like waking up a pit bull terrier by poking a stick in its ear. We should let sleeping dogs lie." Some environmentalists feel that asbestos is a ticking time bomb—that it should be removed as soon as possible, as one can never predict when building insulation will be damaged.

Most of the initial concern about asbestos was related to crocidolite, **blue asbestos.** Evidence implicating this material in the causation of cancer in humans was already well established over two decades ago. It is a material with thin, straight and relatively short fibers that more readily penetrate lung passages and that is a more potent carcinogen than the white form. Crocidolite is mined in South Africa and Australia and has not been used much in North America.

THE DETAILED CHEMISTRY OF THE TROPOSPHERE

Some of the overall reactions involved in the production of secondary air pollutants from primary pollutants have been discussed in previous sections. However, in order to be able to reduce or eliminate specific types of pollution, it is vital to understand the detailed *mechanisms* by which the processes occur in air. In the material that follows, the principles of reactivity in both clean and polluted air are outlined, following which the oxidation of CH_4 and of SO_2 are analyzed in detail.

The hydrogen halides (HF, HCl, HBr) and fully oxidized gases such as carbon dioxide and sulfur trioxide are unreactive (from the oxidation-reduction point of view) because no further oxidation is possible with them; they eventually are deposited on the Earth's surface, often as a result of dissolving in falling raindrops. Thus we shall not consider these substances further in the discussions that follow.

TRACE GASES IN CLEAN AIR

From biological and volcanic sources, the atmosphere regularly receives inputs of the partially oxidized gases carbon monoxide, CO, and sulfur dioxide, SO_2, and of several gases that are simple compounds of hydrogen some of whose atoms are in a highly reduced form (e.g., H_2S, NH_3); the most important of these "natural" substances are listed in Table 3-2.

Although most of these gases are gradually oxidized in air, none of them reacts directly with diatomic oxygen. Rather, their reactions all begin when they are attacked by the hydroxyl free radical, $OH^\bullet$, even though the concentration of this species in air is exceedingly small. In

TABLE 3-2	GASES EMITTED INTO THE ATMOSPHERE FROM NATURAL SOURCES	
		Important natural source
CH_4	methane	anaerobic biological decay
NH_3	ammonia	anaerobic biological decay
H_2S	hydrogen sulfide	anaerobic biological decay
HCl	hydrogen chloride	anaerobic biological decay, volcanoes
CH_3Cl	methyl chloride	oceans
CH_3Br	methyl bromide	oceans
CH_3I	methyl iodide	oceans
CO	carbon monoxide	atmospheric CH_4, fires
SO_2	sulfur dioxide	volcanoes

clean tropospheric air, the hydroxyl radical is produced when a fraction of the excited oxygen atoms resulting from the photochemical decomposition of the trace amounts of atmospheric ozone react with gaseous water to abstract one hydrogen atom from each H_2O molecule:

$$O_3 \xrightarrow{\text{UV-B}} O_2 + O*$$

$$O* + H_2O \longrightarrow 2\,OH^{\bullet}$$

Recall from Problem 2-6 that because the corresponding reaction involving unexcited atomic oxygen is endothermic, its activation energy is high and consequently it occurs far too slowly to be a significant source of atmospheric $OH^{\bullet}$. As we shall see later, the oxidation of methane to carbon dioxide generates additional hydroxyl radicals in the atmosphere.

PROBLEM 3-8

The concentration of $OH^{\bullet}$ in air averages 8.7×10^6 molecules per cubic centimeter. Calculate its molar concentration, and its concentration in parts per trillion, assuming that the total air pressure is 1.0 atm and the temperature is 15°C.

The hydroxyl free radical is reactive toward a wide variety of other molecules, including the hydrides of carbon, nitrogen, and sulfur listed in Table 3-2, and toward many molecules containing multiple bonds (double or triple bonds) including CO and SO_2. Although suspected for decades of playing a pivotal role in air chemistry, the presence of $OH^•$ in the troposphere was only recently confirmed since its concentration is so very small. The great importance of the hydroxyl radical to tropospheric chemistry arises because it, not O_2, initiates the oxidation of *all* the gases in Table 3-2 (other than HCl). Without $OH^•$ and its related reactive species $HOO^•$, these gases would not be efficiently removed from the troposphere, nor would most pollutant gases such as unburnt hydrocarbons emitted from vehicles. Indeed, $OH^•$ has been called the "tropospheric vacuum cleaner." The reactions that it initiates constitute the flameless, low-temperature "burning" of the reduced gases of the lower atmosphere. An example of this process is the net oxidation of methane gas, CH_4, into the completely oxidized product carbon dioxide, CO_2, which eventually becomes deposited on the ocean floor as a carbonate.

$$CH_4 + 2\,O_2 \longrightarrow CO_2 + 2\,H_2O$$

As we shall see below, this overall reaction occurs by a sequence of reactions, the first of which involves the hydroxyl radical. Recall that a sequence of steps (molecular events) by which an overall reaction proceeds is called the reaction's mechanism.

PRINCIPLES OF REACTIVITY IN THE TROPOSPHERE

Most gases in the troposphere are gradually oxidized by a sequence of reactions involving free radicals. For a given gas, the sequence can be predicted from the principles discussed below; these generalizations have much in common with those developed in Chapter 2 for processes in the stratosphere.

   As mentioned above, the usual initial reaction of an atmospheric gas is with the hydroxyl free radical rather than with molecular oxygen, O_2, since a large activation energy is required for the latter reaction to occur. With molecules that contain a multiple bond, the hydroxyl radical usually reacts by *adding* itself to the molecule at the position of the multiple bond. Recall the general principle that spontaneous radical reactions tend to produce stable products—that is, products containing strong bonds. Thus it may be easier to understand that $OH^•$ addition does not occur to an oxygen atom since the O—O bonds that would result are weak, and does not occur to CO *double* bonds since they are very strong relative to the single O—O or C—O bond which would be produced.

For example, the OH$^•$ radical adds to the sulfur atom, forming a strong bond, but not to an oxygen atom in sulfur dioxide:

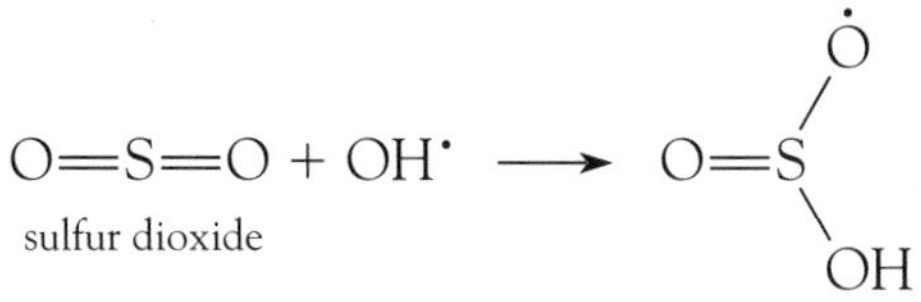

$$O{=}S{=}O + OH^• \longrightarrow \quad O{=}\overset{\displaystyle \overset{•}{O}}{\underset{\displaystyle OH}{S}}$$

sulfur dioxide

(Here and elsewhere in this book, we write Lewis structures that assume electrons can form double bonds.) Hydroxyl radical does not add to carbon dioxide, O$=$C$=$O, since it contains only very strong C$=$O bonds. However, OH$^•$ addition does occur to the carbon atom in carbon monoxide, CO, since the triple bond is thereby converted to the very stable double bond and a new single bond is also formed:

$$^{\ominus}C{\equiv}O^{\oplus} + OH^• \longrightarrow HO{-}\overset{•}{C}{=}O$$

carbon monoxide

This process is exothermic because the third CO bond in carbon monoxide is weak relative to the other two. Generally, OH$^•$ does not add to multiple bonds in any fully oxidized species such as CO_2, SO_3, and N_2O_5, since such processes are endothermic and therefore are very slow to occur at atmospheric temperatures. Similarly, N_2 does not react with OH$^•$ because the component of the nitrogen-to-nitrogen bond that would be destroyed is stronger than the N$-$O bond that would be formed.

For molecules that do not have a reactive multiple bond but that do contain hydrogen, OH$^•$ reacts with them by the abstraction of a hydrogen atom to form a water molecule and a new reactive free radical. For CH_4, NH_3, H_2S, and CH_3Cl, for instance, the reactions are as follows:

$$CH_4 + OH^• \longrightarrow CH_3^• + H_2O$$
$$NH_3 + OH^• \longrightarrow NH_2^• + H_2O$$
$$H_2S + OH^• \longrightarrow SH^• + H_2O$$
$$CH_3Cl + OH^• \longrightarrow CH_2Cl^• + H_2O$$

Because the H$-$OH bond formed in these reactions is very strong, the processes are all exothermic; thus only small activation energy barriers exist to impede these reactions (see Box 2-2).

PROBLEM 3-9

Why aren't gases such as CF_2Cl_2 (a CFC) readily oxidized in the troposphere? Would the same be true for CH_2Cl_2?

PROBLEM 3-10

The abstraction of the hydrogen atom by $OH^{\bullet}$ in HF is endothermic. Comment briefly on the expected rate of this reaction: would it be (at least potentially) fast, or necessarily very slow in the troposphere?

PROBLEM 3-11

The hydroxyl radical does not react with gaseous nitrous oxide, N_2O, even though the molecule contains multiple bonds. What can you deduce about the probable energetics (endo- or exothermic character) of this reaction from the observed lack of reactivity?

A few gases emitted into air can absorb some of either the UV-A or the visible component of sunlight, and this input of energy is sufficient to break one of the bonds in the molecule, thereby producing two free radicals. For example, most molecules of atmospheric formaldehyde gas, H_2CO, react by photochemical decomposition after absorption of UV-A from sunlight:

$$H_2CO \xrightarrow[\text{($\lambda < 338$ nm)}]{\text{UV-A}} H^{\bullet} + HCO^{\bullet}$$
$$\text{formaldehyde}$$

As shown above, the initial reaction of a gas emitted into air usually produces free radicals, almost all of which are extremely reactive. The predominant fate in tropospheric air for most simple radicals is reaction with diatomic oxygen, often by an addition process (one of the oxygen atoms attaches, or "adds on," to the other reactant). For instance, O_2 reacts by addition with the methyl radical:

$$CH_3^{\bullet} + O_2 \longrightarrow CH_3OO^{\bullet}$$

Notice that $CH_3OO^{\bullet}$ itself is a free radical; the terminal oxygen forms only one bond and carries the unpaired electron:

$$H_3C\!-\!\ddot{\underset{\displaystyle\cdot\cdot}{O}}\!-\!\dot{\underset{\displaystyle\cdot\cdot}{O}}: \quad \text{or just} \quad H_3C\!-\!O\!-\!\dot{O}$$

Species such as $HOO^{\bullet}$ and $CH_3OO^{\bullet}$ are called *peroxy* radicals since they contain a peroxide-like $O\!-\!O$ bond; recall that $HOO^{\bullet}$ is called the *hydroperoxy* radical.

As radicals go, peroxy radicals are less reactive than most. They do *not* readily abstract hydrogen since the resulting peroxides would not be very stable energetically; since the transfer of H to the peroxy radical would be endothermic, and thus would possess a large activation energy, such reactions are usually so slow as to be of negligible importance. In contrast to stratospheric $HOO^{\bullet}$, peroxy radicals in the troposphere do not react with atomic oxygen or with UV light because of their extremely low concentrations in this region of the atmosphere. The most common fate of peroxy radicals in tropospheric air, except of the very cleanest type of air such as that over oceans, is reaction with nitric oxide, $NO^{\bullet}$, by the transfer of the "loose" oxygen atom (see Chapter 2), thereby forming $NO_2^{\bullet}$ and a radical which has one fewer oxygen atom:

$$HOO^{\bullet} + NO^{\bullet} \longrightarrow OH^{\bullet} + NO_2^{\bullet}$$

$$CH_3OO^{\bullet} + NO^{\bullet} \longrightarrow CH_3O^{\bullet} + NO_2^{\bullet}$$

It is by this type of reaction that most atmospheric $NO^{\bullet}$ is oxidized to $NO_2^{\bullet}$. Recall that this reaction also is typical of the types encountered in stratospheric chemistry (see Chapter 2).

For radicals that contain non-peroxy oxygen, the reaction with molecular oxygen frequently involves the abstraction by O_2 of an H atom; this process occurs provided that, as a result, a single bond involving oxygen is converted to a double one, or a double bond involving oxygen is converted to a triple one. As examples, consider the three reactions below in which a carbon-oxygen single (or double) bond is converted to a double (or triple) one:

$$CH_3\!-\!\dot{O} + O_2 \longrightarrow H_2C\!=\!O + HOO^{\bullet}$$

$$HO\!-\!\dot{C}\!=\!O + O_2 \longrightarrow O\!=\!C\!=\!O + HOO^{\bullet}$$

$$H\!-\!\dot{C}\!=\!O + O_2 \longrightarrow {}^{\ominus}C\!\equiv\!O^{\oplus} + HOO^{\bullet}$$

In general, such processes will not occur unless a new bond is created in the product free radical, since the strength of the newly created

H—OO bond is not sufficient to compensate for that of the original bond to hydrogen that is broken in the process.

If there is no suitable hydrogen atom for O_2 to abstract, then when it collides with a radical, it instead *adds* to it at the site of the unpaired electron, as was previously discussed for simple radicals. For example, radicals of the type $R—\overset{\bullet}{C}{=}O$, where R is a chain of carbon atoms, add O_2 to form a peroxy radical:

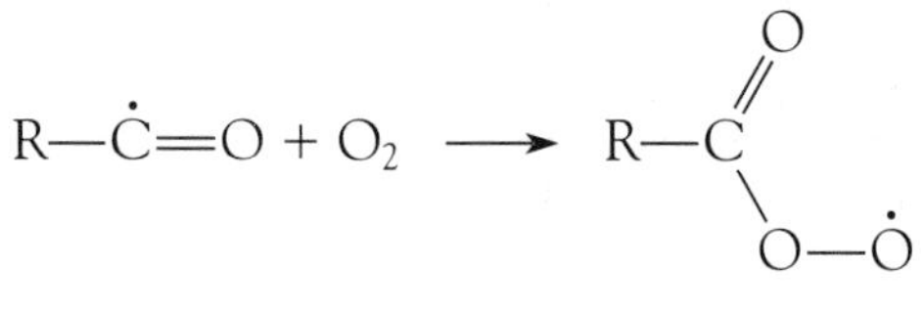

$$R—\overset{\bullet}{C}{=}O + O_2 \longrightarrow R—C\overset{\displaystyle O}{\underset{\textstyle O—\overset{\bullet}{O}}{\Big\backslash}}$$

The only exception to the rule that oxygen-containing radicals react with O_2 occurs when the radical can decompose spontaneously in a thermoneutral or exothermic fashion. An example of this rare phenomenon is discussed in the section on photochemical smog.

The generalizations discussed above are summarized in the form of the "decision trees" diagrammed in Figure 3-6. By using these diagrams, you can deduce the sequence of reactions by which most atmospheric gases in the troposphere are oxidized.

By assuming that the free radicals in atmospheric reactions are in a steady state, it is possible to mathematically relate the concentrations of the various chemicals and rate constants of the various reactions to each other. This analysis is illustrated in Appendix II of this book.

TROPOSPHERIC OXIDATION OF METHANE

Gaseous methane, CH_4, is released into the atmosphere in large quantities as a result of **anaerobic** (i.e., O_2-free) biological decay processes and of the use of coal, oil, and especially natural gas. Details concerning its production, and of the effects on climate of atmospheric methane, are discussed in Chapter 4. Here, however, we shall be concerned with its conversion to carbon dioxide.

The sequence of reactions by which methane is slowly oxidized in the atmosphere can be deduced by applying the principles discussed above. Since CH_4 contains no multiple bonds, the sequence is initiated by a hydroxyl radical abstracting a hydrogen atom from a methane molecule, giving the methyl radical $CH_3^{\bullet}$:

$$CH_4 + OH^{\bullet} \longrightarrow CH_3^{\bullet} + H_2O$$

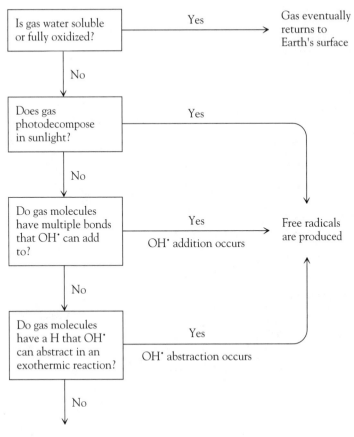

FIGURE 3-6A

Decision tree illustrating the fate of gases emitted into air.

Since the $CH_3^•$ radical contains no oxygen, we deduce that it adds O_2, producing a peroxy radical:

$$CH_3^• + O_2 \longrightarrow CH_3OO^•$$

Further, since $CH_3OO^•$ is of the peroxy type, we deduce that it reacts with $NO^•$ molecules in air to oxidize them by transfer of an oxygen atom:

$$CH_3OO^• + NO^• \longrightarrow CH_3O^• + NO_2^•$$

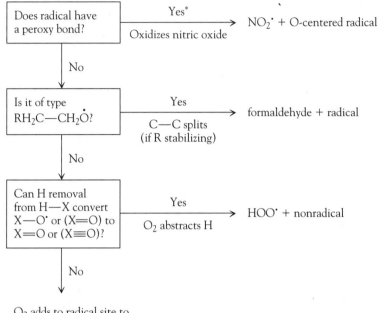

*Under conditions with significant nitric oxide present and before radical + radical reactions become important.

FIGURE 3-6B
Decision tree illustrating the fate of airborne free radicals.

The radical $CH_3O^{\cdot}$ contains a C—O bond that can become C=O upon loss of one hydrogen, so we conclude from our principles that in the next step O_2 abstracts a hydrogen (H) atom, thereby producing the nonradical product formaldehyde, H_2CO:

$$CH_3O^{\cdot} + O_2 \longrightarrow H_2CO + HOO^{\cdot}$$

Thus methane is converted to formaldehyde during its oxidation. Since the latter substance is reactive as a gas in the atmosphere, the mechanism is not complete at this point. After several days (on average), most formaldehyde molecules decompose photochemically by the absorption of UV-A in sunlight, resulting in the cleavage of a C—H bond, and the consequent formation of two radicals:

$$H_2CO \xrightarrow[\text{(λ < 338 nm)}]{\text{UV-A}} H^{\cdot} + HCO^{\cdot}$$

A minority of formaldehyde molecules react with $OH^{\cdot}$ by H atom abstraction, yielding the same $HCO^{\cdot}$ radical; see Problem 3-16 for the implications of this alternative route.

The hydrogen atom from formaldehyde photolysis is a simple radical, and therefore it reacts by addition to O_2 to yield $HOO^{\cdot}$:

$$H^{\cdot} + O_2 \longrightarrow HOO^{\cdot}$$

Meanwhile, the $H{-}\dot{C}{=}O$ radical reacts by yielding a H atom to O_2 to produce carbon monoxide and $HOO^{\cdot}$:

$$HCO^{\cdot} + O_2 \longrightarrow CO + HOO^{\cdot}$$

Thus carbon monoxide also is an intermediate in the oxidation of methane; indeed most of the CO in a clean atmosphere is derived from this source. Since CO is not a radical and does not absorb visible or UV-A light, we can deduce that it reacts ultimately by hydroxyl radical addition to its triple bond:

$$^{\ominus}C{\equiv}O^{\oplus} + OH^{\cdot} \longrightarrow H{-}O{-}\dot{C}{=}O$$

This radical can convert its O—C bond to O=C by loss of $H^{\cdot}$, so we can deduce that O_2 readily abstracts the hydrogen:

$$H{-}O{-}\dot{C}{=}O + O_2 \longrightarrow O{=}C{=}O + HOO^{\cdot}$$

Carbon in its fully oxidized form of carbon dioxide is ultimately produced from methane by this sequence of numerous steps, which are summarized in Figure 3-7. If we add up the 9 steps involved, then after cancellation of common terms, the overall reaction is seen to be

$$CH_4 + 5\,O_2 + NO^{\cdot} + 2\,OH^{\cdot} \xrightarrow{\text{UV-A}} CO_2 + H_2O + NO_2^{\cdot} + 4\,HOO^{\cdot}$$

If to this is added the conversion of the four $HOO^{\cdot}$ radicals back to $OH^{\cdot}$ by reaction with 4 $NO^{\cdot}$ molecules, the revised overall reaction is

$$CH_4 + 5\,O_2 + 5\,NO^{\cdot} \xrightarrow{\text{UV-A}} CO_2 + H_2O + 5\,NO_2^{\cdot} + 2\,OH^{\cdot}$$

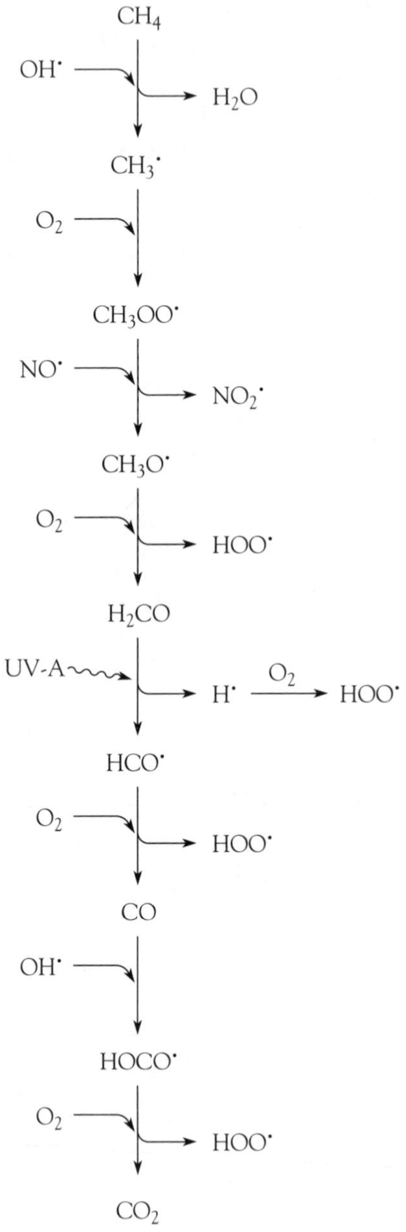

FIGURE 3-7
Steps in the atmospheric oxidation of methane to carbon dioxide.

We conclude that $NO^\bullet$ is oxidized to $NO_2^\bullet$ synergistically, that is, in a mutually cooperative process, when methane is oxidized to carbon dioxide. Note also that the number of $OH^\bullet$ free radicals is increased as a result of the process, due to the photochemical decomposition of formaldehyde. The initial step—the abstraction by $OH^\bullet$ of a hydrogen atom from methane—is a relatively slow process, however, requiring about a decade to occur on average. Once this has happened, however, the subsequent steps leading to formaldehyde occur very rapidly. The slowness of the initial step in methane oxidation, and the increasing amounts of the gas released from the surface of the Earth, have led to an increase in the atmospheric concentration of CH_4 in recent times, as discussed further in Chapter 4.

In general, during the atmospheric oxidation of any of the hydrides (simple hydrogen-containing molecules such as CH_4, H_2S, and NH_3) one or more stable species are encountered along the reaction sequence before the totally oxidized product is formed. These intermediates are also formed independently by various pollution processes. Figure 3-8

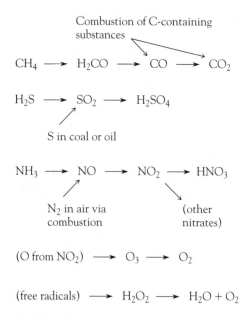

FIGURE 3-8

Stable species (i.e., nontransients) and their additional sources in atmospheric oxidation reactions.

summarizes the sequences for hydrides and partially oxidized materials from the viewpoint of the stable species; close reflection will persuade the reader that the net result is the OH˙-induced oxidation of the reduced and partially oxidized gases emitted into the air from both natural and pollution sources. In a few cases, for instance, methane and methyl chloride, the initiation reaction is sufficiently slow that a few percent of these gases survive long enough to penetrate to the stratosphere due to the upward diffusion of tropospheric air.

PROBLEM 3-12

Using the reaction principles developed above (the decision trees in Figure 3-6 will help here), predict the sequence of reaction steps by which atmospheric H_2 gas will be oxidized in the troposphere. What is the overall reaction?

PROBLEM 3-13

Using the reaction principles and/or decision trees, deduce the series of steps by which H_2S in air is oxidized to SO_3, which then combines with water vapor to yield H_2SO_4. What is the overall reaction?

PROBLEM 3-14

Deduce two short series of steps by which molecules of methanol, CH_3OH, are converted to formaldehyde, H_2CO, in air; the mechanisms should differ according to which hydrogen atom you consider to react first, that of CH_3 or that of OH.

PROBLEM 3-15

Write equations showing the reactions by which atmospheric carbon monoxide is oxidized to carbon dioxide. Then, by adding the process by which HOO˙ is returned to OH˙, deduce the overall reaction.

Deduce the series of steps, and the overall reaction as well, for the oxidation of a formaldehyde molecule to CO_2, assuming that for the particular H_2CO molecule involved the initial reaction is abstraction of H by OH·, rather than photochemical decomposition. Overall, is there any increase in the number of free radicals as a result of the oxidation?

PHOTOCHEMICAL SMOG: THE OXIDATION OF HYDROCARBONS

Notwithstanding the great complexity of the process, the most important features of the photochemical smog phenomenon can be understood by considering only its few main categories of reactions; these do not differ much in type from those occurring in clean air.

We shall restrict our attention to the most reactive VOCs, namely hydrocarbons that contain a C=C bond. Since each carbon in a double bond forms two bonds to the other carbon, it is free to form only two additional bonds. The simplest example is ethene (ethylene), C_2H_4, the structure of which can be written as $H_2C=CH_2$; its full structural formula is shown below:

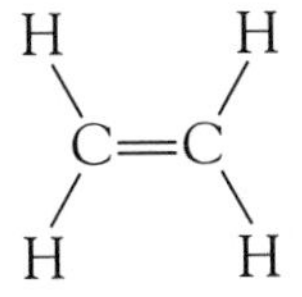

$$\begin{array}{ccc} H & & H \\ \diagdown & & \diagup \\ & C=C & \\ \diagup & & \diagdown \\ H & & H \end{array}$$

ethene

In other, similar, hydrocarbons, one or more of the four hydrogens are substituted by other atoms or groups, often a chain of carbon atoms such as CH_3- or CH_3CH_2- (longer chains are possible), which will be designated in the usual notation simply as R, since it is generally not the chain, but rather the C=C part of the molecule that is the reactive site. (This chain, R, of carbons, however short—it may consist of only one carbon with its attendant hydrogens—is called an **alkyl group.**)

Consider a general hydrocarbon RHC=CHR. In air, it reacts with hydroxyl radical by addition to the C=C bond, which is a faster process (due to a lower activation energy) than is the abstraction of H:

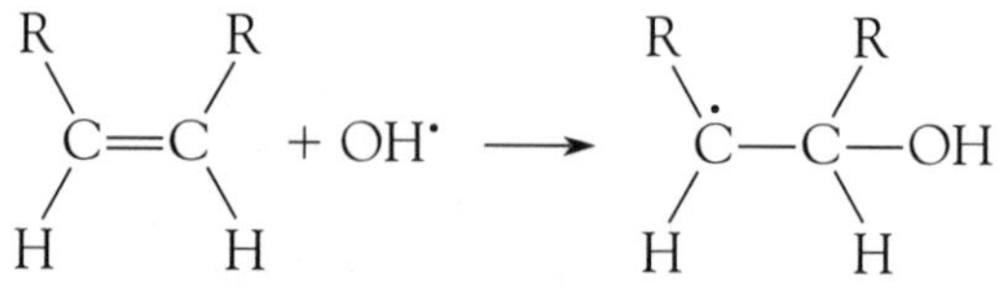

A small fraction of such hydrocarbons react initially with ozone or with atomic oxygen, but we shall ignore these processes as they lie beyond the scope of this book.

As anticipated from the reaction principles for clean air, the carbon-based radical produced from the reaction of hydroxyl radical with the hydrocarbon adds O_2 to yield a peroxy radical, which in turn oxidizes $NO^{\cdot}$ to $NO_2^{\cdot}$:

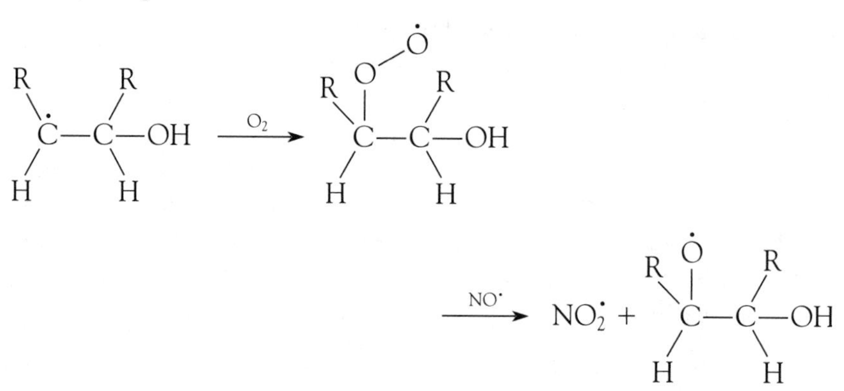

Once much of the $NO^{\cdot}$ has been oxidized to $NO_2^{\cdot}$, photochemical decomposition by sunlight of the latter gives $NO^{\cdot}$ plus O, and the latter quickly combines with molecular oxygen to give ozone. Indeed, it is a characteristic of air pollution driven by photochemical processes that ozone from $NO_2^{\cdot}$ photodecomposition builds up to much higher levels than are found in clean air. The other reactions involving atomic oxygen that were encountered in the stratosphere cannot compete owing to the fast rate of atomic oxygen's reaction with the very abundant O_2 in the troposphere.

$$NO_2^{\cdot} \xrightarrow{\text{UV-A}} NO^{\cdot} + O$$
$$O + O_2 \longrightarrow O_3$$

Nitrogen dioxide is the only significant source of the atomic oxygen from which ozone can form. The reaction of "urban ozone" does not occur to a significant extent until most $NO^{\cdot}$ has been converted to

$NO_2^{\bullet}$, since $NO^{\bullet}$ and O_3 mutually self-destruct if both are present in significant concentrations:

$$NO^{\bullet} + O_3 \longrightarrow NO_2^{\bullet} + O_2$$

Thus the concentration of ozone never builds up substantially as a result of this sequence, since the above three reactions alone corresponds to a null cycle, that is, one in which no net reaction takes place. It is only after most $NO^{\bullet}$ has been oxidized to $NO_2^{\bullet}$ as a result of reactions with peroxy free radicals that the characteristic buildup of ozone occurs.

PROBLEM 3-17

Show that the net effect of the last three reactions is indeed a null cycle.

Ozone destruction via the action of atomic chlorine or bromine, that is, by the mechanism discussed in Chapter 2 for the stratosphere, does not occur in the troposphere due to the lack of free halogen atoms in air at low altitudes. Any chlorine atoms formed in the troposphere would be quickly transformed to HCl, which is highly soluble in water droplets and is readily rained out.

One might anticipate from our reactivity principles (Fig. 3-6) that the two-carbon radical mentioned above ($RCH\overset{\bullet}{O}HOHR$) would lose H by abstraction by O_2, but instead it decomposes spontaneously by cleavage of the C—C bond to give a nonradical molecule containing a C=O bond and another radical, $RHCOH^{\bullet}$:

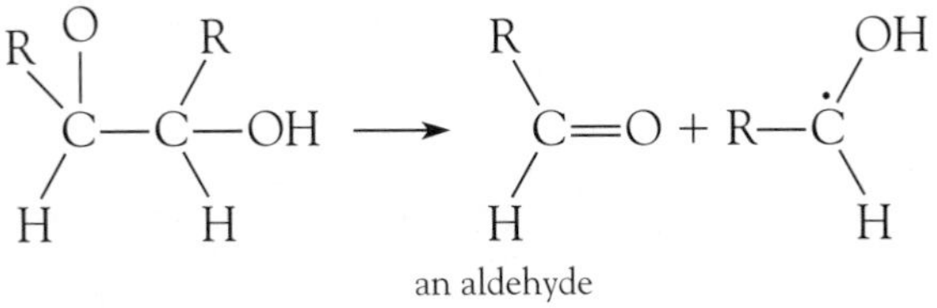

an aldehyde

It happens that the reaction requires no energy input, that is, ΔH is close to zero, because in this case the formation of a C=O bond from C—O compensates energetically for loss of the C—C bond. Since the decomposition of this radical is not endothermic, its activation energy is small and thus the process occurs spontaneously in air.

Molecules of the RHCO type shown above are called "aldehydes." The simplest example is formaldehyde, H_2CO, which was encountered previously in the process of methane oxidation. The carbon-based radical RHCOH• produced in the reaction above subsequently reacts with an O_2 molecule; since loss of the hydroxyl hydrogen allows the C—O bond to become C=O, the oxygen molecule abstracts the H atom:

$$R-\underset{H}{\overset{OH}{\underset{|}{\overset{|}{\dot{C}}}}} \;+\; O_2 \longrightarrow \; HOO• + \underset{H}{\overset{R}{\underset{/}{\overset{\backslash}{C}}}}=O$$

Thus the original RHC=CHR pollutant molecule is converted into two aldehyde molecules, each possessing half the number of carbon atoms. Indeed, as shown in Figure 3-9, by about noon on a smoggy day in Los Angeles most of the reactive hydrocarbons emitted into air by morning rush-hour traffic have been converted to aldehydes. By

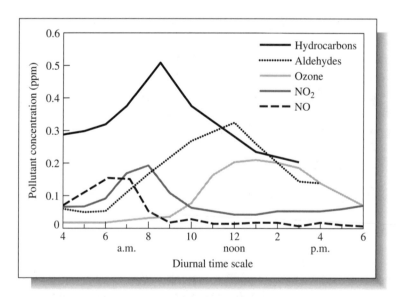

FIGURE 3-9

Time-of-day (diurnal) variation in the average concentrations of gases during days of marked eye irritation in Los Angeles (Source: Redrawn from D. J. Spedding. *Air Pollution*. Oxford: Oxford University Press, 1974.)

midafternoon, most of the aldehydes have disappeared since they have been photochemically decomposed into HCO˙ and R˙ (alkyl) free radicals.

$$RHCO \xrightarrow{\text{sunlight}} R˙ + HCO˙$$

The sunlight-induced decomposition of aldehydes and of ozone leads to a huge increase in the number of free radicals in the air of a city undergoing photochemical smog, although in absolute terms the concentration of radicals is still very tiny. The entire process of conversion of the original RHC=CHR molecule into aldehydes is summarized in Figure 3-10.

PROBLEM 3-18

Assume that the alkyl group (the "R") in the aldehyde RHCO produced by photochemical smog is a simple methyl group, CH_3, and that when it undergoes photochemical decomposition by sunlight the radicals $CH_3˙$ and HCO˙ are produced. Using the air reactivity principles, deduce the sequence of reactions by which these radicals are oxidized to carbon dioxide, and determine the overall reaction of conversion of RHCO to CO_2.

PHOTOCHEMICAL SMOG: THE FATE OF FREE RADICALS

In later stages of photochemical smog, reactions that occur between two radicals can no longer be neglected as they were in clean air. Because their rates are proportional to the *product* of two radical concentrations, these processes are important when the radical concentrations are high. That is, they will occur quickly under such conditions. Generally, the reaction of two free radicals yields a stable, nonradical product:

$$radical + radical \longrightarrow nonradical\ molecule$$

One important example of a radical-radical reaction is that of the combination of hydroxyl and nitrogen dioxide to yield nitric acid, a process that also occurs in the stratosphere:

$$OH˙ + NO_2˙ \longrightarrow HNO_3$$
$$\text{nitric acid}$$

FIGURE 3-10
Mechanism of the
RHC=CHR reaction in
photochemical smog.

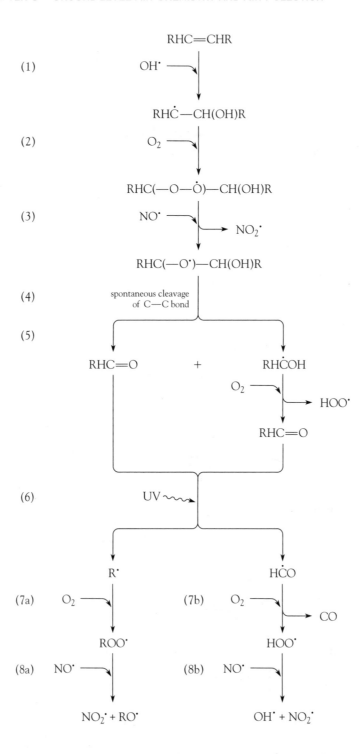

This reaction is the main tropospheric sink for hydroxyl radicals. The average lifetime for an HNO_3 molecule is several days; by then it either has dissolved in water and been rained out or it has been photochemically decomposed back into its components.

Similarly, combination of $OH^{\cdot}$ with $NO^{\cdot}$ gives nitrous acid HONO, also written HNO_2. In sunlight, the nitrous acid is almost immediately photochemically decomposed back to $OH^{\cdot}$ and $NO^{\cdot}$, but at night it is stable. The observed gigantic increase at dawn in the concentration of $OH^{\cdot}$ radicals in the air of smog-ridden cities, which serves to start the oxidation of hydrocarbons, is due to decomposition of the HONO which had been created the previous evening.

$$OH^{\cdot} + NO_2^{\cdot} \longrightarrow \underset{\substack{\text{nitrous} \\ \text{acid}}}{HONO} \xrightarrow{\text{sunlight}} OH^{\cdot} + NO^{\cdot}$$

It is a characteristic of the later stages in the day of a smog episode that oxidizing agents such as nitric acid and ozone are formed in substantial quantities (see Figure 3-9). The reaction of two $OH^{\cdot}$ radicals, or of two hydroperoxy radicals $HOO^{\cdot}$, produces another atmospheric oxidizing agent, hydrogen peroxide, H_2O_2:

$$2\,OH^{\cdot} \longrightarrow \underset{\text{hydrogen peroxide}}{H_2O_2}$$

$$2\,HOO^{\cdot} \longrightarrow H_2O_2 + O_2$$

The latter reaction occurs also in clean air when the concentration of NO_x is especially low, and was encountered in stratospheric chemistry in Chapter 2.

The combination of $R\!-\!\overset{\cdot}{C}\!=\!O$ radicals, produced by H atom abstraction by $OH^{\cdot}$ from aldehydes in the ways discussed above, occurs in the presence of O_2 to produce the following free radical:

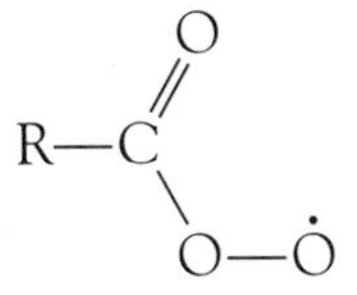

$$R\!-\!\overset{\displaystyle O}{\underset{\displaystyle O\!-\!\dot{O}}{\overset{\displaystyle \|}{C}}}$$

When NO˙ is plentiful, this complex species as expected behaves as a peroxy radical and oxidizes nitric oxide. In the afternoon, when the concentration of NO˙ is very low, the radical reacts instead in a radical-radical process by *adding* to NO_2˙ to yield a nitrate. For the common case for which the alkyl group, R, is CH_3, the nitrate product formed is *peroxyacetylnitrate*, or PAN for short, which is a potent eye irritant in humans and is also toxic to plants.

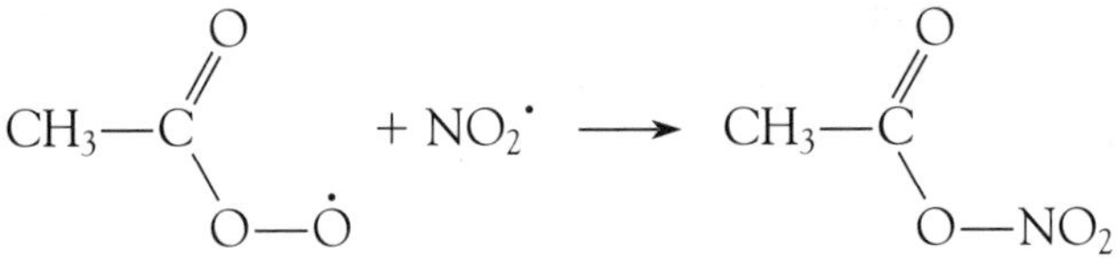

$$CH_3-\overset{\displaystyle O}{\overset{\|}{C}}\diagdown_{O-\dot{O}} \quad + NO_2\text{˙} \quad \longrightarrow \quad CH_3-\overset{\displaystyle O}{\overset{\|}{C}}\diagdown_{O-NO_2}$$
$$\text{PAN}$$

Overall, then, the afternoon stage of a photochemical smog episode is characterized by a build-up of oxidizing agents such as ozone, hydrogen peroxide, nitric acid, and PAN.

Another important species that is present in the later stages of smog episodes is the nitrate radical, NO_3˙, produced when high concentrations of NO_2˙ and ozone occur simultaneously:

$$NO_2\text{˙} + O_3 \longrightarrow NO_3\text{˙} + O_2$$

Although NO_3˙ is photochemically dissociated to NO_2˙ and O rapidly during the daytime, it is stable at night and plays a role similar to OH˙ in attacking hydrocarbons in the hours following sundown:

$$NO_3\text{˙} + RH \longrightarrow HNO_3 + R\text{˙}$$
$$\text{nitric acid}$$

Thus at night NO_3˙, rather than OH˙, initiates the oxidation of reduced gases in the troposphere. The similarity between OH˙ and NO_3˙ is not surprising since both react as —O˙ radicals and form very stable O—H bonds when they abstract hydrogens. The nitrate radical also combines with NO_2˙ to yield N_2O_5, which subsequently adds a water molecule to yield two molecules of nitric acid.

$$NO_2\text{˙} + NO_3\text{˙} \longrightarrow N_2O_5$$
$$\text{dinitrogen pentoxide}$$

$$N_2O_5 + H_2O \longrightarrow 2\ HNO_3$$

PROBLEM 3-19

Some formaldehyde molecules photochemically decompose to the molecular products H_2 and CO rather than to free radicals. Deduce the mechanism and overall reaction for the oxidation to CO_2 for formaldehyde molecules that initially produce the molecular products.

PROBLEM 3-20

Deduce the series of steps by which ethylene gas, $H_2C{=}CH_2$, is oxidized to CO_2 when it is released into an atmosphere undergoing a photochemical smog process. (Assume in this case that aldehydes react completely by photochemical decomposition rather than by $OH^{\bullet}$ attack.)

PROBLEM 3-21

What reaction, of all those discussed in this chapter to this point, might result in a reduction rather than an increase in the level of daytime ozone when additional NO_x is added to air? Would you expect the ozone-reducing reaction to be dominant when the concentration of $HOO^{\bullet}$ is particularly (a) high or (b) low?

OXIDATION OF ATMOSPHERIC SO₂:
THE HOMOGENEOUS GAS-PHASE MECHANISM

When the sky is clear or when clouds occupy only a few percent of the tropospheric volume, the predominant mechanism for the conversion of SO_2 to H_2SO_4 is a homogeneous gas-phase reaction; the mechanism for the conversion consists of several sequential steps. As usual for atmospheric trace gases, it is the hydroxyl radical that initiates the process. Since SO_2 contains multiple bonds but no hydrogen, then it is expected (see Figure 3-6a) that the $OH^{\bullet}$ will *add* to the molecule at the sulfur atom:

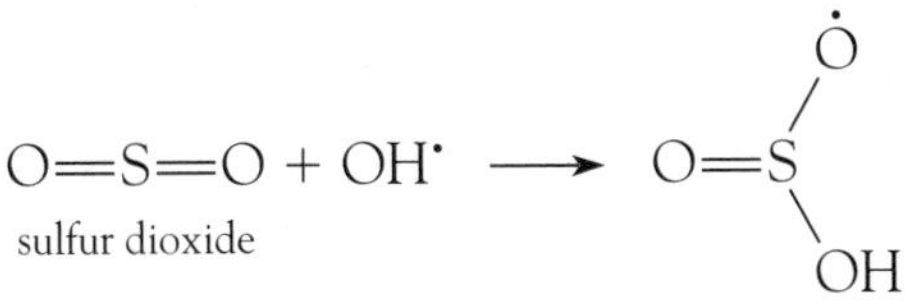

$$O{=}S{=}O + OH^{\bullet} \longrightarrow \overset{\displaystyle \overset{\bullet}{O}}{\underset{\displaystyle OH}{O{=}S}}$$

sulfur dioxide

Since a stable molecule, namely sulfur trioxide, SO_3, can be produced from this radical by the removal of the hydrogen atom, the reaction principles with which we have become familiar predict that the next reaction in the sequence is that between the radical and an O_2 molecule to abstract H:

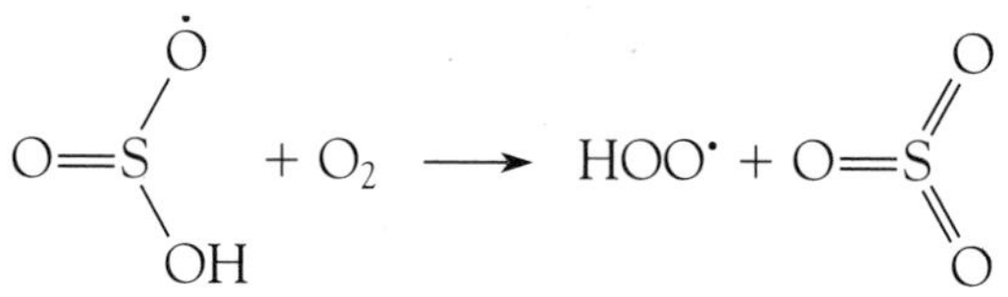

sulfur trioxide

The sulfur trioxide molecule rapidly combines with a gaseous water molecule to form sulfuric acid. Finally, the H_2SO_4 molecules react with water, whether in the form of water vapor or as a mist, to form an aerosol of droplets, each of which is an aqueous solution of sulfuric acid. The sequence of steps from gaseous SO_2 to aqueous H_2SO_4 is summarized below:

$$SO_2 + OH^\bullet \longrightarrow HSO_3^\bullet$$

$$HSO_3^\bullet + O_2 \longrightarrow SO_3 + HOO^\bullet$$

$$SO_3 + H_2O \longrightarrow H_2SO_4(g)$$

$$H_2SO_4(g) \xrightarrow{\;H_2O\;} H_2SO_4(aq)$$

When we include the return of $HOO^\bullet$ to $OH^\bullet$ via reaction with $NO^\bullet$, the overall reaction is seen to be $OH^\bullet$-catalyzed co-oxidation of SO_2 and $NO^\bullet$:

$$SO_2 + NO^\bullet + O_2 \xrightarrow{\;H_2O\;} NO_2^\bullet + H_2SO_4(aq)$$

For representative concentrations of the $OH^\bullet$ radical in relatively clean air, a few percent of atmospheric SO_2 is oxidized per hour by this mechanism. The rate is much faster for air masses undergoing photochemical smog reactions since the concentration of $OH^\bullet$ there is much higher. However, generally only a minority of sulfur dioxide is oxidized in air; the rest is removed by dry deposition before the reaction has time to occur.

AQUEOUS-PHASE OXIDATION OF SULFUR DIOXIDE

Since sulfur dioxide is somewhat soluble in water, a fraction of the atmospheric SO_2 exists in dissolved aqueous form if there is a significant cloud, fog, or mist content to the air. Under these circumstances, most of its oxidation to sulfuric acid occurs in the liquid phase (an aqueous solution) rather than in the gas phase, since the process is inherently faster in aqueous solutions.

Some of the sulfur dioxide dissolved in the aqueous phase occurs as sulfurous acid, H_2SO_3; the process by which the gas dissolves to yield this acid is

$$SO_2(g) + H_2O(aq) \rightleftharpoons H_2SO_3(aq)$$
$$\text{sulfur} \qquad\qquad\qquad \text{sulfurous}$$
$$\text{dioxide} \qquad\qquad\qquad \text{acid}$$

The relative concentrations of SO_2 and of H_2SO_3 are related by the equilibrium constant for this reaction. When gases are dissolved in liquids, the equilibrium constant is expressed as the Henry's Law constant K_H which equals the *equilibrium* molar concentration of the substance in the liquid phase divided by its partial pressure in the gas phase; for the reaction above, K_H is therefore equal to

$$K_H = [H_2SO_3(aq)]/P$$

Here the quantity P is the atmospheric partial pressure of gaseous SO_2. Typically the concentration of gaseous SO_2 is about 0.1 ppm, which at 1 atmosphere total air pressure corresponds to a partial pressure of 0.1 $\times 10^{-6}$ atm; that is, $P = 10^{-7}$ atm. Since at 25°C, $K_H = 1$ M atm^{-1}, we can write

$$[H_2SO_3(aq)] = PK_H = 1 \times 10^{-7} M$$

This value of about 10^{-7} M for the equilibrium concentration of $H_2SO_3(aq)$ is deceptive since it by no means represents all the dissolved sulfur dioxide. Although it is a weak acid, H_2SO_3 has a K_a which is sufficiently large (1.7×10^{-2}) that in dilute solutions—such as are encountered in atmospheric droplets—most of the sulfurous acid that forms initially subsequently ionizes to bisulfite (or hydrogen sulfite) ion, HSO_3^-:

$$H_2SO_3 \rightleftharpoons HSO_3^- + H^+$$
$$\text{sulfurous} \qquad\qquad \text{bisulfite}$$
$$\text{acid} \qquad\qquad\qquad \text{ion}$$

Thus $[H_2SO_3]$ represents the concentration only of the H_2SO_3 that does *not* ionize; its value is held fixed at 10^{-7} M owing to the equilibrium with gaseous SO_2.

If we assume that the above reaction is the only significant source of acidity, then from its stoichiometry it follows that $[HSO_3^-] = [H^+]$, and thus the K_a expression

$$1.7 \times 10^{-2} = [HSO_3^-][H^+]/[H_2SO_3]$$

becomes

$$1.7 \times 10^{-2} = [HSO_3^-]^2/10^{-7}$$

After solving this equation, it is found that

$$[HSO_3^-] = 4 \times 10^{-5} \text{ M}.$$

Thus the ratio of bisulfite ion to sulfurous acid is 400:1 and the total dissolved sulfur concentration is about 4×10^{-5} M, rather than just the 1×10^{-7} M which represents only the contribution from the unionized acid. Since the concentration of hydrogen ion produced by the reaction also is 4×10^{-5}, the pH of the raindrops is 4.4. Rain does not become highly acidic if only weak acids are dissolved in it.

PROBLEM 3-22

Confirm by calculation that the pH of CO_2-saturated water at 25°C is 5.6, given that the CO_2 concentration in air is 350 ppm, i.e., 0.00035 atm, and that for carbon dioxide the Henry's Law constant $K_H = 3.4 \times 10^{-2}$ mol L^{-1} atm^{-1} at 25°C. Furthermore, the ionization constant, K_a, for H_2CO_3 has a value of 4.5×10^{-7} mol L^{-1} at this temperature. [Hint: use the same techniques previously employed to calculate the pH of sulfur dioxide dissolved in water.] Given that at 15°C, $K_H = 0.047$ mol L^{-1} atm^{-1} and $K_a = 3.8 \times 10^{-7}$, calculate the pH of natural rainwater at this temperature.

PROBLEM 3-23

Calculate the pH of rainwater in equilibrium with SO_2 in a polluted air mass for which the sulfur dioxide concentration is 1.0 ppm.

PROBLEM 3-24

Calculate the concentration of SO_2 that must be reached in polluted air if the dissolved gas is to produce a pH of 4.0 in raindrops without any oxidation of the gas.

If strong acids are already present in the water droplet, they, rather than sulfurous acid, control its pH. Under these conditions, the bisulfite ion concentration can be calculated through rearrangement of the K_a expression to give

$$[HSO_3{}^-] = 1.7 \times 10^{-2} \times 10^{-7} / [H^+]$$
$$= 1.7 \times 10^{-9} / [H^+]$$

Here the H^+ concentration is overwhelmingly due to the release of H^+ from the strong acids, since the additional amount contributed from the H_2SO_3 ionization will be negligible in comparison. Thus the bisulfite ion concentration, $[HSO_3{}^-]$, is inversely proportional to $[H^+]$. At a pH of 2, which is about the lowest pH found in acid rain, the bisulfite concentration is calculated to be only about 10^{-7} M, which is comparable to that of the unionized acid, H_2SO_3; at higher pHs the bisulfite form dominates. The reduction in bisulfite concentration with increasing acidity is consistent with the application of Le Chatelier's Principle to the H_2SO_3 dissociation reaction; that is, the concentrations in the reaction previously at equilibrium shift so as to reduce the influence of the added H^+ from the strong acids in establishing the new position of equilibrium.

Dissolved sulfur dioxide, SO_2, is oxidized to sulfate ion, $SO_4{}^{2-}$, by trace amounts of the well-known oxidizing agents hydrogen peroxide, H_2O_2, and ozone, O_3, that are present in the airborne droplets. Indeed, these reactions currently are thought to constitute the *main* oxidation pathways for atmospheric SO_2, except under clear sky conditions when the gas-phase, homogeneous mechanism predominates. The ozone and hydrogen peroxide result mainly from sunlight-induced reactions in photochemical smog; thus SO_2 oxidation occurs most rapidly in air that has also been polluted by reactive hydrocarbons and nitrogen oxides. Since the smog reactions occur predominantly in summer, rapid oxidation of SO_2 to sulfate also is characteristic of the summer season.

Hydrogen peroxide is highly soluble in water, since it readily forms hydrogen bonds with water molecules. Its Henry's Law constant K_H is

7.4×10^4 M atm^{-1}. Since its atmospheric gas-phase concentration is 1 ppb, its partial pressure P' is 1×10^{-9} atm. Further, since

$$K_H = [H_2O_2(aq)] / P'$$

then
$$[H_2O_2(aq)] = P' K_H$$
$$= 1.0 \times 10^{-9} \text{ atm} \times 7.4 \times 10^4 \text{ M atm}^{-1}$$
$$= 7 \times 10^{-5} \text{ M}$$

Although this value of 7×10^{-5} M—70 μM, when expressed in micromolar units—seems tiny by comparison to the concentration values usually encountered in laboratory solutions, it is sufficient to oxidize dissolved atmospheric sulfur dioxide at an appreciable rate. Measured concentrations of H_2O_2 in clouds and rainwater range from about 0.01 μM to more than 200 μM. Because gas-phase concentrations are highest in the afternoons and in summer, and lowest at night and in the winter, one may conclude that sunlight is a dominant factor in determining the H_2O_2 concentration.

Although dissolved hydrogen peroxide oxidizes dissolved sulfur dioxide by attacking all three of the species H_2SO_3, HSO_3^-, and SO_3^{2-}, the fastest and dominant reaction occurs with the bisulfite ion:

$$H_2O_2 + HSO_3^- \longrightarrow H_2O + HSO_4^- \qquad \text{overall}$$

hydrogen   bisulfite   bisulfate
peroxide   ion   ion

This overall reaction is acid-catalyzed, so the attacking species presumably is the $H_3O_2^+$ ion

$$H_2O_2 + H^+ \rightleftharpoons HO\!-\!OH_2^+$$

Given that O—O bonds are weak, cleavage of the O—O single bond should occur without much difficulty, thereby producing an OH$^+$ ion and water molecule. Indeed, the observed reaction with bisulfite ion consists of the transfer of OH$^+$ from $H_3O_2^+$ to HSO_3^-; the oxygen bonds to the sulfur, resulting in the formation of an H_2SO_4 molecule and a water molecule, as illustrated in Figure 3-11.

As proven by the analysis in Box 3-2, the rate of SO_2 oxidation by hydrogen peroxide is independent of pH; although an increase in the H$^+$ concentration does increase the amount of protonated peroxide, it also decreases the concentration of the bisulfite ion, and thus there is no *net* effect on the rate.

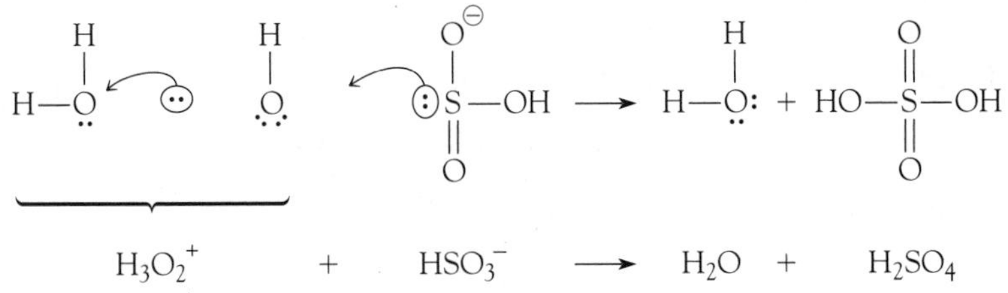

FIGURE 3-11

Oxidation of HSO_3^- to H_2SO_4 by $H_3O_2^+$. Curved arrows indicate formal movement of electron pairs during the reaction.

In contrast, the oxidation of bisulfite by *ozone* is *not* catalyzed by hydrogen ion since O_3 molecules do not readily add H^+; thus the reaction of interest is simply the transfer of an oxygen atom from O_3 to bisulfite:

$$HSO_3^- + O_3 \longrightarrow HSO_4^- + O_2 \quad \text{(overall reaction)}$$

Thus the net effect of increasing H^+ is to reduce the rate of the overall reaction, since the equilibrium concentration of HSO_3^- is thereby decreased by the effect of H^+ on the H_2SO_3 dissociation reaction. For these reasons, oxidation of bisulfite in rather acidic droplets (pH < 5) proceeds mainly by H_2O_2 whereas only in less acidic droplets does oxidation occur mainly by ozone. Since oxidation of HSO_3^- by ozone is faster than that of H_2SO_3 or of SO_3^{2-}, the oxidation by ozone of dissolved SO_2 is "turned off" once the water droplet becomes fairly acidic.

The uncatalyzed oxidation of dissolved SO_2 by molecular oxygen, O_2, is thought to proceed at too slow a rate to make a significant contribution to the overall speed of the reaction. However if catalysts such as ions of iron or other metals are present in the droplets (due to their release into air from pollution sources or even just from dust), this generalization may not be valid.

PROBLEM 3-25

Prove that for the oxidation of dissolved SO_2 in atmospheric water droplets that occurs by reaction of HSO_3^- with ozone, the rate is proportional to the concentration of the ozone times the partial pressure, P, of SO_2, and is inversely proportional to $[H^+]$.

3-2   # THE RATE LAW FOR
HETEROGENEOUS SO_2 OXIDATION

Since the reactants in the slow step of the oxidation mechanism are $H_3O_2^+$ and HSO_3^-, it follows from the Collision Theory of Reaction Rates that the rate of the reaction is proportional to the product of their concentrations (since the collision rate between them is also proportional to this product):

$$rate = k \, [H_3O_2^+] \, [HSO_3^-]$$

The concentration of the positive ion is determined by its equilibrium with H_2O_2 and H^+ (see main text) for which

$$K = [H_3O_2^+] / [H_2O_2] \, [H^+]$$

so by rearranging we obtain

$$[H_3O_2^+] = K \, [H_2O_2] \, [H^+]$$

Substituting this into the previous expression gives the rate of oxidation as

$$rate = k \, K \, [H^+] \, [H_2O_2] \, [HSO_3^-]$$

The concentration of HSO_3^- itself is *inversely* proportional to that of H^+ since the former is controlled by the equilibrium of the H_2SO_3 dissociation process (see main text). From the expression for K_a for this dissociation, we obtain

$$[HSO_3^-] = K_a \, [H_2SO_3] / [H^+]$$

which upon substitution in the rate equation yields

$$rate = k \, K \, K_a \, [H^+] \, [H_2O_2] \, [H_2SO_3] / [H^+]$$
$$= k \, K \, K_a \, [H_2O_2] \, [H_2SO_3]$$

Since $[H_2SO_3] = P K_H$ from Henry's Law, we obtain as our final expression

$$rate = k \, K \, K_a \, K_H \, P \, [H_2O_2]$$

PROBLEM 3-26

Assuming that the rate of oxidation by hydrogen peroxide is always proportional to $[H_3O_2^+]$, establish the dependence on $[H^+]$ of the oxidation of H_2SO_3. Repeat your analysis for oxidation of sulfite ion, SO_3^{2-}.

RADIOACTIVITY: AIR POLLUTION FROM RADON GAS

Radon gas is a substance that occurs naturally, but which is an environmental hazard when it becomes concentrated in the air of enclosed spaces such as houses. The production, accumulation, and health effects of radon are explained below, but can be understood only after a discussion of some fundamental properties of radioactivity.

PRINCIPLES OF RADIOACTIVITY

Although most types of atomic nuclei are stable indefinitely, some are not. The unstable, or **radioactive,** ones decompose by emitting a small particle that usually carries a great deal of energy. In many such processes, atoms are converted from those of one element to those of another as a result of this emission. Very heavy elements are particularly prone to this type of decomposition, the particles emitted in these cases usually being either "alpha" or "beta" in character.

An **alpha (α) particle** is a radioactively emitted particle that has a charge of +2 and a mass number of 4—it has 2 neutrons in addition to two protons—and happens to be identical to a common helium nucleus. Thus, an alpha particle is written as $^4_2$He, where 4 is its mass number, and 2 refers to its nuclear charge (i.e., number of protons). The nucleus that remains behind after an atom has lost an alpha particle has a nuclear charge that is 2 units smaller than the original, and it is 4 units lighter. Thus, for example, when a $^{226}_{88}$Ra (radium-226) nucleus emits an α particle, the resulting nucleus has a mass of $226 - 4 = 222$ units and a nuclear charge of $88 - 2 = 86$; this is a wholly new element that is an isotope of the element radon. The process can be written as a "nuclear reaction":

$$^{226}_{88}\text{Ra} \longrightarrow {}^{222}_{86}\text{Rn} + {}^4_2\text{He}$$

Notice that both the total mass number and the total nuclear charge individually balance in such equations.

A **beta (β) particle** is an electron; it is formed when a neutron splits into a proton and an electron in the nucleus. Since the proton remains behind in the nucleus when the electron leaves it, the nuclear charge (or *atomic number*) *increases* by one unit (you may imagine this effect as "subtracting a negative particle"). There is no change in mass number for the nucleus, since the total number of neutrons + protons remain the same. (Recall that *mass number* means *number* of heavy particles—protons and neutrons—and *not* the actual mass of the nucleus.) Thus, for example, when an atom of lead of the isotope $^{214}_{82}Pb$ (lead-214) decays radioactively by the emission of a β particle, the nuclear charge of the product is $82 + 1 = 83$, corresponding to the element bismuth, and the mass number remains 214:

$$^{214}_{82}Pb \longrightarrow \ ^{214}_{83}Bi + \ ^{0}_{-1}e$$

Notice that the symbol $^{0}_{-1}e$ for the electron here shows its mass number (zero) and its charge so that the equation can readily be seen to be balanced by considering the subscripts and superscripts.

One other type of radioactivity occurs when a nucleus emits a **gamma (γ)** "particle," which is a huge amount of energy concentrated in one photon and possesses no mass as a particle. Neither the nuclear mass number nor the nuclear charge changes when a γ particle is emitted.

PROBLEM 3-27

Deduce the nature of the species whose identity is disguised by a blank for each of the following nuclear reactions:

a. $^{222}_{86}Rn \longrightarrow \ ^{4}_{2}He +$ _____

b. $^{214}_{83}Bi \longrightarrow \ \beta +$ _____

c. $^{214}Po \longrightarrow \ ^{214}Pb +$ _____

d. _____ $\longrightarrow \ ^{234}_{90}Th + \alpha$

The radioactive decomposition of the atoms in a sample of an isotope does not occur all at once. For example, in a sample of uranium-238, ^{238}U, just large enough to be visible, there are about 10^{20} atoms; only about 10,000,000—10^7—of them decompose in a given second, so it requires billions of years for the decomposition process to be complete for the sample as a whole. A measure often used to express such decomposition rates is the time period required for half the nuclei in a sample to disintegrate, or its **half-life,** $t_{1/2}$. For example, the half-life of uranium-238 is about 4.5 billion years. Thus about half of this type of uranium existing when the Earth was formed (about 4.5 billion years ago) has now disintegrated; half the *remaining* uranium-238, amounting to one-quarter of the original, will disintegrate in the next 4.5 billion years, leaving one-quarter of the original still intact. After three half-lives have passed, only one-eighth of the original will remain, and only one-sixteenth is still there after four half-lives. This type of decay of the members of a population (nuclei, in this case) is *exponential*; it is in contrast to *linear* decay (a simple process not relevant to radioactivity) in which if 50% of a sample disappears in x years, all of it is gone at the end of 2x years.

RADON

Many rocks and granite soils contain uranium, and so the radioactive decomposition process takes place under our feet each day, with radon gas being one of its unwelcome products. Each $^{238}_{92}U$ nucleus eventually emits an α particle, and thereby an atom of the thorium isotope $^{234}_{90}Th$ is formed:

$$^{238}_{92}U \longrightarrow {}^{234}_{90}Th + {}^{4}_{2}He$$

This process is the first of 14 sequential radioactive decay processes that such a nucleus undergoes; the last of these reactions produces $^{206}_{82}Pb$, which is a nonradioactive (stable) isotope of lead that therefore stops the sequence.

Of particular interest is that portion of the 14-step sequence of ^{238}U radioactive decay which involves radon, since this element is the only one, other than the helium from the alpha particles, that is gaseous and which therefore is mobile. The immediate precursor of the radon is radium-226, which has a half-life of 1,600 years and which decays by emission of an α particle:

$$^{226}_{88}Ra \longrightarrow {}^{222}_{86}Rn + {}^{4}_{2}He$$

The radon isotope has a half-life of 3.8 days—time enough to diffuse through the solid rock or soil in which it is initially formed. Most radon escapes directly into outdoor air since the surface of the Earth at which it appears is not covered by a building. The very small background concentration of radon in air nevertheless yields about half our exposure to radioactivity, the other half arising from small contributions from a variety of sources such as gamma rays from outer space.

Most radon that seeps into homes comes from the top meter of soil below and around the foundation; radon produced much deeper than this will probably decay to a nongaseous and therefore immobile element before it reaches the surface. Loose, sandy soil allows the maximum diffusion of radon gas, whereas frozen, compacted, or clay soil inhibits its flow. Radon enters basements of homes through holes and cracks in their concrete foundations. The intake is increased significantly if the air pressure in the basement is low. The material used to construct the homes, and water from artesian wells, are other potential sources of radon in homes. For example, when well water is heated and exposed to air, as occurs when it exits from a shower head, radon is released to the air.

Some scientists have pointed out that radon gas accumulates to unhealthy levels in caves, including some that are often used for recreational purposes.

Radon, the heaviest member of the noble gas group, is chemically inert under ambient conditions and remains a monatomic gas. As such, it becomes part of the air that we breathe once it enters our homes. Because of its inertness, physical state, and low solubility in body fluids, radon *itself* does not pose much of a danger to us; the chance that it will disintegrate during the short time it is present in our lungs is small, and the range of α particles in air before they lose most of their energy is less than 10 cm. The danger arises from the radioactivity arising from the three elements produced in sequence by the disintegration of radon—namely polonium, lead, and bismuth; such "descendants" are termed **daughters** of radon. In macroscopic amounts, these elements are solids, and when formed in air from radon they all quickly adhere to dust particles. Some dust particles adhere to lung surfaces when inhaled, and it is in this condition that these elements pose a health threat. In particular, both the ^{218}Po which is formed directly from ^{226}Rn, and the ^{214}Po that is formed later in the sequence illustrated in Figure 3-12, emit energetic α particles that can cause radiation damage to the bronchial cells near which the dust particles reside. This damage can eventually lead to lung cancer, and indeed "radon" (or rather its daughters) is the second-leading cause of such cancers, though it follows smoking by a wide margin.

Notice that the sequence (see Figure 3-12) following radon decay to ^{210}Pb formation takes less than a day on average to occur; in contrast, disintegration of this lead to bismuth has a half-life of 22 years, and in fact most of the lead will have been cleared from the body before this occurs.

Although some radon daughters in the above sequence disintegrate by β particle emission, the deleterious health effects of such particles are considered to be negligible because the α particles carry much more energy than do the β, and it is the disruption of cell molecules by the burst of high energy that initiates cancer. (The energy breaks bonds within the molecules.)

The greatest exposure to α particles from radon disintegration is experienced by those miners who work in poorly ventilated underground uranium mines. Their rate of lung cancer is indeed higher than that of the general public, even after corrections have been made for smoking. From statistical data relating their excess incidence of lung cancer to their cumulative level of exposure to radiation, a mathematical relationship between cancer incidence and radon exposure has been developed. Scientists have attempted to extrapolate this relationship in order to determine the risk to the general population from the generally lower levels of radon to which the public is exposed. Based upon linear extrapolation, one recent estimate concluded that the increased chance of contracting lung cancer is 1 in 250, and therefore that about 10,000 lung cancer deaths per year in the United States are due to this cause. However this estimate may be too high, since miners work in much dustier conditions than are found in homes and their breathing during hard labor is much deeper than normal; consequently there is much more chance that radon daughters will find their way deep into the lungs of miners compared to those of us in the general population.

Two epidemiological studies published in 1994, one from Sweden and the other from Canada, reach contradictory conclusions about the risk of radon to householders. In the Swedish report, the rate of lung cancer in nonsmokers and especially in smokers was found to increase with increasing levels of radon in their homes. The Canadian study focused upon residents of Winnipeg, Manitoba, which has the highest

$$^{226}_{88}\text{Ra} \xrightarrow[1600\ y]{-\alpha} {}^{222}_{86}\text{Rn} \xrightarrow[3.8\ d]{-\alpha} {}^{214}_{82}\text{Po} \xrightarrow[3\ m]{-\alpha} {}^{214}_{82}\text{Pb}$$

$$\xrightarrow[27\ m]{-\beta} {}^{214}_{83}\text{Bi} \xrightarrow[20\ m]{-\beta} {}^{214}_{84}\text{Po} \xrightarrow[<1\ s]{-\alpha} {}^{210}_{82}\text{Pb} \xrightarrow[22\ y]{-\beta} {}^{210}_{83}\text{Bi}$$

FIGURE 3-12

The radium-radon radioactivity series. An "$-\alpha$" above an arrow signifies (for example) the emission of an alpha (α) particle. The time period indicated below an arrow is the half-life of the unstable isotope.

average radon levels in Canada; no linkage between radon levels and lung cancer incidence was found.

The "radon problem" has received greatest attention in the United States, where currently there are programs in place to test the air in the basements of a large number of homes for significantly elevated levels of radon. Once radon is identified, the owners can then alter the air circulation patterns in these dwellings to reduce future radon levels in living areas, and thereby reduce the additional risk to the residents of contracting lung cancer.

REVIEW QUESTIONS

1. In the "micrograms per cubic meter" concentration scale, to what substances do micrograms and cubic meters refer?

2. In general terms, what is meant by "photochemical smog"? What are the initial reactants in the process? Why is sunlight required?

3. What is meant by a "primary" pollutant and by a "secondary" pollutant? Give examples.

4. What is the chemical reaction by which most atmospheric $NO^{\bullet}$ initially is produced? From which two sources do most urban $NO^{\bullet}$ arise?

5. Describe the strategies by which reduction of urban ozone levels have been attempted. What difficulties have been encountered in these efforts?

6. Describe the processes by which the "three-way catalyst" works to transform air pollutants released by an automobile engine.

7. What is "acid rain"? What two acids predominate in it?

8. What are the main anthropogenic sources of sulfur dioxide? Describe the strategies by which these emissions can be reduced.

9. What species are included in the measure "total reduced sulfur"?

10. Explain why the predominant acid in acid rain differs in eastern and western North America.

11. Using chemical equations, explain how acid rain is neutralized by limestone.

12. Describe the effects of acid precipitation upon a. dissolved levels of aluminum, b. fish populations, and c. trees.

13. Write a balanced equation illustrating the reactions that occur between one molecule of ammonia with a. one molecule of nitric acid, and b. with one molecule of sulfuric acid.

14. Define the term *aerosol*, and differentiate between "coarse" and "fine" particulates. What are the usual origins of these three types of atmospheric particles?

15. List the reasons why coarse particles usually are of less danger to human health than are fine particles.

16. What are the usual concentration units for suspended particulates? What would the designation PM_{40} mean?

17. Discuss the relationship between atmospheric particulates and haze.

18. Describe the major health effects of air pollutants.

19. What are the main sources of formaldehyde in indoor air? What are its effects?

20. What are the main sources of nitrogen dioxide and of carbon monoxide in indoor air?

21. What are the two forms of asbestos? Why is asbestos of environmental concern?

22. What are the names and formulas for six of the gases that are released into our atmosphere from biological or volcanic sources? What chemical species initiates their oxidation?

23. What is the two-step mechanism by which the hydroxyl free radical is produced in clean air?

24. Explain why $OH^{\cdot}$ reacts more quickly to abstract hydrogen from other molecules than does $HOO^{\cdot}$.

25. What are the two different initial steps by which atmospheric formaldehyde, H_2CO, is decomposed in air?

26. What is the fate of $OH^{\cdot}$ radicals that react with $NO^{\cdot}$? with $NO_2^{\cdot}$? with other $OH^{\cdot}$?

27. What is the fate of $NO_2^{\cdot}$ molecules that photodissociate? that react with ozone? that react with $RC(O)OO^{\cdot}$ radicals?

28. Why does the production of high concentrations of $NO_2^{\cdot}$ lead to an increase in ozone levels in air? Why does this not occur if much $NO^{\cdot}$ is present?

29. Describe both the homogeneous and the heterogeneous mechanisms by which sulfur dioxide is oxidized in the atmosphere. What are the oxidizing agents? Are any of the processes affected by pH?

30. What is the nature of the radioactive emissions α and β particles? Why are the former more dangerous to health than the latter?

31. What is meant by the term *half-life*?

32. Explain the origin of radon in buildings.

33. Explain why the "daughters" of radon are more dangerous to health than is radon itself.

SUGGESTIONS FOR

FURTHER READING

1. Smet, J. M., and J. D. Spengler, eds. 1991. *Indoor Air Pollution: A Health Perspective*. Baltimore, MD: Johns Hopkins University Press.

2. Selinger, B. 1986. *Chemistry in the Marketplace*, 3rd ed. Sydney, Australia: Harcourt Brace Jovanovich.

3. Chameides, W. L., and D. D. Davis. 1982. Chemistry in the troposphere. *Chemical and Engineering News* Oct. 4, 1982: 38–52.

4. Pearce, F. 1987. Acid Rain. *New Scientist* Nov. 5, 1987 (supplement): 1–4.

5. Lippmann, M. 1991. Health effects of tropospheric ozone. *Environmental Science and*

Technology 25 (12): 1954–1961.

6. Finlayson-Pitts, B. J., and J. N. Pitts, Jr. 1986. *Atmospheric Chemistry*. New York: Wiley. [A comprehensive guide to the detailed chemistry of the atmosphere]

7. Lents, J. M., and W. J. Kelly. 1993. Clearing the air in Los Angeles. *Scientific American* October 1993: 32–39.

8. Pearce, F. 1990. Whatever happened to acid rain? *New Scientist* Sept. 15, 1990: 57–60

9. Calvert, J. G., et al. 1985. Chemical mechanisms of acid generation in the troposphere, *Nature*, 317: 27-35.

10. Schulze, E. -D. 1989. Air pollution and forest decline in a spruce forest, *Science*, 244: 776–783.

11. Anon. 1994. Air pollution in the world's megacities. *Environment*, 36(2); 4–13.

12. Smith, W. H. 1991. Air pollution and forest damage. *Chemical and Engineering News*, Nov. 11: 30.

CHAPTER 4

THE GREENHOUSE EFFECT AND GLOBAL WARMING

By now, everyone has heard the prediction that the "greenhouse effect" will affect climates around the world in the twenty-first century and beyond. The term "greenhouse effect" in ordinary usage simply means that average air temperatures will increase by several degrees as a result of the buildup of carbon dioxide and other "greenhouse" gases in the atmosphere. Indeed, many scientists believe that such global warming has been under way already for some time, and is largely responsible for the temperature increase of about two-thirds of a degree Celsius that has occurred since 1860.

Some of us, particularly those who currently suffer through severe winters each year, may look forward to a warmer climate. After all, in the eleventh and twelfth centuries, an increase of only 0.5°C beyond today's average worldwide temperature was sufficient to allow farming on the coast of Greenland, for vineyards to flourish extensively in England, and for the Vikings to travel the North Atlantic and settle in Newfoundland. However, the climate changes predicted for the twenty-first century do not present such a pleasant prospect.

According to computer projections made in 1992 by the Intergovernmental Panel on Climate Change (IPCC), a group sponsored by the United Nations Environmental Plan, if no additional steps are taken to reduce emissions of CO_2 and the other problematic gases, then by about the year 2040 the average global air temperature will be 1°C higher than at present. By 2100 it will increase by still another 1.5°C. The total amount of global rainfall will increase, although there will be regions that will receive less rainfall than before these changes took place. Annually, the number of days having intense rain showers or very high temperatures both will increase. Sea levels will rise by about 18 centimeters by 2040 and by 48 centimeters by 2100, effects due mainly to the thermal expansion of seawater and the melting of glaciers. Although this increase in sea level may seem small, there are countries such as Bangladesh in which much of the population currently lives on land that would be flooded by a rise in sea level of only about 50 centimeters. Temperature and moisture changes will occur quickly compared to those that have taken place in the past, and consequently some ecosystems will be destabilized. However, in contrast to some fears expressed in the 1980s, it now seems unlikely that temperature increases in the twenty-first century will be enough to cause melting of the Antarctic and Greenland ice sheets, both of which sit above sea level on land and whose transfer by the breaking off of icebergs into the oceans would cause a major increase in sea levels.

Although scientists who work with hypothetical models of world climate feel fairly confident about their predictions for the *average* extent of global warming that will result from an unabated greenhouse effect, it is much harder to make specific, reliable predictions for individual regions. The temperature increase will probably be greater close to the polar regions. There may well be enough melting of ice in the Arctic region for the Northwest Passage to be used for commercial transport. In subtropical areas, the monsoon rains will likely be heavier. But in the midwest regions of the United States and the areas just north of them in Canada, the soil probably will become much less moist because of increased rates of evaporation in the warmer air; the continued suitability of this land for the growing of grain is open to question. There probably will be more extreme heat waves in summers and fewer prolonged cold snaps in winters. There will be longer frost-free growing seasons at northern latitudes, but increased chances that heat stress will affect crops grown there. The boreal forest of central Canada could be eliminated by fire by 2050; indeed the rate of fires in these woodlands is already climbing.

In his recent book *Global Warming: Are We Entering the Greenhouse Century?*, climatologist Stephen Schneider speculated on what the coming "greenhouse century" might look like to residents of the world. He asked will the sea level rise in New York, the lake level fall in Chicago, the river traffic dry up on the Mississippi, the bayous continue to disappear in Louisiana, the temperatures repeatedly sear the inhabitants of Dallas and Phoenix, smoke clouds fill the skies near Denver and San Francisco, and water shortages in the Rockies turn the lawns and swimming pools of Los Angeles brown and dry? Will northern Canada and Siberia become the new breadbaskets with shiploads of their grain plying ice-free waters of the now-frozen North? Will there be more frequent devastating floods in Indonesia, India or (as there were in 1988) Bangladesh and threats to the security of developing nations?

There may be both positive and negative effects associated with any significant increase in the average global temperature. Indeed, the phenomenon of rapid **global warming**—with its demands for such large-scale adjustments—is generally considered to be our most crucial worldwide environmental problem. Unlike stratospheric ozone depletion, which has manifested itself in spectacular fashion in the form of the ozone hole, the phenomenon of global warming due to the greenhouse effect has yet to be observed in convincing fashion. No one currently is sure of the extent or timing of future temperature increases, nor is it likely that reliable predictions for individual regions will ever be available much in advance of the events in question. If current models of the atmosphere are correct, however, significant global warming will occur in coming decades. Thus it is important that we understand the factors that are driving this increase in global temperatures so that we can take steps now to avoid potential catastrophes caused by rapid change in the future.

In this chapter, the mechanism by which global warming could arise is explained, and the nature and sources of the chemicals that are responsible for the effect are analyzed.

THE MECHANISM OF THE GREENHOUSE EFFECT

THE EARTH'S ENERGY BALANCE

The Earth's surface and atmosphere are kept warm primarily by energy from the Sun. Maximum solar output (see dashed portion of the curve in Figure 4-1) lies in the range of visible light—that of wavelengths

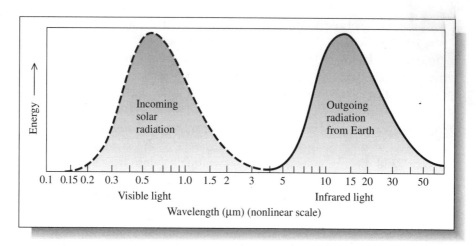

FIGURE 4-1

Wavelength distributions for light emitted by the Sun (dashed curve) and by the Earth's surface and troposphere (solid curve). (Source: J. Gribbin. Inside science: The greenhouse effect. *New Scientist*, supplement to Oct. 22, 1988 edition.)

between 0.40 and 0.75 μm (i.e., 400–750 nm). Visible light falls between the violet (400 nm) and red (750 nm) limits of sunlight. As explained in Chapter 2, much of the ultraviolet light (wavelengths < 0.4 μm) from the sun is filtered out in the stratosphere and warms the air there rather than at the surface of the Earth. Beyond the "red limit," that is, the maximum wavelength for visible light, we receive in sunlight some **infrared** (IR) light in the 0.8–4 μm region. Of the total incoming light covering all wavelengths that impinges upon the Earth, about 50% reaches the surface and is absorbed by it. A further 20% of the incoming light is absorbed by gases—UV by stratospheric ozone, and IR by CO_2 and H_2O—and by water droplets in air; the remaining 30% is reflected back into space by clouds, ice, snow, sand, and other reflecting bodies, without being absorbed.

Like any other warm body, the Earth emits energy; indeed the amount of energy that the planet absorbs and the amount that it releases must be equal if its temperature is to remain constant. The emitted energy (see the solid portion of the curve in Figure 4-1) is neither visible nor UV light but rather it is infrared light having wavelengths ranging from 4 to 50 μm; this is called the **thermal infrared** region since the energy is a form of heat, the same kind of heat energy a heated iron pot would radiate.

Some gases in air can temporarily absorb thermal infrared light of specific wavelengths, and so, not all the IR emitted from the Earth's surface and atmosphere escapes directly to space. Shortly after its absorption by airborne molecules such as CO_2, this infrared light is re-emitted in all directions—completely randomly. Thus some thermal IR is redirected back toward the Earth's surface, is reabsorbed, and consequently further heats both the surface and the air. This phenomenon, the redirection of thermal IR towards the Earth, as shown in Figure 4-2, is called the **greenhouse effect,** and is responsible for the average temperature at the Earth's surface being $+15°C$ rather than $-15°C$, the temperature it would be if there were no atmosphere. The very fact that our planet is not entirely covered by a thick sheet of ice is due to the natural operation of the greenhouse effect. The surface is as much warmed by this mechanism as it is by the solar energy it receives directly! The atmosphere operates in the same way as a blanket, retaining within the immediate region some of the heat released by a body and thereby increasing the local temperature.

The phenomenon that worries environmental scientists is that increasing the concentration of the trace gases in air that absorb thermal infrared light (piling on more "blankets," so to speak) would result in the redirection of even more of the outgoing thermal infrared energy and would thereby increase the average surface temperature beyond 15°C. This phenomenon will be referred to as the **enhanced greenhouse effect,** to distinguish its effects from the one that has been operating naturally for millennia.

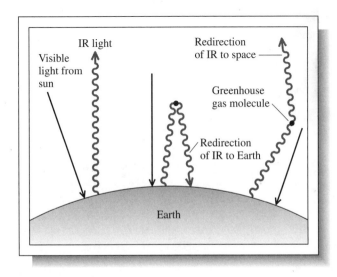

FIGURE 4-2

Schema of the operation of the greenhouse effect in Earth's troposphere.

The principal constituents of the atmosphere, N_2, O_2, and Ar, are *incapable* of absorbing infrared light—the reasons for this fact will be discussed in the following section. The atmospheric gases that in the past have produced most of the greenhouse warming are water (responsible for about two-thirds of the effect) and carbon dioxide (which is responsible for about one-quarter). Indeed, the absence of water in the dry air of desert areas leads to low nighttime temperatures even though the daytime temperatures are quite high on account of direct absorption of solar energy. More familiar to people living in temperate climates is the crisp chill in winter air on cloudless days and nights.

MOLECULAR VIBRATIONS: ENERGY ABSORPTION BY GREENHOUSE GASES

Light is absorbed most completely when its frequency matches the frequency of an internal motion within a molecule that it encounters. For frequencies in the infrared region, the relevant motions are the **vibrations** of the molecule's atoms relative to each other.

The simplest vibrational motion in a molecule is the oscillatory motion of two bonded atoms X and Y relative to each other. In this motion, called **bond stretching**, the X to Y distance increases beyond its average value R, then returns to R, then contracts to a lesser value, and finally returns to R, as illustrated in Figure 4-3a. Such oscillatory motion occurs in all bonds of all molecules under all temperature conditions, even at absolute zero. A huge number of such vibrational cycles occurs each second. The exact frequency of the oscillatory motion depends

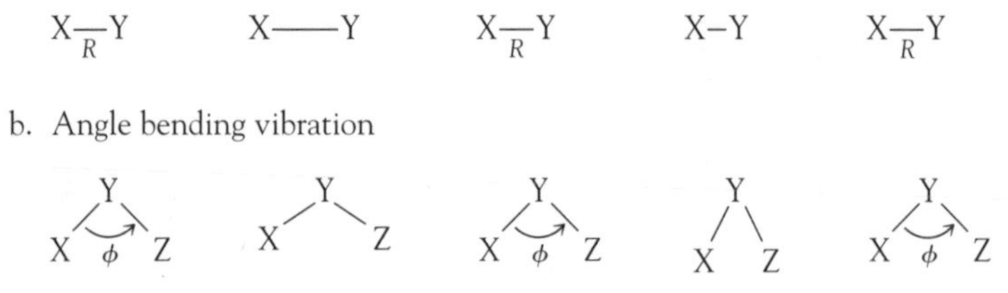

a. Bond stretching vibration

b. Angle bending vibration

FIGURE 4-3

The two kinds of vibrations within molecules. Bond stretching (a) is most simply illustrated by a diatomic molecule XY (e.g., CO, NO, O_2). R represents the average value of the X-Y distance. In (b), angle bending vibration is shown for a molecule XYZ (e.g., CO_2, H_2O—better written as OCO and HOH to make the comparative structure clearer.) The average X-Y-Z angle is indicated by the symbol ø.

primarily upon the type of bond, that is whether it is single or double or triple, and upon the identity of the two atoms involved. For many bond types, for example, with the CH bond in methane and the CO bond in carbon monoxide, the stretching frequency does not fall within the thermal infrared region. The stretching frequency of carbon-fluorine bonds does, however, fall within the thermal infrared range of 4 to 50 μm and thus any molecules in the atmosphere with C—F bonds will absorb outgoing thermal IR light and enhance the greenhouse effect.

The other relevant type of vibration is an oscillation in the distance between two atoms X and Z bonded to a common atom Y but not bonded to each other. Such motion alters the XYZ bond angle from its average value Φ, and is called a **bending vibration**. All molecules containing three or more atoms possess bending vibrations. The oscillatory cycle, of bond angle increase followed by a decrease and then another increase, and so on, is illustrated in Figure 4-3b. The frequencies of many types of bending vibrations occur within the thermal infrared region.

Box 4-1 presents reasons why some atmospheric gases are unable to absorb infrared light, and should be reviewed in connection with the foregoing discussion and Figures 4-3.

THE MAJOR GREENHOUSE GASES

CARBON DIOXIDE: EMISSIONS AND EFFECTS

As stated previously, the absorption of light by a molecule occurs most efficiently when the frequencies of the light and of one of its vibrations match almost exactly. However, light of somewhat lower or somewhat higher frequency than that of the vibration is absorbed to some extent by a collection of molecules. This ability of molecules to absorb infrared light over a short range of frequencies rather than just at a single frequency occurs because it is not only the energy associated with vibration which changes when an infrared photon is absorbed; there can also be a change in the energy associated with the rotation (tumbling) of the molecule about its internal axes. This **rotational energy** of a molecule can be either slightly increased or slightly decreased when IR light is absorbed to increase its vibrational energy. Thus photon absorption occurs not only at the exact frequency of the vibration but to some extent also at slightly higher or lower frequencies; generally the absorption tendency of gas falls off as the light's frequency lies farther and farther in either

4-1   WHY SOME MOLECULES ABSORB IR AND OTHERS DO NOT

In order for infrared light to be absorbed by a molecule during a vibration, there must be a difference in the position in the molecule between its center of positive charge—of its nuclei, in other words—and the center of negative charge—of its electron "cloud." More compactly stated, in order to absorb IR light, the molecule must have a dipole moment. These centers of charge coincide in free atoms and (by definition) in *homonuclear* diatomic molecules like O_2 and N_2. Thus argon gas, Ar, diatomic nitrogen gas, N_2, and diatomic oxygen, O_2, do not absorb IR light. For carbon dioxide, during the vibratory motion in which both CO lengths lengthen and shorten simultaneously, there is at no time any difference in position between the centers of positive and negative charges, since both lie precisely at the central nucleus.* Consequently, during this vibration, called the **symmetric stretch**, the molecule cannot absorb IR light. However, in the **antisymmetric stretch** vibration in CO_2, the contraction of one CO bond occurs when the other is expanding, or vice-versa, so that during the motion the centers of charge no longer necessarily coincide. Consequently, IR light at this vibration's frequency can be absorbed since at some points in the vibration the molecule does have a dipole moment. Similarly, the bending vibration in a CO_2 molecule, in which the three atoms depart from a colinear geometry, is a vibration that can absorb IR light at that frequency.

PROBLEM 4-1

Deduce whether the following molecules will absorb infrared light due to internal vibrational motions:

a. H_2 b. CO   c. Cl_2 d. O_3 e. CCl_4 f. NO·

None of the four diatomic molecules in this list actually absorbs light in the *thermal* infrared region. What does this imply about the frequencies of the bond stretching vibrational motion of those molecules that can in principle absorb IR light?

*Technically there must be a *change* in the distance between the centers before and after the absorption of IR light, but in practice this always occurs if there is a separation between the charges to begin with.

direction from the vibrational frequency. The absorption spectrum for carbon dioxide in a portion of the infrared range is shown in Figure 4-4. For CO_2, the maximum absorption of light in the thermal infrared range occurs at a wavelength of 15.0 μm, which corresponds to a frequency of 2×10^{15} cycles per second; the absorption occurs at this particular frequency since it matches that of one of the vibrations in a CO_2 molecule, namely the OCO angle bending vibration.

The carbon dioxide molecules currently present in air collectively absorb about half of the outgoing thermal infrared light having wavelengths in the 14–16 μm region, together with a sizeable portion of that in the 12–14 and 16–18 μm regions, which are called **shoulders** of the main 15 μm absorption shown in Figure 4-4, and arise from variations in the energy absorbed by the rotational motions. It is for this reason that in Figure 4-5 the solid curve, which represents the amount of IR light that actually escapes from our atmosphere, falls so steeply around 15 μm; the vertical separation between the dashed and solid curves is proportional to the amount of IR of a given wavelength that is being absorbed rather than emitted. Increasing further the CO_2 concentration in the atmosphere will prevent more of the remaining IR from escaping, especially in the shoulder regions, and hence will further warm the air.

Measurements on air trapped in ice core samples indicate that the atmospheric concentration of carbon dioxide in preindustrial times (i.e., before about 1750), was about 280 ppm; it increased by about one-quarter, to 356 ppm, by 1992. Currently it is growing at an annual rate of 0.4% or 1.5 ppm, double that of the 1960s. A plot of the increase in the

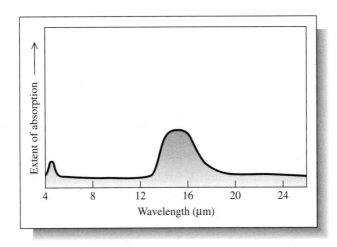

FIGURE 4-4

Infrared absorption spectrum for carbon dioxide. (Source: T. G. Spiro and W. M. Stigliani. *Environmental Issues in Chemical Perspective*. Albany: State University of New York Press, 1980.)

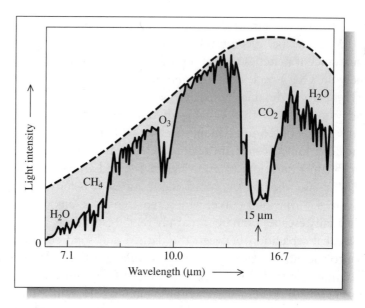

FIGURE 4-5

Actual (experimentally measured) intensity of thermal IR light leaving the Earth's atmosphere (solid curve) compared with the intensity that would be expected without absorption by greenhouse gases (dashed curve). Wavelengths of greatest absorption by gases are indicated on the horizontal axis. (Source: E. S. Nisbet. *Leaving Eden*. Cambridge, England: Cambridge University Press, 1991.)

atmospheric CO_2 concentration over time in recent years is shown in Figure 4-6. Huge quantities of CO_2 are removed from the air each spring and summer because of the process of photosynthesis in plants:

$$CO_2 + H_2O \longrightarrow O_2 + \text{polymeric } CH_2O \text{ (plant fiber)}$$

The term used for the product "polymeric CH_2O" in the above equation is an umbrella word for plant fiber, typically the cellulose that gives wood its mass and bulk. The CO_2 "captured" by the photosynthetic process is no longer free to function as a greenhouse gas—or as any gas— while it is packed away in this polymeric form. The carbon that is trapped in this way is called **fixed carbon.** However, the biological decay of this plant material, the very *reverse* of this reaction, that occurs mainly in the fall and winter replaces the withdrawn carbon dioxide. The annual fluctuations in the CO_2 concentrations visible in Figure 4-6 are due to the growth spurt in spring and summer that removes CO_2 from air and the decay cycle in fall and winter that increases it.

Some scientists suspect that the rate of photosynthesis is speeding up as the level of CO_2 and the air temperature increase, and that the formation of greater amounts of fixed carbon represents an important sink for the gas. Indeed, an increase in the biomass of northern temperate forests is now suspected to be the most likely sink to account for the annual atmospheric CO_2 loss for which scientists had previously been unable to assign a cause. In fact, anthropogenic releases of CO_2 amount only to about 4% of the enormous amounts produced by nature, and thus a very small variation in the rate at which carbon is absorbed into biomass could have a large effect on the residual amount of CO_2 that accumulates in the atmosphere. Unfortunately, scientists still do not completely understand the global carbon cycle. As Figure 4-6 indicates, however, there is no doubt that the atmospheric CO_2 concentration is increasing.

Much of the exponential increase in anthropogenic contributions to the increase in carbon dioxide concentration in air is due to the combustion of **fossil fuels**—chiefly coal, oil, and natural gas—which were formed eons ago when plant and animal matter was covered by geological deposits before it could be more thoroughly broken down by air oxidation. On average, each person in the industrial countries is responsible for the release of about 5 metric tons (a metric ton is 1,000 kg, i.e., 2,200 lb) of CO_2 from carbon-containing fuels each year! Some of this per capita output is direct, for example that released in the exhaust gases when vehicles are driven and homes are warmed by burning a fossil fuel.

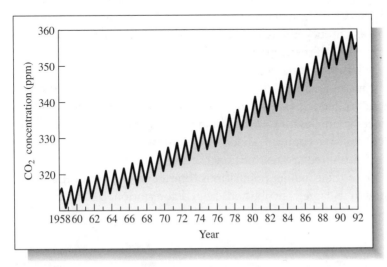

FIGURE 4-6

Yearly variation in atmospheric CO_2 concentrations. The sawtooth variations are due to seasonal effects (see text). (Source: Redrawn from B. Hileman. 1992. *Chemical & Engineering News* (April 27): 7–19.)

The remainder is indirect, and arises when energy is used to produce and transport goods, heat and cool factories, classrooms, and offices, produce and refine oil—in fact, to accomplish virtually any constructive economic purpose in an industrialized society.

The annual per capita release of carbon dioxide in developing countries is about one-tenth as large as it is in developed countries, but it is increasing. A significant amount of carbon dioxide is added to the atmosphere when forests are cleared and the wood burned in order to provide land for agricultural use. This sort of activity occurred on a massive scale in temperate climate zones in past centuries (consider the immense deforestation that accompanied the settlement of the United States and southern Canada) but has now shifted largely to the tropics. The greatest single *amount* of current deforestation occurs in Brazil, and involves both rain forest and moist deciduous forest, but the annual *rate* of deforestation on a percentage basis is actually greater in South East Asia (1.6%) and in Central America (1.5%) than in South America (0.6%). In some temperate areas, such as the Canadian province of British Columbia, where forestry is an important industry, more CO_2 is released by burning shrubs and stumps after logging than is generated by all other activities combined! Notwithstanding forestry operations, the total amount of carbon contained in the forests of the Northern Hemisphere apparently is increasing, though the annual increment does not equal the decreases discussed above in Asia and South and Central America. Overall, deforestation accounts for about one-quarter of the annual anthropogenic release of CO_2, the other three-quarters originating mainly with the combustion of fossil fuels.

A plot of the total annual emissions, in terms of the mass of carbon, over time for carbon dioxide from fossil fuel combustion and from cement production (see Problem 4-2) is shown in Figure 4-7. The top 20 producer countries for carbon dioxide, as of 1987, are listed in Table 4-1; notice that per capita, Canada and Australia emit almost as much as does the United States, although their totals are not as impressive because of their smaller populations. Brazil comes close to the top of the list due to its large contributions from deforestation.

PROBLEM 4-2

Some carbon dioxide is released into the atmosphere when calcium carbonate rock (limestone) is heated to produce the quicklime, i.e. calcium oxide, used in the manufacture of cement:

$$CaCO_3(s) \longrightarrow CaO(s) + CO_2(g)$$

Calculate the mass, in metric tons, of CO_2 released per metric ton of limestone used in this process.

About half of the anthropogenic CO_2 emissions currently find a sink; much of it is removed from the atmosphere by dissolution in sea water. However, the top few hundred meters of sea water mixes very slowly with deeper waters; thus carbon dioxide that is freshly dissolved in surface water requires hundreds of years to penetrate to the ocean depths and to be precipitated as insoluble calcium carbonate. Consequently, although the oceans will ultimately dissolve much of the increased CO_2 in the air, the time scale involved is very long, and so the gas continues to accumulate in the atmosphere. At present, the average carbon dioxide molecule has an atmospheric residence time of more than a century—a considerable span of time. Furthermore, in the short run, the ability of the upper ocean layers to absorb carbon dioxide may decrease if the water warms appreciably, since the solubility of gases in water decreases with increasing temperature.

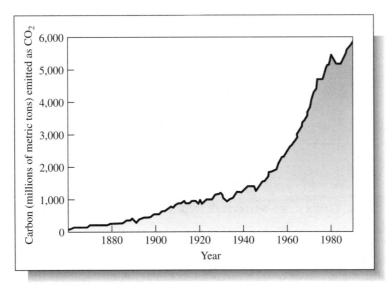

FIGURE 4-7
Annual carbon content of CO_2 emissions from fossil fuel combustion and cement production. (Source: H. Hengeveld. *Understanding Atmospheric Change*. State of Environment Report 91-2, Minister of Supply Services Canada, 1991.)

TABLE
4-1   THE TOP 20 PRODUCERS OF CARBON DIOXIDE FOR 1987

Rank	Country	Total net additions to the atmosphere (per 1,000 metric tons of carbon equivalent)	Percentage share of greenhouse gas increases over CO_2 for previous year	Net per capita additions to the atmosphere (kilograms of carbon equivalent)
1	United States	920,000	17.0	3.8
2	USSR	700,000	13.1	2.5
3	Brazil	460,000	8.5	3.3
4	China	410,000	7.6	0.4
5	India	250,000	4.6	0.3
6	Japan	200,000	3.7	1.6
7	Indonesia	150,000	2.7	0.9
8	FRG	140,000	2.7	2.4
9	United Kingdom	130,000	2.5	2.3
10	Italy	100,000	1.9	1.8
11	France	99,000	1.8	1.8
12	Canada	88,000	1.6	3.4
13	Mexico	77,000	1.4	0.9
14	Poland	76,000	1.4	2.0
15	Colombia	67,000	1.2	2.2
16	Thailand	67,000	1.2	1.3
17	Nigeria	61,000	1.1	0.6
18	Spain	59,000	1.1	1.5
19	GDR	58,000	1.1	3.5
20	Australia	58,000	1.1	3.6
	WORLD	5,400,000	100.0	1.1

Source: From A. L. Hammond, E. Rodenburg, and W. Moomaw. (1990) *Nature* 347 (Oct. 25): 705–706.

CARBON DIOXIDE: MINIMIZING FUTURE EMISSIONS

In devising strategies to minimize the amount of carbon dioxide emitted to the atmosphere in the future, scientists and policymakers take into account the fact that fossil fuels differ in the amount of the gas that they emit per unit amount of energy that they produce. To a good approximation, the amount of heat released when a carbon-containing substance burns is directly proportional to the amount of oxygen it consumes.

Consider the reactions of coal (mainly carbon), oil (essentially polymers of CH_2) and natural gas (essentially CH_4) with atmospheric oxygen:

$$C + O_2 \longrightarrow CO_2$$

$$CH_2 + 1.5\, O_2 \longrightarrow CO_2 + H_2O$$

$$CH_4 + 2\, O_2 \longrightarrow CO_2 + 2\, H_2O$$

If a little thought is given to the stoichiometry of these reactions, it can be understood that, per mole of O_2 consumed and thus per joule of energy produced, natural gas generates less carbon dioxide than does oil, which in turn is superior to coal.

Ton for ton, the potential of a fossil fuel to generate CO_2 and so to cause atmospheric warming depends on its carbon content, and so presumably **carbon taxes**, that is, taxes based on the amount of carbon contained in a fuel rather than upon its total mass, should be instituted as a disincentive to use the less desirable fuels. The possibility of switching to fuels that would emit no carbon dioxide upon their formulation or combustion, such as hydrogen gas produced by solar energy, is explored in Chapter 10.

PROBLEM 4-3

Methanol, CH_3OH, is an "oxygenated" liquid fuel that can be used instead of gasoline in suitably modified cars.

a. By writing the balanced chemical equation for its combustion in air, determine whether it is more similar to coal, oil, or natural gas in terms of the joules of energy released per mole of CO_2 produced.

b. Determine the balanced equation by which methanol can be produced by reacting elemental carbon (coal) with water vapor, given that CO_2 is the only other product in the reaction.

c. Does the combined scheme of parts (a) and (b) above represent a way of using coal but producing less carbon dioxide per joule than through its direct combustion? Explain your answer.

In the future, CO_2 might be removed chemically from the exhaust gas of power plants that burn fossil fuels instead of being released to the atmosphere. The carbon dioxide gas so recovered would then be buried, for example in the deep oceans where it would dissolve, or in empty oil and natural gas wells. Near the sea floor, the carbon dioxide would react with the solid calcium carbonate formed from seashells to produce soluble calcium bicarbonate:

$$CO_2(g) + H_2O(aq) + CaCO_3(s) \longrightarrow Ca(HCO_3)_2 \ (aq)$$

For practical purposes, the CO_2, now chemically trapped in the bicarbonate form, would remain indefinitely in the dissolved state.

Recently it has been conjectured by Norwegian scientists that even relatively shallow injection of the gas in the ocean, at about 200 meters depth, would produce a satisfactory result, provided that the sea floor there is slanted sufficiently to allow the dense, CO_2-rich water to be transported by gravity to greater depths. Over a period of centuries, the excess carbon dioxide would eventually return to the atmosphere, but presumably by that time alternative energy sources would have replaced fossil fuels and the atmospheric CO_2 problem would then be less serious.

An alternative to burying the carbon dioxide itself would be to first combine it with some other abundant substance and then to dispose of the product. For example, it has been proposed to pump CO_2 into a slurry of calcium silicates (a cheap, abundant mineral) that would thereby transform it to aqueous calcium bicarbonate, which could be drained into ocean depths:

$$2CO_2 + H_2O + CaSiO_3 \longrightarrow SiO_2 + Ca(HCO_3)_2 \ (aq)$$

Yet another recent suggestion (proposed by scientists at the University of Stuttgart in Germany) to prevent the immediate release of carbon dioxide to the atmosphere would involve the creation of giant balls of solid carbon dioxide, or "dry ice," and their insulated storage below $-79°C$ (the sublimation point of the solid CO_2) at the Earth's surface. If a very thick blanket of glass wool insulation is used, the release of gaseous carbon dioxide from the 400-meter-diameter balls would not occur for many centuries.

The energy input required for the CO_2-concentrating phase of any of these so-called **carbon sequesterization** schemes would represent a substantial fraction, from one-third to one-half, of the power plant's output. In addition, only a small fraction of CO_2 comes from such sources. It is not clear how carbon dioxide could be extracted from automobile exhaust and other less centralized but collectively important sources.

WATER VAPOR

Water molecules, always abundant in air, absorb thermal IR light due to the H-O-H bending vibration; the peak in the spectrum for this absorption occurs at about 6.3 μm (see Figure 4-8). (Also note that the peaks at wavelengths less than 4 μm fall outside the thermal IR region.) Thus almost all the relatively small amount of outgoing IR in the 5.5–7.5 μm region is intercepted by water vapor (see Figure 4-5). There are other water vibrations that remove thermal infrared light of wavelength 12 μm and longer (Figures 4-5 and 4-8). In fact, water is the most important greenhouse gas in the Earth's atmosphere, though on a per molecule basis it is a less efficient absorber than is CO_2.

The equilibrium vapor pressure of liquid water, and consequently the maximum concentration of water vapor in air, increases exponentially with temperature. Thus the amount of thermal IR redirected by water vapor will rise as a result of any global warming induced by the other greenhouse gases and will amplify the temperature increase. Since it comes about as an indirect effect of increasing the levels of other gases, the warming increment due to water is usually incorporated without further comment, so to speak, with the direct warming effects of the other gases. Consequently, water is not usually listed explicitly among gases whose increasing concentrations are enhancing the greenhouse effect.

Water in the form of liquid droplets in clouds also absorbs thermal IR. However, such clouds also reflect some incoming sunlight, both UV

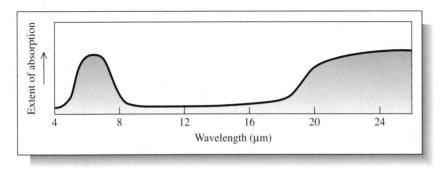

FIGURE 4-8

The infrared absorption spectrum for molecules of water in the vapor state. The largest amount of absorption in the thermal IR (4–50 μm) occurs at and around 6.3 μm. (Source: T. G. Spiro and W. M. Stigliani. *Environmental Issues in Chemical Perspective*. Albany: State University of New York Press, 1980.)

and visible, back into space. It is not yet clear whether the additional cloud cover produced by increasing the atmospheric water content will have a net positive or a net negative contribution to global warming. Clouds over tropical regions are known to have a zero net effect, but those in northern latitudes produce a net cooling effect since their ability to reflect sunlight outweighs their ability to absorb IR. Thus, if increased air temperatures produce more of this latter type of cloud, the enhancement of global warming would be damped. However, no one is sure that *additional* northern cloud cover would occur at the same altitudes and act in the same manner as do the currrent clouds.

As a result of absorption mainly by carbon dioxide and water, it is essentially only infrared light from 8 to 13 μm that escapes the atmosphere efficiently (see Figure 4-5). Since light of these wavelengths passes unimpeded, this portion of the spectrum is called a **window**.

OTHER SUBSTANCES THAT AFFECT GLOBAL WARMING

The injection into the atmosphere even in trace amounts of gases that can absorb thermal IR light will lead to additional global warming, that is, the *enhanced* greenhouse effect. Particularly serious are pollutant gases that absorb thermal IR in the window region, since the absorption by H_2O and CO_2 in other regions is already so great that there remains little such light for trace gases to absorb. In considering which potential pollutants might contribute to global warming, recall that we can dismiss free atoms and homonuclear diatomic molecules as they cannot absorb IR light (see Box 4-1). Heteronuclear diatomic molecules such as CO and NO are also not of direct concern since their only vibration—bond stretching—has a frequency that lies outside the thermal IR region. In general, however, most gases consisting of molecules with three or more atoms are of concern since they possess many vibrations that absorb IR, one or more of which usually fall in the thermal infrared region. The *important* trace greenhouse gases, that is, ones whose concentration is small in absolute terms but whose ability even at these levels to warm the air is substantial, are detailed below, following a discussion of their average lifetimes in the atmosphere. The nature, abundance, rate of increase, and relative effectiveness in promoting global warming of these gases are summarized in Table 4-2.

TABLE 4-2 SUMMARY OF INFORMATION ABOUT GREENHOUSE GASES

Gas	Current abundance	Rate of increase	Relative warming effectiveness*
CO_2	356 ppm	+0.4%	1
CH_4	1.74 ppm	+0.6%	23
N_2O	0.31 ppm	+0.25%	270
CFC-11 ($CFCl_3$)	0.26 ppb	n/a	14,000
CFC-12 (CF_2Cl_2)	0.47 ppb	n/a	19,500
Tropospheric O_3	0.03 ppm	n/a	———

* Per molecule, over a 20-year integrated time horizon.

TRACE GASES: ATMOSPHERIC RESIDENCE TIME

In order to assess the impact of any substance upon the enhanced greenhouse effect, it is necessary to know how long the substance can be expected to remain in the atmosphere, since the longer its atmospheric lifetime, the greater its total effect. Every atmospheric gas that is present in or near a steady state can be characterized by a **residence time** T_{avg}, which represents the average amount of time one of its molecules exists in air before it is removed by one means or another. In terms of the total atmospheric amount C and the average rate R of input or output of a substance per unit time

$$T_{avg} = C/R$$

Instead of using amounts of gases, concentrations may also be employed for both C and R in this formula. Thus if the atmospheric concentration of a gas currently is 6.0 ppm, and if its global rate of input, as determined by dividing the yearly amount of input by the volume of the atmosphere, is 2.0 ppm per year, its average residence time is 6.0 ppm/2.0 ppm yr^{-1}, or 3.0 years.

The average residence time of the greenhouse gases carbon dioxide, nitrous oxide, and the CFCs are more than a century, so the influence of gases now being emitted into the atmosphere will extend over very long periods of time. In contrast, methane has a residence time of only about a decade.

PROBLEM 4-4

If the average, steady-state residence time of a trace atmospheric gas is 50 years and its input rate is 2.0×10^6 kilograms per year, what is the total amount of it in the atmosphere?

PROBLEM 4-5

The steady-state concentration of an atmospheric gas of molar mass 42 g mol^{-1} is 7.0 μg per gram of air, and its residence time is 14 years. What is the annual total release of the gas into the atmosphere as a whole? Note that the mass of the entire atmosphere is 5.1×10^{21} g; air's average molar mass = 29.0 g mol^{-1}.

METHANE

Following carbon dioxide and water, methane, CH_4, is the next most important greenhouse gas. A methane molecule contains four C—H bonds. Although C—H bond stretching vibrations occur well outside the thermal IR region, HCH bond angle bending vibrations absorb at 7.7 μm, at the edge of the thermal IR window, and consequently methane absorbs IR in this region. Per molecule, increasing the amount of methane in air causes 23 times the warming effect as does adding more carbon dioxide since CH_4 molecules absorb a greater fraction of the thermal IR photons that pass through them than do CO_2 molecules. However, the current 80-fold greater increase in CO_2 molecules means that at present CH_4 is much less important in global warming. To date, methane is estimated to have produced about one-third as much global warming as has carbon dioxide.

The atmospheric concentration of methane has almost doubled as compared with its preindustrial value; almost all of this increase has occurred in the twentieth century. Historically (i.e., before 1750), the methane concentration was constant at 0.75 ppm, but in the 1980s it rose by about 0.6% per year and reached 1.7 ppm. In the 1970s, the rate of increase was twice as large as this; it is not known with certainty why the rate decreased and is now almost zero.

The rise in the atmospheric CH_4 level is presumed to be the consequence of such human activities as increased food production, fossil fuel use, and forest clearing. Methane is produced biologically in the **anaerobic** decomposition of plant material. Such processes occur on a huge scale where the decay of plants occurs under water-logged conditions, for example in natural wetlands such as swamps and bogs, and in rice paddies. Indeed, the original names for methane were "swamp gas" and "marsh gas." The expansion of wetlands that occurs by the flooding of land to produce more hydroelectric power could add to this total. The production of methane from natural and artificial wetlands is currently considered to be the largest single source of the gas.

Ruminant animals—including cattle, sheep, and certain wild animals—produce huge amounts of methane as a byproduct in their stomachs when they digest the cellulose in their food. The animals subsequently belch the methane into the air. The decrease in population of some methane-emitting wild animals (e.g., buffalo) has been far exceeded by the huge increase in the population of cattle and sheep, the net result being a large increase in emissions of methane. Some of the reduction in the annual rate of increase of atmospheric methane may be due to the slowdown in the acceleration in worldwide cattle production.

Anaerobic decomposition of the organic matter in garbage in landfills is another important source of methane in air. In some communities, methane from landfills is collected and used to generate heat rather than being allowed to escape into the air. The burning of biomass, such as forests and grasslands in tropical and semitropical areas, releases methane to the extent of about 1% of the carbon consumed, along with larger amounts of carbon monoxide (both compounds are products of incomplete—poorly ventilated—combustion).

In summary, there are six different significant sources of atmospheric methane: in order of importance they are wetlands, fossil fuels, landfills, ruminant animals, rice paddies, and biomass burning. All but the first are subject to such large uncertainties that even their relative ordering is subject to revision, and individual percentages for each will not be quoted here.

The carbon in all living matter contains a small, constant fraction of a radioactive isotope, carbon-14 (^{14}C), taken in via the carbon cycle when photosynthesis captures atmospheric CO_2, and when animals in turn feed off plant matter. This fact underlies the radiocarbon dating methods used by archaeologists and anthropologists: when an organism dies, its ^{14}C decays at a known rate that makes the date of its death calculable. (The assumptions justifying these methods are that biotic carbon and atmospheric carbon in CO_2 are balanced—in equilibrium with one another—and that the level of atmospheric ^{14}C is constant.)

However, in the case of atmospheric methane, the average fraction of ^{14}C is less than the value found in living tissue, indicating that some of the CH_4 escaping into air must be "old carbon" that has been trapped in the ground for so long that its ^{14}C content has diminished to zero as a result of radioactive decay through the ages. Most methane containing old carbon is released into air as a byproduct of the mining, processing, and distribution of fossil fuels. Methane trapped in coal is released into the atmosphere when this material is mined, as is methane in oil when it is pumped from the ground. The transmission of natural gas, which is almost entirely methane, involves losses into air due to leakage from pipelines, and is the largest of the sources of old carbon. Recent measurements of the methane levels in the air of various cities indicates that much of the loss from pipelines occurs in Eastern Europe. Because of the losses to air of methane that occur during the transmission of natural gas, the *net* greenhouse-enhancing effect of using natural gas for combustion can be several times that of oil, even after the advantage of methane in producing less CO_2 per joule is considered! (See page 163 and Problem 4-6.) Finally, there is probably a small contribution to the old carbon source from methane trapped in permafrost in far northern latitudes; the methane was formed by the decay of plant matter that lived there many thousands of years ago when the polar climate was much warmer than it is today.

PROBLEM 4-6

The replacement by natural gas of oil or coal used in power plants has been proposed as a mechanism by which CO_2 emissions can be reduced. However, much of the advantage of switching to gas can be offset since methane escaping into the atmosphere from gas pipelines is 23 times as effective, on a molecule-per-molecule basis, in causing global warming as is carbon dioxide. Calculate the maximum percent of CH_4 that can escape if replacement of oil by natural gas is to reduce the rate of global warming. (Hint: Recall that the heat energy output of the fuels are proportional to the amount of O_2 they consume.)

The dominant sink for atmospheric methane, accounting for about 90% of its loss from air, is reaction with the hydroxyl free radical, $OH^{\bullet}$:

$$CH_4 + OH^{\bullet} \longrightarrow CH_3^{\bullet} + H_2O$$

The other sinks for methane gas are reaction with soil and loss to the stratosphere. As discussed previously in Chapter 3, the reaction with hydroxyl radical is but the first step of a sequence that transforms methane ultimately to CO and then to CO_2. The average lifetime of atmospheric CH_4 is about 15 years. The yearly loss of methane by this reaction (see Problem 4-7 below) is about 480 Tg (where 1 Tg, or Teragram, is 10^{12} grams—one million metric tons) and thus the net sink from all sources amounts to about 530 Tg per year. Some scientists speculate that the concentration of $OH^{\bullet}$ in the atmosphere may be decreasing because of the increasing concentrations of CO and CH_4, the two gases which constitute its main sinks; thus some of the observed increase in methane concentration may be due to a reduction in the rate of its reaction with hydroxyl radical.

PROBLEM 4-7

The 1992 concentration of atmospheric methane was 1.74 ppm, and the newly determined (1992) rate constant for the reaction between CH_4 and $OH^{\bullet}$ is 3.6×10^{-15} cm^3 molecule^{-1} s^{-1}. Calculate the rate, in Tg per year, of methane destruction by reaction with hydroxyl radical, the concentration of which is 8.7×10^5 molecules cm^{-3}. Note that the atmospheric mass $= 5.1 \times 10^{21}$ g; air's average molar mass $= 29.0$ g mole^{-1}.

PROBLEM 4-8

The total amount of methane in the atmosphere in 1992 was about 5,000 Tg, and was increasing by about 0.6% annually due to the fact that the annual input rate exceeded the annual output rate of 530 Tg yr^{-1}. Calculate the percent by which anthropogenic releases of methane, which account for two-thirds of the total, had to be reduced if the atmospheric concentration of this gas was to be stabilized in 1992.

A minor sink for methane exists in the stratosphere, to which a few percent of it eventually rises. It reacts there with OH$^{\bullet}$, or atomic chlorine or bromine, or excited atomic oxygen; reaction with the latter produces hydroxyl radicals and eventually water molecules:

$$O^* + CH_4 \longrightarrow OH^{\bullet} + CH_3^{\bullet}$$
$$OH^{\bullet} + CH_4 \longrightarrow H_2O + CH_3^{\bullet}$$

Stratospheric water vapor acts as a significant greenhouse gas. About one-quarter of the total global warming caused by increased methane emissions is thus not brought about directly: it is due to this effect in the stratosphere by which the region's water content is increased.

There has been speculation among some scientists that the rate of release of methane into air could greatly increase in the future as a result of temperature rises from the enhanced greenhouse effect. (For instance, anaerobic biomass decay, as in a common landfill, would be accelerated at higher temperatures.) In turn, the additional release of methane would itself cause a further rise in temperature. This is an example of **positive feedback**: the operation of a phenomenon produces a result that itself *further* enhances (amplifies) the result. *Feedback* is a reaction to change; with positive feedback, this reaction accelerates the pace of future change. On the other hand, a system whose output reduces the subsequent level of output displays **negative feedback**. An example of negative feedback from daily life is the attempt by a business to raise its profits by increasing its prices; however, the rise in price often results in a reduction in demand for the item of concern, and the rise in profits is less than anticipated. (No value judgment as to the desirability of the effect is implied by the term *positive* and *negative*; only the increase or decrease in the pace of change is meant.)

Methane release from biomass decay among the extensive bogs and tundra in Canada, Russia, and Scandinavia could increase with increasing air temperature and would constitute positive feedback. However, the rate of biomass decay, and thus of CH_4 production, depends also on soil moisture and therefore on rainfall, which probably would be affected by climate change in an as yet uncertain direction, so the net feedback from this source could be positive or negative.

There is much methane currently frozen in the permafrost of far northern regions; it was produced from the decay of plant materials during warm spells in the region, but became trapped due to glaciation as temperatures became lower and lower at the start of the last Ice Age. Melting of the permafrost due to global warming could release large amounts of this methane. Melting would also allow the decomposition of organic matter currently frozen in the permafrost, with the consequent release of more methane. As well, there are monumental amounts of methane trapped at the bottom of the oceans on continental shelves in the form of "methane hydrate." This substance has the approximate formula $CH_4 \cdot 6H_2O$, and is an example of a **clathrate compound,** that is, a rather remarkable structure that forms when small molecules occupy vacant spaces ("holes") in a cagelike polyhedral structure formed by other molecules. In the present case, methane is "caged" in a 3-D ice lattice structure formed by the water molecules. Clathrates form under conditions of high pressure and low temperature (such as found in cold waters and under ocean sediments). If sea water warmed by the enhanced greenhouse effect penetrates to the bottom of the oceans, the clathrate compounds could decompose and release their own methane, as well as reservoirs of pure methane currently trapped below them, to the air above. Methane trapped far below the permafrost in northern areas and in offshore areas in the Arctic also exists in the form of clathrates; it would be released eventually if the Arctic warmed sufficiently. Although the uncertainties concerning methane feedback are large, the stakes are higher than with any other gas. Measurements made thus far do not indicate any significant emissions from these sources. It has even been suggested by some scientists that CH_4 released from clathrates may be oxidized to CO_2 before it reaches the air. But a few scientists believe in the possibility that several positive feedback mechanisms, including those involving methane, could combine to trigger an unstoppable warming of the globe. (This worst-case scenario is called the "runaway greenhouse effect.") Such climate change would threaten all life on Earth, as the temperature would rise markedly, ocean currents probably would shift and rainfall patterns would be very different from those we know.

PROBLEM 4-9

Calculate the mass of methane gas trapped within each kilogram of methane hydrate.

NITROUS OXIDE

Another significant greenhouse trace gas is nitrous oxide, N_2O, "laughing gas," the structure of which is NNO rather than the more symmetric NON. Its bending vibration absorbs IR light in a band at 8.6 μm, that is, within the window region, and in addition one of its bond stretching vibrations is centered at 7.8 μm, on the shoulder of the window and at the same wavelength as one of the absorptions for methane. Per molecule, N_2O is 270 times as effective as CO_2 in causing global warming. Like that of methane, the atmospheric concentration of nitrous oxide was constant until about 300 years ago, at which time it began to increase, though the level now has grown only by a total of 9% and the present yearly growth rate is about 0.25%—a rate that seems quite low, but that will yield a greater increase in the level of N_2O in less than forty years than the last three centuries have witnessed! The increased amounts of nitrous oxide that have accumulated in air since preindustrial times have produced about one-third the magnitude of the additional warming that methane has induced.

PROBLEM 4-10

Given that the atmospheric concentration of N_2O currently is 310 ppb, and its annual increase amounts to 0.25%, calculate the extent to which the input rate exceeds the output rate.

The greater part of the natural supply of the gas comes from release by the oceans and most of the remainder is contributed by processes occurring in the soils of tropical regions. The gas is a byproduct of the biological **denitrification** process in aerobic environments and in the biological **nitrification** process in anaerobic environments; the chemistry of both processes is illustrated in Figure 4-9. In denitrification, fully oxidized nitrogen in the form of the nitrate ion, NO_3^-, is reduced mostly to molecular nitrogen, N_2. In nitrification, reduced nitrogen in the form of ammonia or the ammonium ion is oxidized mostly to nitrite (NO_2^-) and nitrate ions. Chemically, the existence of the nitrous oxide byproduct in both processes is simple to rationalize: nitrification (oxidation) under oxygen-limited conditions yields some N_2O, which has less oxygen than the "intended" nitrite ion, and denitrification (reduction) under oxygen-rich conditions yields some N_2O, which has more oxygen than the "intended" nitrogen molecule. Recently it has been established that nitrification is more important than denitrification as a global source of N_2O. Normally, about 0.001 mole of N_2O is emitted per mole of nitrogen oxidized, but this value increases substantially when the ammonia or ammonium concentration is high and relatively little oxygen is present.

Apparently nitrous oxide released from new grasslands is particularly significant in the years following the burning of a forest. Some portion of the nitrate and ammonium fertilizers used agriculturally, particularly in tropical areas, is similarly converted (an unintended effect, to be sure) to nitrous oxide and released into the air.

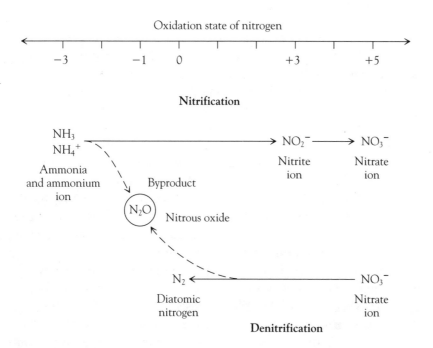

FIGURE 4-9
Nitrous oxide production during the biological cycling of nitrogen.

Previously it was believed that fossil fuel combustion released nitrous oxide as a byproduct of the chemical combination of the N_2 and O_2 in air, but this belief was based upon faulty experiments. It is now known that only when the fuel itself contains nitrogen, as does coal and biomass (but not gasoline or natural gas), does N_2O form; apparently N_2 from air does not enter into this process at all. However, some of the $NO^•$ produced from atmospheric N_2 during fuel combustion in automobiles is unavoidably converted to N_2O rather than to N_2 in the three-way catalytic converters currently in use and is subsequently released into air. Some of the newer catalysts developed for use in automobiles do not suffer from this flaw of producing and releasing nitrous oxide during their operation.

As mentioned in Chapter 2, there are no sinks for nitrous oxide in the troposphere. Instead, all of it rises eventually to the stratosphere where each molecule absorbs UV light and decomposes, usually to N_2 and atomic oxygen, or reacts with atomic oxygen.

CFCs AND THEIR REPLACEMENTS

Gaseous compounds consisting of molecules with carbon atoms bonded exclusively to fluorine and/or chlorine atoms have perhaps the greatest potential among trace gases to induce global warming, since they are both very persistent and absorb strongly in the 8–13 μm window region. Absorption due to the C—F bond stretch is centered at 9 μm, and the C—Cl stretch and various bond-angle bending vibrations involving carbon atoms bonded to halogens also occur at frequencies that lie within the window region. As discussed in Chapter 2, the chlorofluoro-carbons $CFCl_3$ and CF_2Cl_2 have already been released into the atmosphere in large quantities and have long residence times. Due to this persistence, and to their high efficiency in absorbing thermal IR in the window region, each CFC molecule has the potential to cause the same amount of global warming as do tens of thousands of CO_2 molecules. The *net* effect of CFCs on global temperature is approximately zero, however. The heating that the CFCs produce by the redirection of thermal infrared is virtually cancelled by a separate effect, the cooling that they induce in the stratosphere by their destruction of ozone. (Recall from Chapter 2 that the stratosphere is heated when oxygen atoms, recently detached photochemically from ozone molecules, collide with O_2 molecules to produce an exothermic reaction.) It should be realized however that the cooling and heating effects of CFCs occur at very different altitudes, so that their net effect on the Earth's *weather* (as opposed to the net temperature of the atmosphere) may not be zero.

Ironically, the use of CFCs in insulating freezers, refrigerators, and air conditioners has reduced the energy requirements of this equipment and so has reduced CO_2 emissions resulting from electricity production.

The influence of CFCs on climate in the future will be reduced as a result of the requirements of the Montreal Protocols, which ban further production after the year 1995, as discussed in Chapter 2. The proposed HCFC and HFC replacements for CFCs have shorter atmospheric lifetimes and absorb less efficiently in the center of the window region, and thus on a molecule-for-molecule basis they pose less of a greenhouse threat. However, if their levels of production and release become high in future decades because of expanding world population and increasing affluence, they will make significant contributions to global warming if they are released into the air. For this reason, many people feel that these substances must be used only in closed systems from which leakage to the atmosphere does not occur, and that these substances must be recovered from equipment before its eventual disposal. Indeed, prevention of the chronic release of long-lived gases of all types to the atmosphere is a principle now agreed to by many scientific, business and governmental groups.

PROBLEM 4-11

Fully fluorinated compounds such as tetrafluoromethane and hexafluoroethane are released as byproduct wastes into the air in the production of aluminum. They were also briefly considered as CFC replacements. Will such molecules have a sink in the troposphere? Will they act as greenhouse gases? Would your answers be the same for monofluoro-methane and -ethane?

OZONE

Like methane and nitrous oxide, tropospheric ozone, O_3, is a "natural" greenhouse gas, but one which has a short tropospheric residence time. One of ozone's OO bond stretching vibrations occurs between 9 and 10 μm, that is, in the window region. Indeed the "bite" near 9 μm from the outgoing thermal IR displayed in Figure 4-5 is due to ozone. Its bending vibration occurs near that for CO_2, and thus it does not contribute much to the enhancement of the greenhouse effect since atmospheric carbon dioxide already removes much of the outgoing light at this frequency.

As explained in Chapter 3, ozone is formed in the troposphere from oxygen atoms produced by the photochemical dissociation of nitrogen dioxide:

$$NO_2^{\bullet} \xrightarrow{\text{UV-A}} NO^{\bullet} + O$$

$$O + O_2 \longrightarrow O_3$$

Because of pollution from power plants and motor vehicles, the levels of nitrogen dioxide and consequently of ozone in the troposphere probably have increased since preindustrial times. The best guess is that approximately 10% of the increased global warming potential of the atmosphere results from increases in tropospheric ozone, though this value is very uncertain. The amount of thermal IR absorbed by *stratospheric* ozone has probably dropped slightly due to its recent decline there.

AEROSOLS

In Chapter 2, we saw that the initial neglect by scientists of the effects of atmospheric aerosol particles, specifically ice crystals in the stratosphere, led to a large underestimation of the amount of ozone that would be destroyed by chlorine. Recently it has been realized that a similar neglect of the effect of aerosols led scientists to *overestimate* the amount of short-term global warming to be expected.

Specifically, the sulfate-rich tropospheric aerosols produced from air pollutants, especially over urban areas in the northern hemisphere, reflect sunlight back into space more effectively than they absorb it. Consequently, their existence means that less sunlight is available to be absorbed by the surface and in the lower troposphere and converted to heat. The sulfate aerosols are not particularly effective in trapping thermal IR emissions from the Earth's surface. Thus the net effect of the aerosols is to cool the air near ground level, and thereby to offset some of the effects of global warming induced by greenhouse gases. Indeed, two effects have been observed that are consistent with the proposed effect of aerosols. In the first place more warming has been observed in the relatively unpolluted southern hemisphere than in the northern, and, secondly, more warming has occurred with nighttime temperatures than with daytime.

About 20 to 40% of the current temperature-enhancing effect of the greenhouse gases probably is negated by aerosol deflection of sunlight. Ironically, the introduction of controls on SO_2 emissions from power plants in the United States by the Clean Air Act of 1990 could diminish this effect, since, as discussed in Chapter 3, the sulfate aerosol—sulfuric acid neutralized to varying extents by ammonia—originates from oxidized sulfur dioxide. However, the anthropogenic sulfate aerosol concentrations over southern Europe, the Middle East, and parts of Russia and China are considerably higher than the current maximum values in North America, and will not be affected by this legislation. Globally, increased sulfate emissions in the future due to expanded coal burning in China might well overcome the effect of controls on SO_2 emissions in the United States.

Naturally produced aerosols from massive volcanic eruptions can also influence climate to a noticeable extent. The 1991 eruption of Mount Pinatubo in the Philippines, the largest anywhere in the world in the twentieth century, injected huge quantities of SO_2 directly into the stratosphere, yielding a sulfate aerosol which remained there for several years. Because this aerosol efficiently reflected incoming sunlight, a decrease of 0.5°C in average global temperature was projected and observed for 1992 and 1993. Any signs of global warming due to the buildup of greenhouse gases would have been masked for these years by this effect.

Aerosols also result from the oxidation of the gas dimethylsulfide (DMS), $(CH_3)_2S$, which is produced by marine phytoplankton and subsequently released into the air over oceans. Once in the troposphere, DMS undergoes oxidation, some of it to SO_2, which then can oxidize to sulfuric acid, and some to methanesulfonic acid, CH_3SO_3H. Both these acids form aerosol particles, which in turn lead to the formation of water droplets and hence of clouds over the oceans. The particles and droplets deflect incoming light from the sun. Some scientists believe that increased emissions of dimethylsulfide by oceans will occur when seawater warms as a result of the enhancement of the greenhouse effect, and that this negative feedback will temper global warming.

PROBLEM 4-12

The structure for methanesulfonic acid is that of sulfuric acid with one OH replaced by CH_3. Using the principles of tropospheric gas-phase reactivity discussed in Chapter 3, deduce a) how DMS can be converted to formaldehyde and the radical $CH_3—\dot{S}$ by decomposition of an intermediate radical, and b) how the radical $CH_3—\dot{S}$ is eventually converted to $CH_3SO_3\dot{}$, which then abstracts a hydrogen atom to form methanesulfonic acid.

PREDICTIONS ABOUT GLOBAL WARMING

The combined greenhouse effect enhancement from the increases in concentrations of the trace gases methane, nitrous oxide, ozone and the CFCs is now almost as large as that from the increases in carbon dioxide. In order to conveniently summarize the temperature-enhancing effects of all greenhouse gases by a single number, the concept of an **Effective (or Equivalent) Carbon Dioxide concentration** has been devised. For this scale, one considers the *increases* in the concentration of greenhouse gases (other than carbon dioxide) that have occurred since preindustrial times, calculates the global temperature change that should result from these increases, and deduces the increase in CO_2 concentration that *alone* would have had the same effect. The sum of the *actual* carbon dioxide concentration plus the equivalent contributions from increases in the other gases give the Effective CO_2 concentration, which is now slightly over 400 ppm, since the other gases have added the equivalent of about 50 ppm of CO_2. If current trends continue, the Effective CO_2 concentration will be double the preindustrial CO_2 concentration of 280 ppm by the year 2025, although the real CO_2 concentration itself will not have doubled until about 2100. It should be realized that the global warming induced by greenhouse gases is not necessarily linearly proportional to their concentrations; the reasons for this are explained in Box 4-2.

As discussed at the beginning of this chapter, scientists have noted that the emissions to the atmosphere of carbon dioxide and the trace gases which collectively enhance the greenhouse effect continue to rise with time. In research projects that began in the 1980s and continue today, they attempt by computer modelling to predict the consequences of these increases upon the future climate of the planet. There are some uncertainties in such an endeavor, including the fact that we don't as yet fully understand all the sources and sinks of the gases. More important, the sign and the magnitude of the feedback effect of the additional cloud cover expected from warming of the lakes and oceans is not certain; if additional cloud cover occurs mainly at high altitudes the warming feed-back will be positive whereas if it occurs nearer the surface it will be negative.

Notwithstanding these uncertainties, the potential consequences of the enhanced greenhouse effect are so important to life on Earth that much effort has been devoted to refining these computer-based predictions. The most recent IPCC projection for changes in the next 100 years would result from enhancement of the greenhouse effect is a "business as usual" scenario, that is, with no significant controls placed on gas emissions, is shown in Figure 4-10. The temperature and sea level increases shown there are somewhat smaller than were previous predictions but still are worrisome.

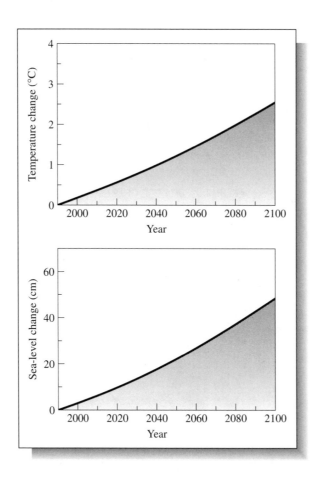

FIGURE 4-10
Predicted increases in the twenty-first century in global average air temperature and sea level. (Source: T. M. L. Wigley and S. C. B. Raper. Implications for climate and sea level of revised IPCC emissions scenarios. *Nature* 357 (1992): 293–300. Copyright 1992 Macmillan Magazines Limited. Redrawn with permission.)

BOX
4-2
THE NONLINEARITY OF
WARMING AND CONCENTRATION

Intuitively, it seems reasonable that the greenhouse warming induced by any gas should be directly proportional to its concentration, but in fact, for reasons developed below, this is true only for gases that are present in near-zero concentrations.

The fraction F of the light of any wavelength that is absorbed by a gas is logarithmically related to the concentration c of the gas and the distance d through which the light travels; this relationship is named the **Beer-Lambert Absorption Law:**

$$\log_e(1 - F) = -Kcd$$

or

$$F = 1 - e^{-Kcd}$$

Here K is a proportionality constant.

A plot of the absorbed fraction F versus the concentration, in the form of the prod-

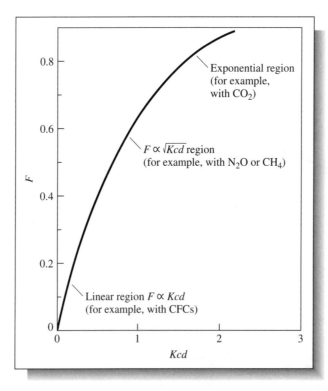

FIGURE 4-11
The dependence of light absorption, F, by a substance upon its atmospheric concentration.

uct Kcd, is illustrated in Figure 4-11. Near zero concentration, F is indeed linearly related to the concentration. Consequently, doubling the concentration of a CFC gas doubles the amount of thermal IR that it absorbs. CFC gases display this behavior since their atmospheric concentrations, zero before the twentieth century, are even now very low, and since they absorb in the window region. Physically, this is equivalent to saying that since only a tiny fraction of the IR light at a wavelength in the window is absorbed, each CFC molecule in the path of IR light emitted from the surface stands an equal chance of absorbing a given outbound photon.

In general, however, doubling the concentration of a gas will not double the amount of thermal IR energy absorbed by it, as illustrated by the nonlinearity of the plot in Figure 4-11. Physically, the reason for the nonlinearity is that for relatively abundant gases such as CO_2, a substantial part of the light at its characteristic absorbing wavelengths already is extracted by existing atmospheric molecules of carbon dioxide. Thus the amount of light for additional carbon dioxide molecules to be exposed to and potentially to absorb is much less than that which leaves the surface of the Earth. Thus doubling the CO_2 concentration does not double the amount of IR absorbed by it. For methane and nitrous oxide, the situation is intermediate between the extremes for CO_2 and CFCs; their absorption and thus their global warming potential rises approximately with the square root of their concentrations.

PROBLEM 4-13

Using Figure 4-5 as a guide, predict the order of enhancement (from greatest to least) on global warming of the introduction to the atmosphere of new, long-lived polyatomic gases having IR absorption vibrations at a) 7; b) 12; c) 16 micrometers (μm). At which one of these three wavelengths would the behavior of F versus concentration for the new gas be most nonlinear?

PROBLEM 4-14

Calculate the ratio of the fractions F of light absorbed by a gas under two conditions specified only as follows: first, after it has doubled its atmospheric concentration, and, second before the doubling occurs, assuming that initially $Kcd = 0.001$. Does the ratio indicate the gas belongs to the region where F is linear with c, or where F is proportional to the square root of c? Repeat the calculation and analysis for an initial Kcd value of 0.9.

An increase in the average atmospheric temperature means that more energy is contained in the air and water at the Earth's surface, and that more violent weather disturbances could result. A recent study by Australian researchers predicts that tropical storms, including hurricanes, will be stronger and more frequent if global warming occurs.

Ironically, an increase in greenhouse gases is predicted to cause a *cooling* of the stratosphere. This phenomenon occurs for two reasons. In the first place, most thermal IR is absorbed at low altitudes (the troposphere), and little is left over to warm the stratosphere. Secondly, at stratospheric temperatures CO_2 emits more thermal IR to space than it absorbs—most of the absorption at these altitudes is due to water vapor and ozone—and so increasing its concentration cools the stratosphere. Indeed, the observed cooling of the stratosphere has been taken to be a signal that the greenhouse effect is undergoing enhancement.

Some scientists have speculated that human health will be affected adversely by global warming. For example, the expected doubling in the annual number of very hot days in temperate zones will affect people who are especially vulnerable to extreme heat, such as the very young, the very old, and those having chronic respiratory diseases, heart disease, or high blood pressure. Domestic violence and civil disturbances could also increase, as they tend to occur more frequently in hot weather. Less directly, global warming may extend the range of insects carrying diseases such as malaria into regions where people have developed no immunity. Food production in temperate areas will probably also be affected by the attack of insects that in the past have been killed off in large measure during the winters but that could survive and flourish under warmer conditions. Animal health could also be affected by the spreading of disease by parasites.

Although the world overall will become more humid as a result of global warming, some areas will become drier. To make matters worse, most areas of the world that currently suffer from drought are predicted to become even drier.

The question of whether any global warming from enhancement of the greenhouse effect has yet occurred is a controversial one. Periodic cycles of heating and cooling have occurred naturally in the past, as is clear from Figure 4-12 (top and middle). According to computer simulations, the increases in carbon dioxide and other gases that have occurred since preindustrial times should have produced a measurable increase in the average air temperature by now. The variations in the average

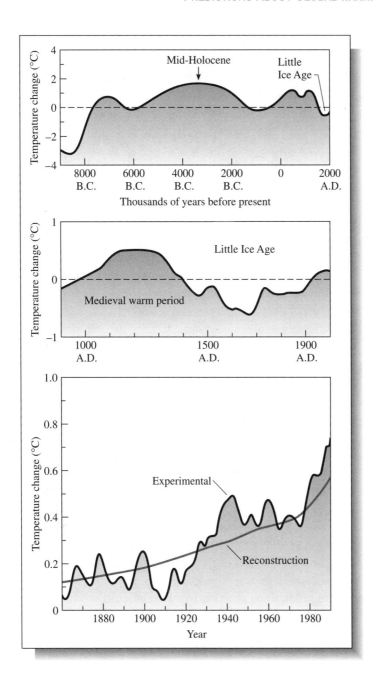

FIGURE 4-12
Global average air temperatures variations (*top*) in past 11,000 years, (*middle*) in past 1,100 years, and, (*bottom*) from 1860–1990. For the last period experimental changes are shown by the thin, fluctuating curve, and predicted changes due to green-house gas emissions and aerosols are shown by the thicker curve. (Source: Top and middle: redrawn with permission from B. Hileman, *Chemical and Engineering News*, April 27, 1992. Bottom: from T. M. L. Wigley and S. C. B. Raper. Implications for climate and sea level of revised IPCC emissions scenarios. *Nature* 357 (1992): 293–300. Copyright 1992 Macmillan Magazines Limited. Redrawn with permission.)

worldwide air temperature for the period 1860–1990 are illustrated by the thin, oscillating line in Figure 4-12, bottom. Clearly an increase of about one-third of one degree Celsius occurred between 1910 and 1940, that is, before fossil fuel combustion underwent its dramatic increase. There also have been increases of about the same magnitude since the mid-1970s. Since the natural temperature variations observed in the past are of the same order of magnitude as those that have occurred recently, it is not absolutely certain that all or indeed that any part of the observed increases in the last 100 years are attributable to the enhancement of the greenhouse effect. The thicker curve in Figure 4-12, bottom represents the reconstruction, that is, the "retrospective prediction," from computer simulations of the temperature trends that should have occurred due to the increase of greenhouse gases and atmospheric aerosols in the 1860–1990 period. The agreement between the calculated and the observed temperature trends is not very quantitative owing perhaps to the influence of the natural fluctuations. Although the year-to-year oscillations are unexplained, the magnitude of the overall observed temperature increase, and the temporary flattening of the rise from the 1950s to the 1970s due to the influence of industrial-based aerosols, are both accounted for in semi-quantitative fashion by the simulations. The cooling of average temperatures in 1992 and 1993 (compared to those for 1990) that resulted from the eruption of Mount Pinatubo interrupted the warming trends of the 1980s, as illustrated for 1992 in Figure 4-13.

We conclude by commenting upon the paradox that faces humanity today concerning the enhancement of the greenhouse effect. On the one hand, there exists the possibility that doubling or quadrupling the Effective CO_2 concentration will have no measurable effect on climate, and that efforts taken to prevent such an increase not only would represent an economic burden for both the developed and the developing worlds, but would perhaps be wasted in the outcome. On the other hand, if the predictions of scientists who model the Earth's climate turn out to be realistic, but we do nothing to prevent further buildup of the gases, both present and future generations will collectively suffer from rapid and perhaps cataclysmic changes to the Earth's climate.

The call to action on the issue of potential climate change was stated perhaps most effectively by the Vice-President of the United States, Al Gore, in his 1992 book *Earth in the Balance: Ecology and the Human Spirit:*

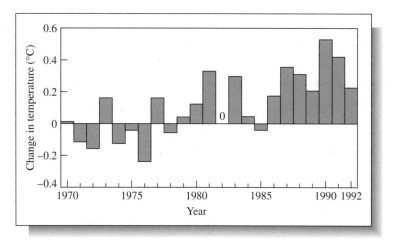

FIGURE 4-13

Cumulative changes in global surface temperatures in recent years (relative to the 1951–1980 average, marked "0" on the vertical axis). (Source: B. Hileman. Global warming trend interrupted in 1992. *Chemical and Engineering News* (Jan. 25, 1993): 7.)

even after highly publicized warnings from virtually the entire global scientific community that the current pattern of our civilization is creating dramatic changes in global climate patterns, likely to be many times larger than any experienced in the last 10,000 years, we are doing virtually nothing to address the principal causes of this catastrophe in the making. We know from the history of climate changes that they can cause unprecedented social and political upheavals, especially in fragile, densely populated societies.

REVIEW QUESTIONS

1. What are the three principal types of incoming light from the sun?

2. What is the wavelength range, in μm, for infrared light? In what portion of this range does the Earth receive IR from the sun? What are the wavelength limits for the "thermal IR" range?

3. Explain in terms of the mechanism involved what is meant by the *greenhouse effect*. Explain what is meant by the *enhancement of the greenhouse effect*.

4. Explain the relationship between the frequency of vibrations in a molecule and the frequencies of light it will absorb.

5. What are the two main sources of carbon dioxide in the atmosphere? What is its main sink?

6. Where could CO_2 extracted from power plant emissions be buried instead of being released to the atmosphere?

7. Why is natural gas considered to be an environmentally superior fuel to oil or coal? What phenomenon involved in its transmission by pipeline might offset this advantage?

8. Is water vapor a greenhouse gas? If so, why is it not usually present on lists of such substances?

9. What is meant by the term "window" as applied to the emission of IR from the Earth's surface? What is the range of wavelengths of this window?

10. What are four important trace gases that contribute to the greenhouse effect?

11. Why don't N_2 and O_2 absorb thermal IR? Why don't we consider CO and NO to be trace gases which could contribute to enhancing the greenhouse effect?

12. What is meant by the "residence time" of a gas in air? How is it related to the gas's rate R of input/output and to its total concentration C?

13. What are the six most important sources of methane?

14. What are the three most important sinks for methane in the atmosphere? Which one of them is dominant?

15. Explain what is meant by *positive* and *negative* feedback. Give an example of each as it affects global warming.

16. Is the enhancement of the greenhouse effect by extra water vapor in the atmosphere due to increased temperature an example of feedback? If so, is it positive or negative feedback? Would an increase in the rate and amount of photosynthesis with increasing temperatures and CO_2 levels be a case of positive or negative feedback?

17. Explain in chemical terms what is meant by "nitrification" and "denitrification." What are the conditions under which nitrous oxide production is enhanced as a byproduct of these two processes?

18. What are the main sources and sinks for N_2O in the atmosphere?

19. Are the proposed CFC replacements themselves greenhouse gases? Why is their emission considered to be less of a problem in enhancing the greenhouse effect than was that of the CFCs themselves?

20. Explain why absorption of thermal IR by CFCs is directly proportional to their concentrations, whereas this linearity of relationship is not true for carbon dioxide.

21. Explain how the Effective CO_2 concentration differs conceptually and numerically from the real CO_2 concentration in air.

SUGGESTIONS FOR

FURTHER
READING

1. Houghton, J. T., et al. *Climate Change. The IPCC Scientific Assessment* (1990) and the Supplementary Report (1992). Cambridge, U.K.: Cambridge University Press.

2. Schneider, S. H. 1989. *Global Warming.* San Francisco: Sierra Club Books.

3. Hileman, B. 1989. Global Warming. *Chemical and Engineering News* March 13, 1989:25–44.

4. White, R. M. 1990. The great climate debate. *Scientific American* July 1, 1990: 36–43.

5. Jones, P. D., and T. M. L.Wigley, 1990. Global warming trends, *Scientific American* August, 1990: 84–91.

6. Pearce, F. 1989. Methane: the hidden greenhouse gas, *New Scientist* May 6, 1989: 37–41.

7. Post, W. M., et al. 1990. The global carbon cycle. *American Scientist* 78: 310–326.

8. Wigley, T. M .L., and S. C. B. Raper, 1992. Implications for climate and sea level of revised IPCC emission scenarios, *Nature* 357: 293–300.

9. Crutzen, P. J. 1991. Methane's sinks and sources. *Nature* 350: 380-381.

10. Charlson, R. J., et al. 1992. Climate forcing by anthropogenic aerosols. *Science* 255: 423–429.

11. Schneider, S. H. Detecting climatic change signals: Are there any fingerprints? *Science* 263: 341–347.

12. Dixon, R. K., et al. 1994. Carbon pools and flux of global forest ecosystems. *Science* 263: 185–190.

INTERCHAPTER:
THE GREENHOUSE EFFECT

AN INTERVIEW WITH STEPHEN H. SCHNEIDER

Stephen H. Schneider received his Ph.D. from Columbia University and is currently a Professor in the Department of Biological Sciences and Senior Fellow at the Institute for International Studies at Stanford University and Senior Scientist at the National Center for Atmospheric Research. Editor of the scientific journal *Climatic Change* and a frequent witness at Congressional hearings, he was the recipient of a MacArthur Foundation Fellowship in June 1992. His current research interests include climatic change and other environmental/science public policy issues, and advancing public understanding of science.

Are you optimistic that developed countries will stabilize or reduce their CO_2 emissions in coming decades? What about rapidly developing countries?

I'm quite optimistic that developed countries can not only stabilize but substantially reduce their CO_2 emissions in the decades ahead, and do so at negligible cost and even make money in the process! The state of the art in engineering with regard to energy efficiency is far ahead of personal, corporate, and governmental energy production and use practices. It's possible to realize savings of 10 percent at worst and more than 50 percent at best on energy expenditures for energy services, ranging from transportation to heating to industrial machinery processes—without sacrificing cost-effectiveness. Although initial investments to replace inefficient existing technologies will carry up-front capital costs, the savings on reduced energy bills over the lifetime of these products or services justify the expectation that the resulting CO_2 emissions reductions can be made at below zero net cost. Of course, this does sometimes involve confronting political interests, since energy efficiency could involve shifts from poor fuels such as coal to cleaner, more efficient fuels such as natural gas, or to renewable power generated by windmills or biomass, for example. My optimism regarding large cost-effective energy reductions diminishes considerably in view of the political difficulties that stand in the way of substantially altering the current energy practices of developed countries—even though it is economically and technically and environmentally advantageous to replace current inefficient practices.

    With regard to developing countries, the situation is even more difficult. Even though developing countries are plagued with far less efficient energy systems than developed countries, so that simply replacing existing technologies with state-of-the-art technologies would produce an even bigger benefit than it would in already efficient countries such as Japan or Italy, developing countries often don't possess the technical skill or sufficient capital to proceed very far along this line. Furthermore, since per capita consumption is so low in developing countries and since developing countries are rightfully demanding improved standards of living, it will be difficult to stabilize or reduce their CO_2 emissions in the coming decades regardless of the efficiency of technology used. What can be done to help the environment, however, is to have the expected *increases* in developing country's emissions reduced by a drastic factor by applying state-of-the-art technology. However, this will take substantial international bargaining, dealing with transfers of technology, financing of such transfers, a reduction in population growth rates, and the deployment of more efficient energy-production and end-use systems in developing countries. Developed countries will have to contribute capital and know-how.

Is it possible that recent increases in the incidence of unusual weather such as flooding and violent storms are due in part at least to the enhancement of the greenhouse effect?

It is possible that human activity may put enough pressure on weather systems to cause them to depart from their normal state and so create the sort of unusual weather conditions that we have recently experienced. Indeed, actual forcing to the climate system by humans has already come about as a result of global-scale heating due to increased levels of greenhouse gases, combined with regional-scale, even more intense, cooling associated with industrial aerosols and smoke aerosols arising from the burning of biomass. The very patchy nature of aerosol cooling versus the more globally distributed greenhouse heating could generate unusual weather patterns, but to date climate modeling has not determined whether the unusual events are part of the random perverseness of nature or forced by human activities—or both. It will require a decade or two of research to make that determination more reliably.

Will replacements for CFCs cause an enhancement of the greenhouse effect?

Yes. Although these replacements are themselves greenhouse chemicals for the most part, they have a shorter lifetime in the atmosphere than CFCs so their net contribution to the greenhouse effect will most likely be comparatively less severe. At the same time, we can also look forward to a degree of relief from the problem of stratospheric ozone depletion. However, if they tempt us by their supposed harmlessness to use them on a scale sufficiently enormous to overwhelm the system, they would turn out to be a cure even worse than the original disease. What is needed is a long-term substitute that has vastly less potential for environmental destruction. Current replacements are interim strategies on the way toward such long-term solutions.

On what grounds do some scientists still doubt predictions of global warming? Are most scientists agreed that it will occur?

This is a very complicated question and difficult to answer concisely. Most scientists agree that global warming has already occurred, pointing to the fact that the current global average surface air temperature is about 0.5°C (1°F) warmer than it was a century ago. The doubts primarily center around the question of whether this change is a natural fluctuation or whether it has been induced by special "forcing" factors such as the doubling of methane levels and the 25% increase in CO_2 levels that humans have caused since the industrial revolution. A warming or cooling trend of one-half degree Celsius could be an "act of nature" with a probability of 10 or 20 percent. These odds are obtained by looking over the last 5,000 years of proxy climate records (for example, tree ring widths) which suggest that such trends occur perhaps once or twice a millennium by the operation of natural causes. However, the vast majority of the knowledgeable scientific community does not doubt the fact that increasing levels of greenhouse gases cause heat to be trapped in the lower part of the earth's atmosphere and at the surface, and that this effect is very likely to lead to some degree of warming. The most contested question is whether processes within the climate system will tend to stabilize global temperatures, limiting any changes to a relatively modest warming (on the order of one degree Celsius or less in the next century), or rather cause a substantial and catastrophic degree of warming to occur (on the order of five or more degrees over the next century). It will probably require a decade or two more of debate and measurement and calculation before this question can be resolved. Unfortunately, while the scientists debate, the real world is "performing an experiment" of its own, one which will determine whether we will be lucky or face potentially serious environmental consequences in the decades ahead.

What specific personal actions can readers of this book take to make a difference about global warming?

There is a whole range of things that individuals can pursue to leave a "lighter footprint on the face of the earth." We can see to it that we do not waste energy or materials, since the mobilization of energy and the use of materials (for example, fuels or paper) both cost us money and degrade the environment. Therefore, exercising simple prudence and common sense in buying efficient light bulbs, cars, well-insulated houses, better industrial machines, in not wasting paper, in recycling materials, and so forth, can make a marked difference both to the economy and to the environment. We might alter our lifestyles to use bicycles more, or do more walking or other things which provide the added benefit of reducing costs and improving our health. However, over the course of the next century the bulk of emissions increases will come not so much from developed countries, but rather from the developing world as it presses its legitimate claims for a larger share of the energy and material resources. We have to reduce the expansiveness of lifestyles in developed countries to allow the "environment space" that will give the developing countries some freedom for growth.

BACKGROUND
ORGANIC CHEMISTRY

Organic chemistry is the chemistry of carbon compounds. In the late nineteenth century and throughout the twentieth, chemists devised special techniques of synthesis by which millions of new organic compounds could be prepared and studied. Thousands of these compounds have found uses and have become items of commerce, even though many of them are toxic to living organisms. Environmental problems are posed whenever living beings—we humans ranking not among the least of these—become exposed to them, even at low concentrations, while they are in active use or as a result of their final disposal. In Chapter 6, the most important environmental problems caused by toxic organic chemicals are discussed in detail. As a prelude, however, in this chapter we provide some necessary background in organic chemistry for those students whose previous education has not included this subject. As part of this material, we shall also include a discussion of some simple chlorine-containing organic compounds that are of environmental concern.

The organic compounds of interest environmentally are mostly electrically neutral molecules containing covalent bonds. Stable compounds of this type inevitably involve the formation of *four* bonds by carbon. Conceptually at least, chemists view *all* organic chemicals as "derived" from those simple organic compounds that contain only carbon and hydrogen, that is **hydrocarbons.** We shall follow this convention, and divide our discussion into several sections, each one of which deals with a different type of hydrocarbon.

ALKANES

The simplest hydrocarbons are those which contain strings of carbon atoms, each one singly bonded to their closest neighboring carbon atom(s) and to several hydrogen atoms. Such hydrocarbons are called **alkanes,** of which the simplest are methane, CH_4, ethane, C_2H_6, and propane, C_3H_8. Commercial supplies of all three are readily available from natural gas wells. Structural formulas for these three alkanes are shown below.

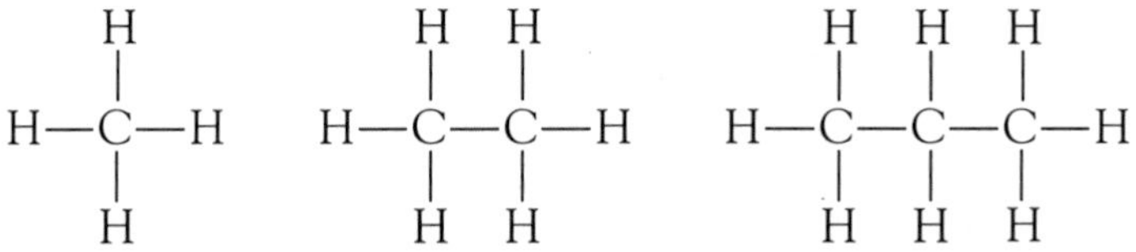

The simplest hydrocarbons are those which contain strings of carbon

For convenience, chemists often write down the formulas for such species by gathering together in one unit all the hydrogens bonded to a given carbon and displaying only the carbon-carbon bonds; thus ethane is represented as CH_3—CH_3 and propane as CH_3—CH_2—CH_3. In another common representation called a **condensed formula,** the C—C single bonds are not shown; rather the formula lists each carbon and the atoms attached to it. For example, the condensed formula for ethane is CH_3CH_3. Each carbon atom in an alkane molecule forms four equiangular single bonds, so the geometry about each carbon is tetrahedral. Thus all alkanes are three-dimensional, nonplanar molecules, even though, for the sake of clarity, their structural formulas seem to represent them as planar molecules involving bond angles of 90° and 180°.

Table 5-1 lists the alkanes having one through twelve carbon atoms in a continuous string (more complex alkanes contain branches, as we shall see shortly). The alkane with four carbon atoms is called butane and is a gas; longer alkanes are liquids or solids under ordinary conditions. When five or more carbons are present in the alkane, the Latin-based abbreviation for that number (*pent* for 5, *hex* for 6, *hept* for 7, *oct* for 8, etc.) is employed as the prefix to the ending *-ane* in its name. Thus, for example, the molecule CH_3—CH_2—CH_2—CH_2—CH_3, which can be written more simply as $CH_3(CH_2)_3CH_3$, is called pentane since it has 5 carbon atoms; its formula is C_5H_{12}. When all the carbons lie in one continuous chain (without branches), the molecule is said to be straight-chained or unbranched, and often the prefix *n-* is added

TABLE
5-1   SOME OF THE SIMPLEST UNBRANCHED ALKANES

Molecular formula	Name	Condensed formula	Boiling point (°C)
CH_4	methane	CH_4	-164
C_2H_6	ethane	CH_3CH_3	-89
C_3H_8	propane	$CH_3CH_2CH_3$	-42
C_4H_{10}	butane	$CH_3CH_2CH_2CH_3$	-0.5
C_5H_{12}	pentane	$CH_3(CH_2)_3CH_3$	36
C_6H_{14}	hexane	$CH_3(CH_2)_4CH_3$	69
C_7H_{16}	heptane	$CH_3(CH_2)_5CH_3$	98
C_8H_{18}	octane	$CH_3(CH_2)_6CH_3$	126
C_9H_{20}	nonane	$CH_3(CH_2)_7CH_3$	151
$C_{10}H_{22}$	decane	$CH_3(CH_2)_8CH_3$	174
$C_{11}H_{24}$	undecane	$CH_3(CH_2)_9CH_3$	196
$C_{12}H_{26}$	dodecane	$CH_3(CH_2)_{10}CH_3$	216

to the name; thus the pentane molecule mentioned above is called
n-pentane.

The constituent atoms of many organic compounds may be "reshuf-
fled" in chemical reactions to yield new structures—the ingredient atoms
remain exactly the same, but the way in which they are linked together
(their order of linkage) can be altered. And so we get distinctly different
compounds. This set of different compounds with the same molecular
formula, but different structures, are **structural isomers** (there are other
types of **isomerism**, too, which we need not consider here). (Sometimes
the difference between isomers is slight: a matter of small differences in
physical properties; sometimes it is enormous: the biological activity of
two isomers may differ profoundly.) Alkanes with four or more carbon
atoms have isomers in which the chain of carbon atoms is branched: not
all the carbons are part of an unbroken path of bonded atoms. An example is the isomer of *n*-pentane illustrated below.

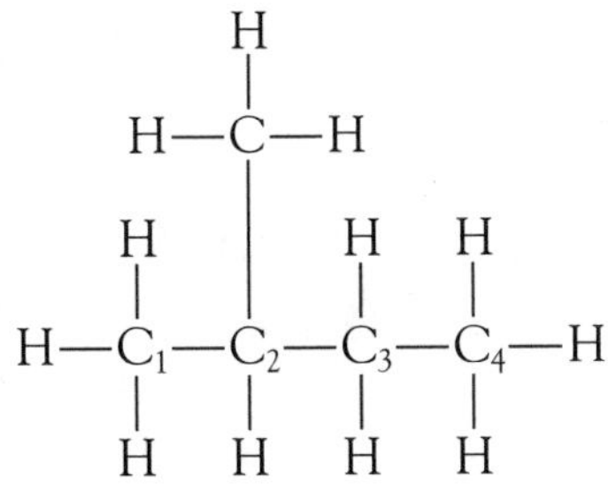

2-methylbutane

In naming such alkanes and other organic molecules, the short chains of carbon atoms which comprise the branches are assigned group names ending in -*yl* which are derived by deleting the -*ane* ending from the name of the alkane hydrocarbon that has the same length (in terms of linked carbon atoms). Thus the CH_3— group is called the methyl group, CH_3CH_2— is called ethyl, and so on. The names of these groups are listed as prefixes to the name for the longest continuous chain of carbon atoms, and each is preceded by a number which indicates the carbon atom of the chain to which the groups are attached. For example, the molecule shown above is called 2-methylbutane, since butane is the alkane consisting of four carbons in an unbranched chain, at the second atom of which the $-CH_3$ group is bonded.

Compounds are known in which one or more of the hydrogen atoms in hydrocarbons such as the alkanes have been substituted by another atom such as fluorine, chlorine, or bromine. Things that substitute for hydrogen atoms are called **substituents.** Indeed, we have already encountered such compounds in Chapter 2; examples included the substituted methanes CF_2Cl_2 (dichlorodifluoromethane) and CHF_2Cl (chlorodifluoromethane) and CF_3Br (bromotrifluoromethane), and the substituted ethane CHF_2-CH_2F which is called 1,1,2-trifluoroethane, where the numbers refer to the carbon numbers to which the fluorines are bonded and whose structure is shown below:

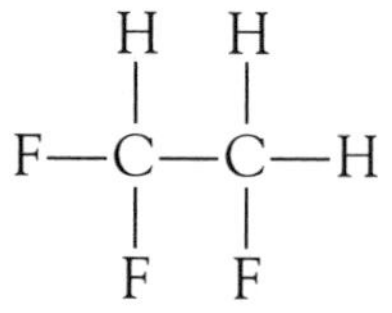

$$\begin{array}{ccc} & H & H \\ & | & | \\ F-&C-&C-H \\ & | & | \\ & F & F \end{array}$$

1,1,2-trifluoroethane

PROBLEM 5-1

Write out the structural formula and the condensed formula for each of the following alkanes: a. *n*-pentane; b. 3-ethylbenzene; c. 2, 3-dimethylbutane.

ALKENES AND THEIR CHLORINATED DERIVATIVES

In some organic molecules, one or more pairs of the carbon atoms are joined by double bonds; since each carbon atom forms a total of four bonds, there are only two additional bonds formed by such carbon atoms. The simplest hydrocarbon of this type is a colorless gas called *ethene*, usually known by its older name *ethylene*:

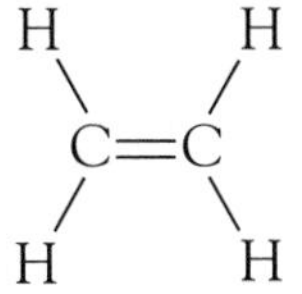

$$
\begin{array}{ccc}
H & & H \\
\diagdown & & \diagup \\
& C = C & \\
\diagup & & \diagdown \\
H & & H
\end{array}
$$

ethene (ethylene)

Notice that the actual planar geometry of the molecule, with bond angles of about 120° around each carbon, can be shown in the structural formula. Condensed formulas normally show the double bond: $CH_2 = CH_2$.

A $C = C$ bond can be a part of a longer sequence of carbon atoms that are joined together by other single or double or triple CC bonds. For example, propene is a three-carbon chain with one adjacent pair of carbons joined by a double bond:

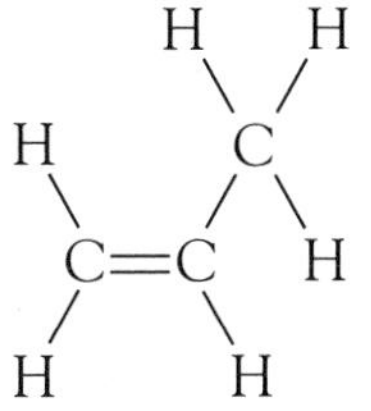

$$
\begin{array}{ccc}
& H \quad H & \\
& \diagdown \diagup & \\
H & C & \\
\diagdown & \diagup \diagdown & \\
C = C & H \quad \text{or} \quad CH_2 = CH - CH_3 \\
\diagup & \diagdown & \\
H & H & \text{propene}
\end{array}
$$

propene

The name for a hydrocarbon chain containing a $C = C$ bond is the same as that used for the alkane of the same length, except that the *-ane* ending of the alkane is replaced by *-ene*. The molecule is numbered such that the $C = C$ unit is part of the continuous chain, and such that the $C = C$ unit is at the lower-numbered end of the chain. Collectively, hydrocarbons containing $C = C$ bonds are called **alkenes.** If there are two $C = C$ bonds in a hydrocarbon, the prefix *di-* is placed before the *-ene* ending; thus the hydrocarbon below is called 1,3-pentadiene:

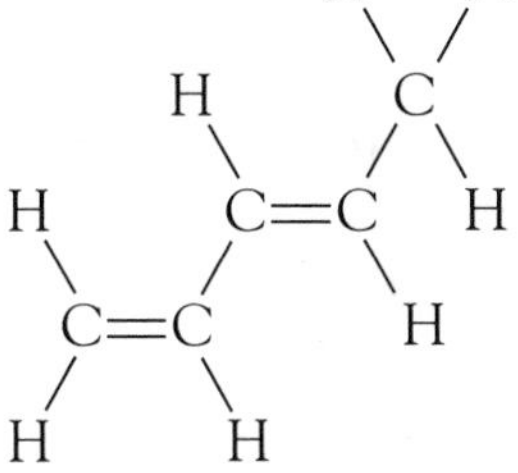

$$
\begin{array}{cccc}
& & H \quad H & \\
& & \diagdown \diagup & \\
& H & C & \\
& \diagdown & \diagup \diagdown & \\
H & C = C & H \\
\diagdown & \diagup & \diagdown & \\
C = C & & H \\
\diagup & \diagdown & \\
H & H &
\end{array}
$$

1,3-pentadiene

The numbers preceding the name are those assigned to the first carbon atom that participates in each of the double bonds. The alternative numbering scheme, that is, assigning the CH_3 carbon on the right to be #1, is not used since the first double bond would then start at carbon #2 and the name would be 2,4-pentadiene and thus the first double bond would not have the lowest possible number.

In some derivatives of ethene, one or more of its hydrogen atoms have been replaced by chlorine atoms. The chloroethenes, like ethene itself, are planar molecules. The simplest example is CH_2=$CHCl$, called *chloroethene* but known in the chemical industry as *vinyl chloride*; it is produced in huge quantities since the common plastic material, polyvinyl chloride (PVC), is subsequently prepared from it.

A number is usually placed in front of the name of the substituent to indicate the specific carbon atom to which it is bonded; thus Cl_2C=CH_2 is called 1,1-dichloroethene to distinguish it from the other isomer, 1,2-dichloroethene, CHCl=CHCl. The molecule 1,1,2-trichloroethene, CCl_2=CHCl, is a liquid solvent that has found extensive uses; these applications and their associated environmental problems are discussed in detail in Box 5-1. Note that the prefix numbers in the compound name here are superfluous since it has no isomers, and thus there is no need to distinguish one isomer from another. The compound is usually referred to by its traditional name trichloroethylene. The structural formulas of a few substitution products of ethene follow:

1,1-dichloroethene   tetrachloroethene

The liquid compound tetrachloroethene, CCl_2=CCl_2, finds use on a large scale as the dry-cleaning solvent used commercially to remove grease spots and other stains on clothing. The prefix 1,1,2,2- is not used as part of its name since it is superfluous (no other arrangements of chlorine being possible), and the traditional name tetrachloroethylene is normally used. Note that when *all* the hydrogens in a molecule have been replaced by a given atom or group, the prefix **per** can be used instead of the actual number; thus tetrachloroethylene is also called perchloroethylene, giving rise to its nickname "perc." Because of its extensive commercial use, and because it can be leached from some PVC water distribution pipes, tetrachloroethylene often contaminates drinking water.

Write structural formulas for each of the following: a. 1,1-dichloropropene; b. perchloropropene; c. 2-butene.

Determine the correct name for each of the following:
a. $CHCl_2CHCl_2$; b. $CH_3—CH_2—CH=CH_2$;
c. $CH_2=CH—CH=CH_2$.

SYMBOLIC REPRESENTATIONS OF CARBON NETWORKS

Organic molecules often contain extensive networks of carbon atoms. Chemists find it convenient to construct shorthand visual representations of such molecules using a symbolic system of lines that indicate only the position of the *bonds* (not including those to hydrogen atoms), rather than writing out a structure in which the C and H atoms are shown explicitly. To indicate the presence of a carbon atom, a "kink" is shown in the chain's representation. For example, the molecule *n*-butane can be represented as (a) or (b) below:

$$CH_3—CH_2—CH_2—CH_3$$

(a)   (b)

The hydrogen atoms are not shown at all in the "stripped down" version; the number of them at any carbon atom can be deduced by subtracting from 4 the number of bonds to that carbon that are displayed explicitly. Thus in the representation below for 2-chloropropane, carbons #1 and #3 must possess 3 hydrogens since they are shown as forming 1 other bond, whereas carbon #2 has 1 hydrogen since it is shown as forming three other bonds:

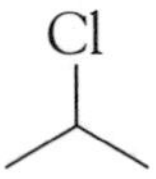

2-chloropropane

TRICHLOROETHYLENE: PROFILE OF A TYPICAL ORGANOCHLORINE SOLVENT

Trichloroethene, often called *trichloroethylene* and abbreviated TCE, is a colorless, oily liquid that has found many uses as a solvent for organic substances that are almost insoluble in water.

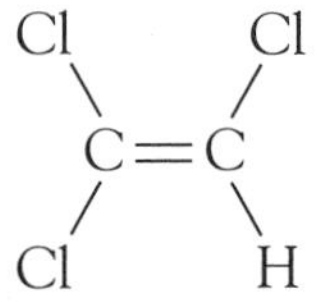

trichloroethylene (TCE)

It is important to have a knowledge at least of the physical properties of such a widespread synthetic material as TCE in order to assess its potential for making its way into air and water. (And into nonaqueous media as well—some pollutants are oil-soluble and can turn up in the fatty tissues of animals.) Thus we shall examine briefly some of the more prominent physical properties of this substance.

TCE boils at 87°C; thus it will remain a liquid even when exposed to the hottest weather. However, its vapor pressure of 8 kPa (i.e., 0.08 atm) at 20°C means that it will evaporate to some extent. Since it is only moderately soluble in water and since its density is 1.47 g/mL—about 50% denser than water—trichloroethylene will form a two-phase system with water (just as "oil and water don't mix") with the TCE at the bottom. Unsubstituted hydrocarbons, in contrast, are less dense than water and form the top layer in such "oil-water" systems.

The solubility in water of organic compounds generally decreases as the number of carbon atoms increases. Solubility also decreases as the proportion of chlorine to hydrogen increases. Trichloroethylene is a small molecule, so even though it has a high Cl/H ratio, it is moderately soluble: 1.1 grams of it dissolves in one liter of water at 25°C. However, its solubility in many organic liquids is greater: when allowed to partition between two layers, one of them water and the other a hydrocarbon, TCE prefers the latter by a large ratio. TCE readily becomes absorbed within organic particles suspended in bodies of water or in sediments. However, in contrast to many other chlorinated hydrocarbons, TCE does not accumulate in large amounts in plants or fish.

Almost all the current usage of trichloroethylene is as an industrial solvent to cleanse grease from metals. Previously it was also used

> as a household and commercial dry-cleaning solvent
>
> to extract constituents from food (e.g., caffeine from coffee)
>
> as a paint stripper
>
> as a fumigant
>
> as an anesthetic agent

For safety and environmental reasons, these other uses have been phased out.

Unfortunately, in the past most TCE was vaporized during its use and thereby became dispersed into the environment. The liquid is still released onto land and into bodies of water as a result of spills and leaks. As a consequence, it is now widely distributed in the environment. TCE is found in about one-third of underground water supplies in the United States, and is the most common chemical detected at "Superfund" toxic dumps (described in more detail in Chapter 7). Near dump sites, its concentration in well water sometimes exceeds 10 mg per liter of water (10 mg/L). The levels in surface waters are usually much less than in underground water since TCE can evaporate into air from the former. Currently the detection limit for TCE in water is about 1 μg/L; levels in surface waters usually vary from less than this value (nondetectable) to about 100 μg/L, that is, 0.1 mg/L. According to the United States Environmental Protection Agency, as little as 22 mg/L of TCE in water leads to chronic health effects in aquatic life. The Canadian government has proposed 0.05 mg/L as the guideline limit for TCE in water intended for drinking by humans and livestock. There is as yet no firm evidence regarding whether or not trichloroethylene causes cancer in humans, although in some studies on rodents, it was discovered to be carcinogenic.

The main sink for TCE becomes available when it evaporates, either from the pure liquid or from water, and enters the atmosphere. Like other molecules containing a $C{=}C$ bond, a hydroxyl radical in air adds to it and initiates a multi-step oxidation process of the type discussed in Chapter 3. Thus TCE released into air contributes to photochemical smog.

PROBLEM 5-4

Write out the full structural formulas for each of the following molecules:

a. 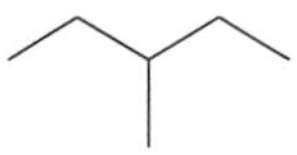   b. 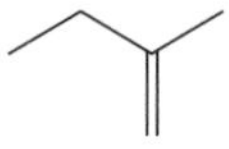   c.

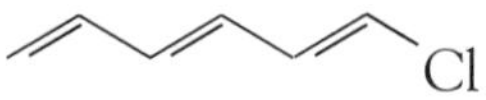

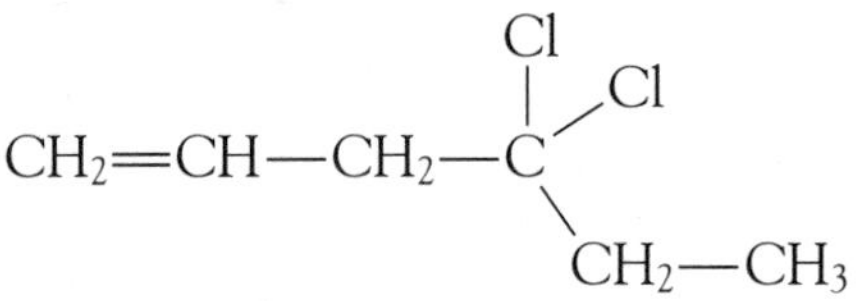

PROBLEM 5-5

Draw symbolic ("kinky") bond diagrams for each of the following molecules:

a. 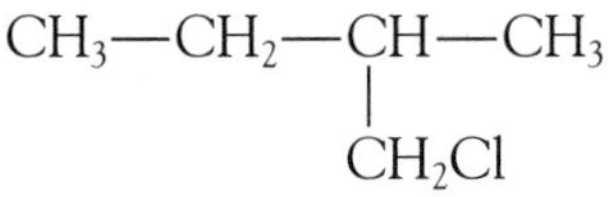 CH_3—CH_2—CH—CH_3 b. 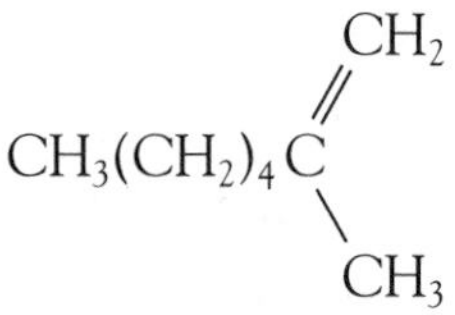 $CH_3(CH_2)_4 C$ \begin{array}{c} CH_2 \\ \| \\ \\ CH_3 \end{array}
             |
             CH_2Cl

c. CH_2=CH—CH_2—C \begin{array}{c} Cl \\ | \\ \diagdown Cl \\ \\ CH_2—CH_3 \end{array}

COMMON FUNCTIONAL GROUPS

In addition to being replaced by simple single-atom substituents like Cl and F, the hydrogen atoms in alkanes and alkenes can be replaced by more complex "attachments" also called **functional groups**—these are typically headed by oxygen or by nitrogen atoms. The common functional groups are listed in Table 5-2. The simplest such polyatomic group is —O—H, usually simply shown as —OH; it is called the **hydroxyl group.** Compounds which correspond to alkanes or alkenes with one hydrogen of a C—H bond replaced by an —OH group are called **alcohols.** Familiar examples are methyl alcohol or methanol (also called wood alcohol), and ethyl alcohol or ethanol (grain alcohol):

TABLE **5-2** SOME COMMON FUNCTIONAL GROUPS

Name of compound type	Functional group
chloride	—Cl
fluoride	—F
alcohol	—OH
ether	—O—
aldehyde	$-C\overset{\displaystyle O}{\underset{\displaystyle H}{}}$
carboxylic acid	$-C\overset{\displaystyle O}{\underset{\displaystyle OH}{}}$
amine	$-N\big\langle$

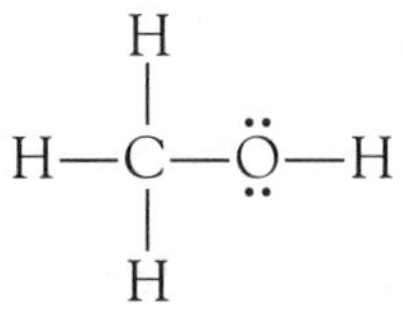

methanol, CH₃OH   ethanol, CH₃CH₂OH

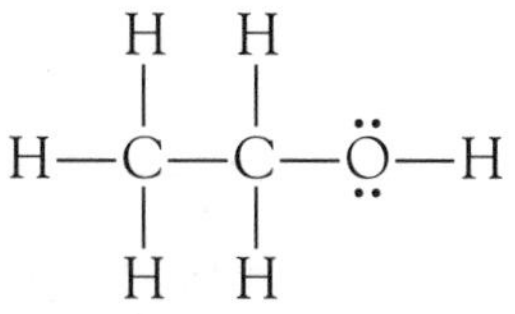

The use of alcohols as fuels is discussed in Chapter 10.

Compounds called **ethers** contain an oxygen atom connected on both sides to a carbon atom or chain:

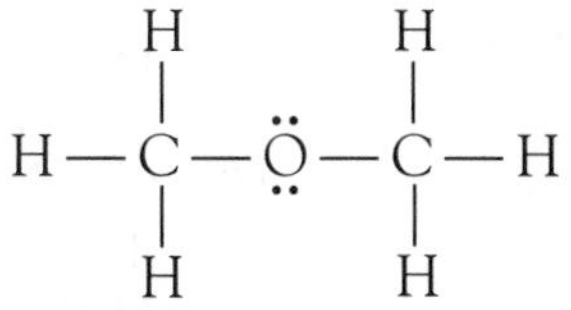

dimethyl ether, $(CH_3)_2O$

In more formal names for such compounds, the —OCH₃ group is known as methoxy, and the —OCH₂—CH₃ group as ethoxy, etc., so that dimethyl ether would be named methoxymethane.

As discussed in Chapter 3, carbon-oxygen double bonds are found in some organic molecules. Molecules that contain the H—C=O group bonded to hydrogen or to a carbon are known as **aldehydes;** the important examples encountered in polluted air are formaldehyde, $H_2C{=}O$, and acetaldehyde, $CH_3C(H){=}O$. (Atoms or groups shown inside parentheses are bonded to the preceding carbon but themselves do not participate in the bond displayed next in the formula.)

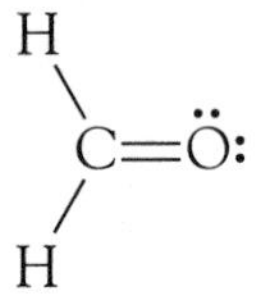

formaldehyde, H_2CO 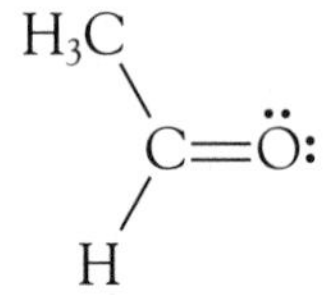   acetaldehyde, CH_3CHO

If the $C{=}O$ group is connected to an —OH group, the system is called a **carboxylic acid;** examples are formic acid and acetic acid.

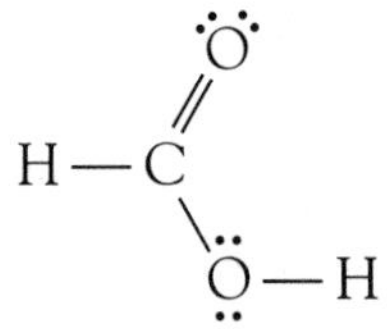

formic acid, HCOOH 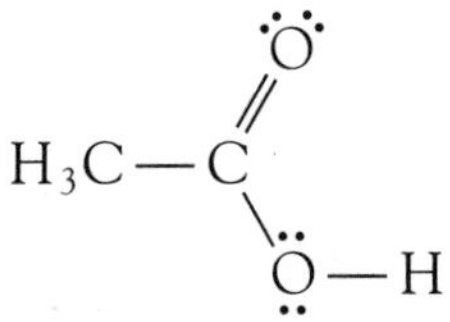   acetic acid, CH_3COOH

Groups headed by nitrogen atoms are known as *amino* groups; they are found to be attached to carbon chains in some organic molecules. Compounds in which the amino group is bonded to a hydrocarbon chain are called **amines.** Note that nitrogen atoms form a total of three bonds, some (or all) of which can be directed to carbons. Two examples follow:

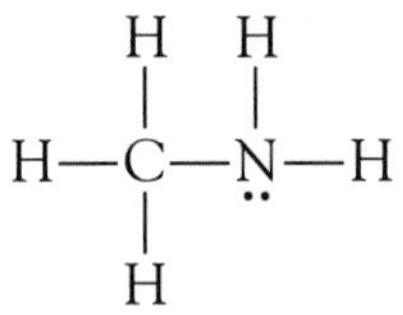

methylamine, CH_3NH_2 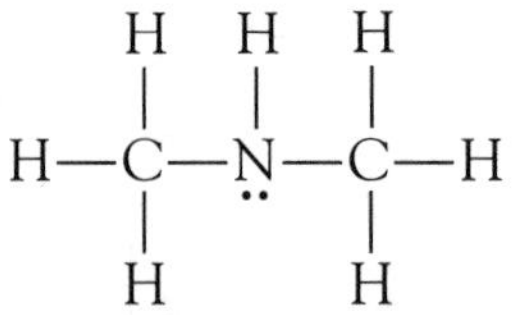   dimethylamine, $(CH_3)_2NH$

Alcohols, acids, and amines that contain short carbon chains are quite soluble in water. The reason for this behavior is that molecules of these three types contain O—H or N—H bonds, each of which possesses a hydrogen atom that is partially depleted of electron density by the highly electronegative atom (O or N) to which it is bonded. The partial positive charge δ^+ of the hydrogen is attracted to regions of unbonded electron density—lone pairs—on atoms of adjacent molecules.

$$:\ddot{O}\!-\!H \boxed{^{\delta+}} \quad :\underset{\displaystyle}{O}\!-\!H$$

with structure: :Ö—H (with H below left O) and H above the right O, hydrogen bond arrow pointing to the bracketed region labeled "hydrogen bond"

Such interactions are called hydrogen bonds; the forces holding the two atoms together—and therefore also holding together the two molecules to which the atoms belong—are not nearly as strong as those of a regular bond within a molecule but are much stronger than the forces that operate between molecules in hydrocarbons. Water molecules are in precisely this situation. They stick together because each hydrogen atom is hydrogen bonded to the lone pair of the closest H_2O molecule to it (see Figure 5-1). The attraction between H_2O molecules from these interactions results in a relatively high boiling point for liquid water, much higher than anticipated for a molecule of its mass. For a (nonionic) substance to be freely soluble in water, these secondary bonds between adjacent water molecules must be replaced by similar interactions between the substance and the water molecules. Consequently, molecules that contain N—H or O—H bonds and a short chain of carbons are soluble in water because the hydrogen bonds they form with H_2O molecules replace those that become broken when the substance is incorporated into the liquid.

Hydrogen atoms bonded to carbon cannot form hydrogen bonds with water molecules, since the carbon is not sufficiently electronegative to produce much of a positive charge on a hydrogen atom bonded to it. In addition, there are no lone pairs on the carbon atoms. Consequently, there is no driving force that can disrupt the extensive network of hydrogen bonding within liquid water in order to incorporate a large number of molecules of hydrocarbons or chlorinated organic molecules. In both hydrocarbon and chlorinated organic molecules all the hydrogens are bonded to carbon, and for this reason, such molecules are not very soluble in water. Even molecules with one O—H or N—H group and many carbon atoms are insoluble in water, since their overall character is dominated by the large number of carbons. The forces of attraction that do exist between organic molecules that contain no hydrogen bonding capacity are quite nonspecific and nondirectional; consequently different hydrocarbons are quite soluble in each other, and organochlorine molecules are soluble in hydrocarbons. We can restate the familiar generalization that "like dissolves like": compounds tend to dissolve in other substances having the same types of intermolecular interactions.

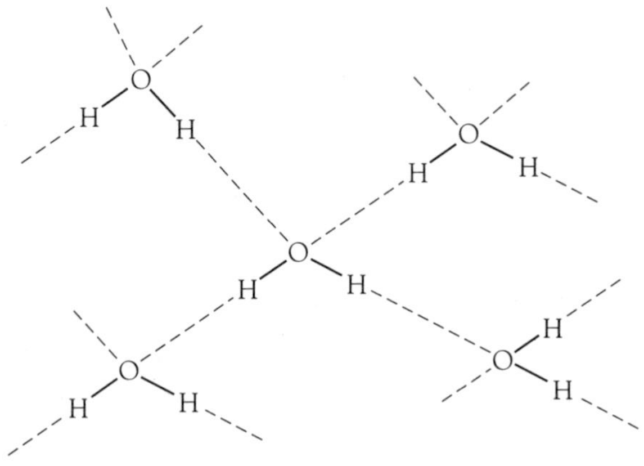

FIGURE 5-1
Hydrogen bonding (dashed lines) between molecules in water.

PROBLEM 5-6

Write the structural formulas and symbolic diagrams for each of the following: a. ethyl alcohol; b. ethylamine; c. acetic acid (the carboxylic acid with a methyl group bonded to the acidic carbon).

RINGS OF CARBON ATOMS

Networks of carbon atoms exist as rings in many organic molecules. The most common rings are those which contain five, six or seven carbon atoms. Molecules containing rings are named by placing the prefix "cyclo" in front of the usual title for the carbon chain of that length. Thus a ring of six carbons, all joined by single C—C bonds, is called cyclohexane. The molecule shown at the right below is called methylcyclopentane.

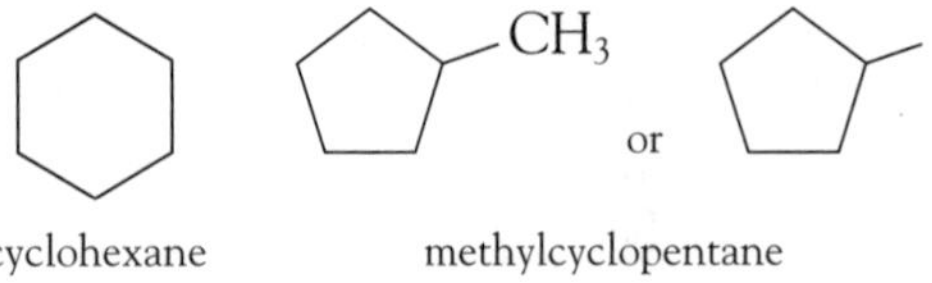

cyclohexane   methylcyclopentane

Some cyclic structures are quite complicated, with some carbon atoms being components of several rings simultaneously; an example is shown below:

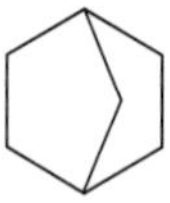

(The jointed line attached at each end to a ring carbon indicates the upward projection from the plane of this page of a "bridge" carbon atom.) We shall not derive systematic names for such complicated molecules, as all extended ring systems of interest to us have "trivial" names—commonly used, handy, traditional, but not descriptive. Unique numbering schemes for such rings also are employed, as we shall see in Chapter 6.

PROBLEM 5-7

Write out both simple and symbolic bond diagrams for the following molecules: a. cyclopropane; b. chlorocyclobutane; c. any isomer of dimethylcyclohexane

BENZENE

One of the most common and most stable organic structural units is the benzene ring, a planar hexagon of six carbon atoms. In the parent hydrocarbon it also contains six hydrogen atoms, one bonded to each carbon, and which also lie in the C_6 plane:

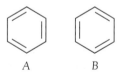

A   *B*

Each carbon in C_6H_6 is bonded to two carbons and to *one* hydrogen, and so in order to form four bonds it must be doubly bonded to one of its neighboring carbons. The two ways of achieving this result are shown in the so-called Kekulé structures (*A* and *B* above). In fact, benzene molecules adopt neither of these two forms, each of which would have alternating short C=C and long C—C bonds; rather they exist in an averaged "resonance" structure in which all CC bonds have the same, intermediate length. (The term "resonance" here alludes to the mathematics of the bond description, and is commonly interpreted as "blend" or "hybrid.") This result is represented by the structure shown below with the hexagon containing an enclosed circle to represent the three double bonds; often however just one of the Kekulé structures is shown, it being understood, at least by chemists, that no actual alternation of bonds is meant.

Since the molecule is planar and has six equal sides, each CCC and CCH angle is 120°. (Note that when benzene occurs as a substituent group in another molecule, it is given the name *phenyl*.)

Physically, benzene is a colorless liquid at room temperature that freezes at 5.5°C and boils at 80°C, that is, just inside the limits for liquid H_2O. Like other hydrocarbons, it is almost insoluble in water and less dense than water. Chronic exposure to occupational levels of benzene has produced leukemia in humans. Benzene's concentration in urban air is appreciable, since it is a component of modern gasoline and is volatilized from it. The environmental consequences of using benzene in gasoline is discussed further in Chapter 10.

Alkenes readily react by addition of molecules such as H_2, HCl, and Cl_2 across the double bonds—that is, with one atom attaching itself to each of the two carbons on the double bond—and thereby convert the C=C units to single bonds. For example, the addition of HCl to ethene produces chloroethane. The corresponding reactions do not readily occur across the bonds in benzene or its derivatives. Benzene can be **hydrogenated,** that is, hydrogen can be added across its double bonds, but only under rather extreme conditions. This difference in behavior of benzene compared to alkenes is an example of the special stability of a six-membered ring containing three sets of alternating double and single bonds. The electrons of the bonds interact with each other in a manner that makes the molecule energetically much more stable than would be expected from adding up single and double bond energies appropriate to alkanes and alkenes. The extra stability disappears if even one of the three double bonds is hydrogenated or otherwise added across by other molecules. Thus the six-membered benzene ring is a unit of great inherent stability, and survives intact in media that would destroy other C=C bonds. Benzene and other molecules that possess this extra stability are said to be **aromatic** systems.

An exception to the rule that benzene does not add across its double bonds occurs when the attacking atom or molecule is a free radical such as a chlorine atom; recall that such species possess an unpaired electron. Thus, whereas molecular chlorine, Cl_2, itself does not add to the double bonds in benzene, a single chlorine atom does. In fact, as early as 1825, Michael Faraday found that benzene would react if chlorine gas was exposed to strong light, which we now realize splits the Cl_2 molecules into free Cl atoms. Once such a reaction starts, it continues until all the carbon atoms have added one chlorine atom, and 1,2,3,4,5,6-hexachlorocyclohexane is produced:

$$C_6H_6 + 3\ Cl_2 \xrightarrow{\text{UV light}} C_6H_6Cl_6$$

benzene   1,2,3,4,5,6,-hexachlorocyclohexane

This technique is used to prepare the insecticide "BHC" which is discussed in Box 5-2.

CHLORINATED BENZENES

Although benzene does not readily undergo addition reactions with molecules or ions that are not free radicals, it does participate in **substitution reactions:** one of its hydrogen atoms can be replaced by a group such as methyl, hydroxyl, and so on, that forms a single bond. Of particular interest is substitution by chlorine, since many compounds of environmental concern contain chlorine-substituted benzene rings. When benzene is reacted with chlorine gas in the presence of a catalyst such as iron (III) chloride, $FeCl_3$, one of the hydrogens (explicitly drawn on the ring below for clarity) is replaced by chlorine, and HCl gas is released:

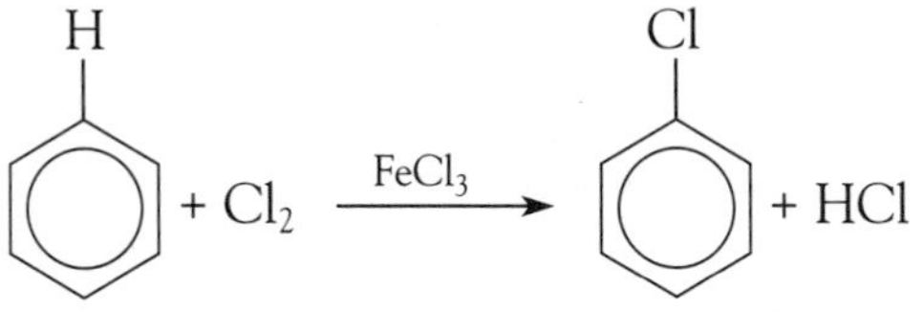

Notice that the aromatic C_6 ring survives intact.

If the reaction is allowed to continue, that is, if excess Cl_2 is available, one or more hydrogen atoms of the chlorobenzene molecules will in turn become replaced by chlorine. There are three isomeric dichlorobenzenes, all of which could in principle be produced in such a reaction:

BOX
5-2
LINDANE AND OTHER
HEXACHLORINATED CYCLOHEXANES

During World War II, the derivative of cyclohexane having one of the two hydrogens on each carbon substituted by chlorine, namely, 1,2,3,4,5,6-hexachlorocyclohexane, was discovered to be an effective insecticide against a wide variety of insects. In fact, there exist eight isomers having this formula; they differ only in the relative orientations of the chlorine atoms bonded to different carbons. Subsequent research indicated that only one of the isomers actually kills insects, the so-called gamma isomer; it is now sold separately under the name Lindane. (The diagrammatic formula below is not intended to illustrate the chlorines' orientations—only their points of attachment.)

1,2,3,4,5,6-hexachlorocyclohexane

Lindane is the active ingredient in several commercial medical preparations used to rid children of lice and scabies, though prolonged use for these purposes should be avoided since the substance can penetrate human skin and subsequently cause serious side effects. For example, it is known that Lindane causes liver damage in laboratory animals.

When 1,2,3,4,5,6-hexachlorocyclohexane is produced commercially, a mixture of five or six of the eight isomers is obtained, including about 13% of the gamma isomer (Lindane). The trivial, somewhat misleading name used for this mixture is BHC ("*benzene hexachloride*"). Like DDT, it was extensively used for mosquito control and in agricultural applications after World War II. However, use of the BHC mixture has been severely restricted since the 1970s on account of its toxicity—it is also a suspected carcinogen—and because some isomers are found to accumulate in biological tissue. Lindane itself is still used to treat seeds and tree seedlings. It is generally considered that the environmental levels of the various hexachlorocyclohexanes are not high enough to pose a health threat. However, a recent study in Finland found that one of the isomers (beta) in BHC was present in significantly higher concentrations in women with breast cancer than in women free of this disease.

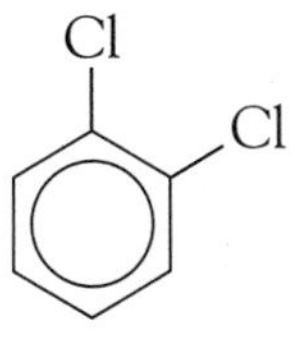

1,2-dichlorobenzene

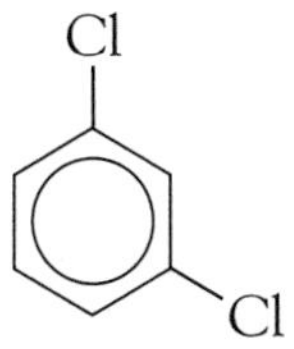

1,3-dichlorobenzene

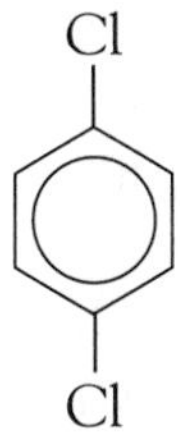

1,4-dichlorobenzene

The numbering scheme begins at one of the "substituted" carbons; the direction of numbering around the ring is chosen to yield the smallest possible number for the second substituent. In older nomenclature, 1,2 disubstituted benzene is called the **ortho-**substituted isomer, 1,3 disubstitution calls for the prefix **meta-,** and the 1,4 isomer is termed **para.** Thus, the compound named 1,4-dichlorobenzene pictured at right is also called paradichlorobenzene or p-dichlorobenzene. When using the Kekulé structure for benzene in which double and single bonds are displayed, it is important to remember that the choice of a structure (A or B on page 207) for the positions of the double bonds is an arbitrary one. This has the consequence that there are only three, not five, isomeric dichlorobenzenes. For example, 1,6-dichlorobenzene is not different from the 1,2 isomer; they represent the same molecule viewed from different perspectives. (To avoid any such complications, many chemists use only the circle-containing hexagon symbol shown on page 208.)

PROBLEM 5-8

Deduce the structures and names for the three chemically different trichlorobenzenes.

The various polychlorobenzenes are used commercially as pesticides and as solvents in the chemical and electrical industries. Some isomeric mixtures were also used in electrical transformer fluids.

The 1,4 isomer of dichlorobenzene is used as an insecticidal fumigant, that is, an airborne insecticide. It is sold as one type of domestic "moth balls," since although it is a crystalline solid, it has an appreciable vapor pressure—it is volatile, even though a solid. Enough of it will vaporize to act as an effective insecticide in the immediate area. The same compound has also been used as a soil fumigant and a pesticide.

However, it is an animal carcinogen and does accumulate to some extent in the environment.

The fully substituted compound, *hexachlorobenzene* or "HCB" (not to be confused with the previously-mentioned compound known as BHC, which still contains all its original hydrogen and which is a cyclohexane rather than a benzene derivative), was formerly used as an agricultural fungicide. Since it is extremely persistent and is still emitted as a byproduct in the chemical industry and in combustion processes, it remains a widespread environmental contaminant. It is of concern because it is known to cause liver cancer in laboratory rodents.

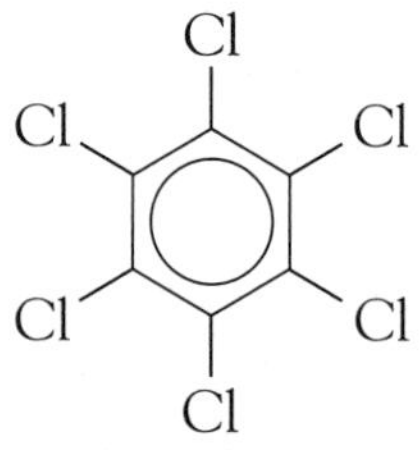

hexachlorobenzene (HCB)

The United States Environmental Protection Agency has recently added HCB to its list of compounds for which drinking water standards have been set; the maximum contaminant level is 0.001 milligrams per liter. Like most organochlorine compounds, hexachlorobenzene is much more soluble in organic media than in water. In bodies of water such as rivers and lakes, organochlorines such as the chlorobenzenes are much more likely to be bound to the surfaces of organic particulate matter suspended in the water and on the muddy sediments at the bottom than to be dissolved in the water itself. From these sources, they enter living organisms such as fish; for reasons discussed in detail in Chapter 6, their concentration in fish is often thousand or millions of times greater than that dissolved in polluted drinking water. For humans, the amount of organochlorines ingested by eating a single Great Lakes fish is generally greater than the total organochlorine content in a lifetime of drinking water from the same Great Lake! Our current daily exposure to HCB is not sufficiently great to pose a significant health hazard even though it is estimated that 99% of Americans have detectable levels of the chemical in their body fat.

REVIEW QUESTIONS

1. What is the name of the hydrocarbon $CH_3(CH_2)_4CH_3$? Is it an alkane? Draw its structural formula.

2. Draw the structural formula for 3-ethylheptane.

3. What would be the name for the substituent group $CH_3CH_2CH_2$—?

4. Draw structural and condensed formulas for
   a. trichloroethene
   b. 1,1-difluoroethene

5. What is the main use for the compound $CCl_2{=}CCl_2$? What two names are used for this compound?

6. Draw structural formulas for each of the following: methyl alcohol, methylamine, formaldehyde, and formic acid.

7. Draw the structural formulas and symbolic representations for cyclopentane and for cyclopentene.

8. Explain what the term *hydrogen bonding* means. Explain why a short-chain alcohol such as methanol is soluble in water whereas a long-chain one such as octanol is not.

9. Is the six-membered ring system in benzene particularly stable or unstable?

10. Do molecules such as H_2 readily add to benzene? Do free radicals such as atomic hydrogen readily add to benzene?

11. What is another name for perchlorobenzene? Has the compound been used as a pesticide? Do you expect it to be more soluble in aqueous or in hydrocarbon media?

SUGGESTIONS FOR FURTHER READING

For more extensive background in organic chemistry, consult a modern introductory textbook such as K. P. C. Vollhardt and N. E. Schore, *Organic Chemistry*, 2d edition, W. H. Freeman and Company, New York, 1994.

CHAPTER

TOXIC ORGANIC
CHEMICALS

The term *synthetic chemicals* is used by the media to describe substances that generally do not occur in nature, but which have been synthesized by chemists from simpler substances.* The great majority of commercial synthetic chemicals are organic compounds, and most use petroleum as the original source of their carbon.

In this chapter, the environmental consequences of the widespread use of synthetic organic chemicals are discussed, with emphasis on those substances whose toxicity has led to concerns regarding human health, especially with respect to cancer and to birth defects, as well as the well-being of the lower organisms.

We shall first discuss pesticides (including insecticides and herbicides), and consider the environmental problems associated with their use. As we shall see, sometimes it is the trace impurities such as dioxins in these commercial substances that are the principal concern. We then consider PCBs, which are industrial chemicals of widespread environmental concern both with respect to their own properties and those of

---

* It may not be an altogether trivial point to note that many artificial, or "man-made"—to borrow the idiom of an earlier era—chemicals are not strictly *synthetic*, or literally "put together" (Greek: *syn* + *thesis*), but may be derived from the *breakdown* of more complex materials. In this book, however, the commonly accepted equivalence between the terms *synthetic* and *artificial* will not be challenged.

215

their contaminants. We conclude with a look at a series of toxic hydrocarbons that are both air and water pollutants.

As implied in Chapters 2 and 5, carbon forms many compounds with chlorine. Due to their toxicity to some plants and insects, many such **organochlorine** compounds, produced by the action of elemental chlorine upon hydrocarbons derived from petroleum, have found extensive use as pesticides. Other organochlorines have been used extensively in the plastics and electronics industries. The carbon-chlorine bond characteristically is difficult to break, and the presence of chlorine also lessens the reactivity of other bonds in organic molecules. For many applications, this lack of reactivity is a distinct advantage. However, this same property means that once organochlorines have entered the environment, they are slow to degrade and tend instead to accumulate. Furthermore, most organochlorine compounds are **hydrophobic:** they do not readily dissolve in water but they are easily soluble in hydrocarbon-like media such as oils or fatty tissue. The lack of an efficient sink for organochlorine compounds, in combination with their hydrophobicity, has led to their accumulation in living organisms, including fish, humans, and other animals. Indeed, the entire planet, including all living things, has undergone low-level contamination by these chemicals. Much of the effort by government agencies and environmental groups in the past decades has involved the documentation of this contamination and the regulation of organochlorine use to prevent concentrations from reaching dangerous levels, particularly in our food supply.

The toxic organic substances discussed in this chapter are mainly, but not exclusively, organochlorine compounds, but we will exclude from the following considerations the chlorofluorocarbons, which were covered in detail in Chapter 2, and the simpler organochlorines such as the chlorobenzenes, which were discussed in Chapter 5.

PESTICIDES

TYPES OF PESTICIDES

Pesticides are substances that kill or otherwise control (for instance, by interfering with the reproductive process) an unwanted organism. The various categories of pesticides are listed in Table 6-1. All chemical pesticides share the common property of blocking a vital metabolic process of the organisms to which they are toxic. We discuss first **insecticides,**

TABLE **6-1** PESTICIDES AND THEIR TARGETS

Pesticide type	Target organism
acaricide	mites
algicide	algae
avicide	birds
bactericide	bacteria
disinfectant	microorganisms
fungicide	fungi
herbicide	plants
insecticide	insects
larvicide	insect larvae
molluscicide	snails, slugs
nematicide	nematodes
piscicide	fish
rodenticide	rodents

which kill insects, and subsequently we consider **herbicides,** which kill plants. Also mentioned in passing are some **fungicides,** substances which are used to control the growth of various types of fungus. Collectively, these three categories represent the great bulk of the one billion kilograms of pesticides that are used annually in North America. Almost half of the usage of pesticides in the United States involves agriculture. Currently, the greatest use of insecticides occurs in the growing of cotton, whereas the majority of herbicide use comes in the growing of corn and soybeans. Most domestic households contain at least one synthetic pesticide; typical examples are weed killers for the lawn and garden, algae controls for the swimming pool, flea powders for use on pets, and sprays to kill insects.

Almost since their introduction, synthetic pesticides have been a concern because of the potential impact on human health of eating food contaminated with these chemicals. About half the foods eaten in the United States contain measurable levels of at least one pesticide. For that reason, many have been banned or restricted in their use. Nevertheless, a 1993 report by the National Academy of Science in the United States pointed out that pesticide regulation to date has not paid enough attention to the protection of human health, especially that of infants and children, whose growth and development are at stake. Currently, the United States Environmental Protection Agency sets maximum acceptable pesticide intakes by dividing by 100 the highest level that causes no adverse effects in test animals; some scientists have suggested that an additional factor of 10 be used in order to protect children.

By way of relieving the immediate urgency of these concerns, other scientists have emphasized recently that plants themselves manufacture insecticides in order to discourage insects and fungi from consuming them, and consequently that we are exposed in our food supply to much higher concentrations of these "natural" pesticides than to synthetic ones.

Traditional Insecticides

Insecticides of one type or another have been used by society for thousands of years. One principal motivation for using insecticides is to control disease: human deaths due to insect-borne diseases through the ages have greatly exceeded those attributable to the effects of warfare. The use of various insecticides has greatly reduced the incidence of diseases transmitted by insects and the rodents which bear them: malaria, yellow fever, bubonic plague, and sleeping sickness scarcely exhaust the list of these scourges. People also try to control insects such as the mosquito and the common fly simply because their presence is annoying. The other principal motivation for insecticide usage is to prevent insects from attacking food crops: even with extensive use of pesticides, about one-third of the world's total crop yield is destroyed by pests or weeds during growth, harvesting, and storage.

The earliest recorded usage of pesticides was the burning of sulfur to fumigate Greek homes around 1000 B.C.; **fumigants** are pesticides that enter the insect as an inhaled gas. The use of SO_2 from the burning of solid sulfur, sometimes by incorporating the element in candles, continued at least into the nineteenth century. Sulfur itself, in the form of dusts and sprays, was also used as an insecticide and as a fungicide; it is still employed in the latter capacity against powdery mildew on plants. Hydrogen cyanide gas has also been used as a fumigant. Its use to prevent damage to specimens in museum cases was recorded in 1877, and a few years later it was used to control insects in fruit trees. It is, of course, very lethal to humans. Inorganic fluorides, such as sodium fluoride, NaF, were used domestically to control ant populations; both sodium fluoride and boric acid were used to kill cockroaches in infested buildings. Various oils, whether from petroleum or living sources such as fish and whales, have found use for hundreds of years as insecticides and as "dormant sprays" to kill insect eggs.

The use of arsenic and its compounds to control insects dates back to at least 900 A.D., and became quite widespread from the late nineteenth century until the Second World War. Paris Green, which is a copper salt containing the arsenite ion, AsO_3^{3-}, was a popular insecticide introduced in 1867. Other salts containing this ion, or the arsenate ion, AsO_4^{3-}, have also been employed; all operate as stomach poisons, and kill insects that ingest them. Arsenic compounds continued to be heavily used as insecticides in the 1930s, 1940s, and early 1950s.

Unfortunately, inorganic and organometallic pesticides are usually quite toxic to humans and other mammals, especially at the dosage levels that are required to make them effective pesticides. Mass poisonings have occurred as a result of the use of some mercury-based fungicides, as discussed in Chapter 9. In addition, heavy metals, such as the arsenic commonly used in such pesticides, are not biodegradable; once released into the environment, they will remain indefinitely in the water, wildlife, soil, or sediments and may enter the food supply if liberated from these sites. During and after the Second World War, many organic insecticides were developed that have largely displaced these inorganic and organometallic substances, as discussed below. Usually only small amounts of the organic compounds are required to be effective against the target pests and thus smaller amounts of chemicals enter the environment. Given a dose of each large enough to act as a pesticide, the organic substances are generally much less toxic to humans than are the inorganic pesticides. Finally, organic pesticides were initially thought to be biodegradable, though as we shall see this has certainly not been found to be true in many cases.

ORGANOCHLORINE INSECTICIDES

In the 1940s and the 1950s, the chemical industries in North America and western Europe produced large quantities of many new pesticides, especially insecticides. The active ingredients in most of these pesticides were organochlorines, many of which share several notable properties:

Stability against decomposition or degradation in the environment

Very low solubility in water, unless oxygen or nitrogen is also present in the molecules

High solubility in hydrocarbon-like environments, such as the fatty material in living matter

Relatively high toxicity to insects but low toxicity to humans

As an example, consider the compound hexachlorobenzene (HCB) mentioned in Chapter 5. It is stable, easy to prepare from chlorine and benzene, and for several decades after World War II it found use as an agricultural fungicide for cereal crops. Although very soluble in organic media such as liquid hydrocarbons, it is almost insoluble in water: only 0.0062 milligrams of HCB dissolve in one liter of water.

As with pollutants in general, means of expressing their "degree of presence" in the environment are fundamental to treating them and their effects on a quantitative basis. The solubilities of trace substances in liquids and solids are often expressed on a "parts per" scale, rather than on a mass or moles per unit volume basis. However, the "parts per" scales for condensed (nongaseous) media express the ratio of the *mass* of the solute to the *mass* of the solution, not the ratio of moles or molecules as is used for gases. Since the mass of one liter of a natural water sample is very close to one kilogram, the HCB solubility quoted above (0.0062 milligrams solute per liter) corresponds to 0.0062 milligrams solute per 1,000 grams of solution. Multiplying both numerator and denominator by 1,000, we conclude that 0.0062 mg/L is equivalent to 0.0062 grams of solute per one million grams of solution, that is, to 0.0062 parts per million. In general, the value for the ppm solubility of any trace substance in water is the same as its value in units of milligrams per liter or micrograms per gram.

PROBLEM 6-1

For aqueous solutions, a. convert 0.04 μg/L to the ppm and ppb scales, and b. convert 3 ppb to the μg/L scale.

But the pollution of aquatic environments is not merely a question of the concentration of pollutant actually in the solution state, and the small values for the solubilities of organochlorines in water may be deceptive on this score. Much greater amounts of these substances are bound to the surfaces of the organic particulate matter suspended in the water or present in the muddy sediments at the bottom of rivers and lakes. From these sources, as well as from the amounts actually dissolved in the water, organochlorines enter both plants and animals living in the natural waters. For reasons that will be discussed in detail below, many organochlorines reach concentrations in living matter that are thousands or millions of times that which is actually dissolved in the water. It is due to this phenomenon that concentrations of organochlorines have often reached dangerous levels in many species. Many organochlorine insecticides have been removed from use as a consequence.

DDT

DDT, or para-dichlorodiphenyltrichloroethane, has had a tumultuous history. It was hailed as "miraculous" in 1945 by Sir Winston Churchill because of its use in the war effort. It was very effective against mosquitoes that carry malaria and yellow fever, against body lice that can transmit typhus, and against plague-carrying fleas. The World Health Organization estimated that malaria reduction programs, one component of which was the use of DDT, saved the lives of more than five million people. Unfortunately, DDT was widely overused, particularly in agriculture, which consumed 80% of its production. As a result, its environmental concentration rose rapidly and it began to affect the reproductive abilities of birds which indirectly incorporated it into their bodies. By 1962 DDT was being called an "elixir of death" by the writer Rachel Carson in her influential book *Silent Spring*, because of its role in decreasing the populations of certain birds such as the bald eagle, whose intake of the chemical in their diet was very high. Further detail concerning the history of DDT is given in Box 6-1.

Structurally, DDT is a substituted ethane. At one carbon, all three hydrogens are replaced by chlorine atoms, while at the other, two of the three hydrogens are replaced by a phenyl (i.e., benzene) ring; each of the rings contains a chlorine atom at the *para* position, that is, directly opposite the ring carbon which is joined to the ethane unit:

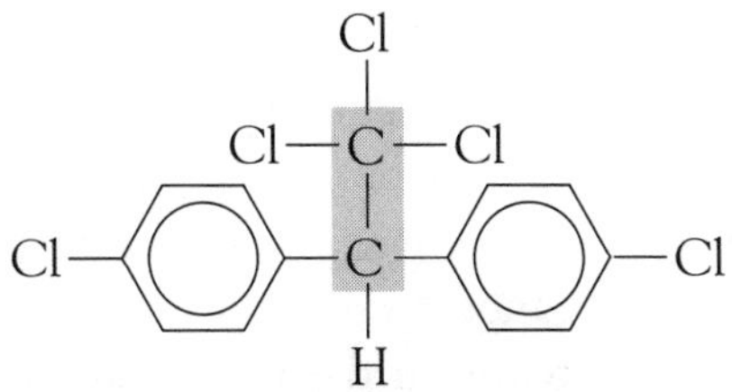

DDT: para-dichlorodiphenyltrichloroethane

DDT's persistence made it an ideal insecticide: one spraying gave protection from insects for weeks to years, depending upon the method of application. Its persistence is due to its low vapor pressure and its consequent slow rate of evaporation, to its low reactivity with respect to

BOX 6-1   THE HISTORY OF DDT

Before DDT, the only insecticides available were those such as arsenic compounds which were very toxic and persistent, and those extracted from plants and which quickly lost their effectiveness once exposed to the elements. Thus DDT seemed at first to be the ideal insecticide: it was not very toxic to humans but highly toxic to insects; the fact that it was persistent represented a further advantage. DDT was discovered to be an insecticide in 1939 by Paul Müller, a chemist working for the Swiss firm Geigy on the development of various chemicals to fight agricultural insects. Müller was awarded the Nobel Prize for medicine and physiology in 1948 in recognition of the many civilian lives DDT saved after the war.

Products containing DDT were marketed within Switzerland beginning in 1941 for a variety of uses. Since Switzerland was a neutral country in World War II, its government informed both the Allies and the Axis countries about the discovery and uses of DDT; however it was only the Western Allies who realized its potential utility for wartime use to combat infestations of disease-carrying insects in hot climates.

During World War I, more than five million deaths had been caused by typhus. To avoid a repetition of such disasters, an incipient epidemic of typhus in Naples, Italy was thwarted by spraying all the civilians and the occupying Allied troops with DDT. Outbreaks of typhus in other parts of Europe, including the concentration camps at Dachau and Belsen, were dealt with in the same way by the Allied troops as they advanced. Aerial spraying with DDT to combat biting insects was carried out by the Allies at Guadalcanal and elsewhere in the Pacific before their troops invaded the islands. DDT was also used to combat mosquitoes that carried malaria in various parts of Europe, both during and after the war.

light and to chemicals in the environment, and to its very low solubility in water. Like other organochlorine insecticides, DDT is soluble in organic solvents and therefore in the fat of animal tissue. We all have some DDT (to the extent of about 3 ppm for North American adults) in our body fat.

Many animal species metabolize DDT by the elimination of HCl; a hydrogen atom is removed from one ethane carbon and a chlorine atom from the other, thereby creating a derivative of ethene called DDE, which stands for dichlorodiphenyldichloroethene:

Once World War II ended, DDT began to be used not only for public health purposes in hot climates but also extensively in developed countries to control insect pests attacking agricultural crops. Initially it was used on fruit trees and on vegetable crops, and subsequently in the growing of cotton. Eventually some insect populations became resistant to DDT, and its effectiveness decreased. This phenomenon led farmers to apply greater and greater amounts of the insecticide, particularly on cotton fields.

Within the scientific community, reservations about DDT as the "perfect insecticide" began to be heard almost as soon as it first went into use. In particular, it was known that DDT in soil persisted for several years and could become magnified in a food chain. The general public became aware of environmental problems associated with DDT upon the publication in 1962 by Rachel Carson of her book *Silent Spring*. In it, she discussed the decline in certain regions of the United States of the American robin, due to its consumption of earthworms that were laden with the DDT used in massive amounts to combat Dutch elm disease. Carson's book stimulated widespread concerns in the public about DDT and other pesticides. Through a series of legal hearings in the United States instigated by lawyers and scientists working with the Environmental Defense Fund, DDT was eventually banned or severely restricted in its use in most states. In 1973, the Environmental Protection Agency banned all DDT uses except those essential to public health. Similar bans were instituted by Sweden in 1969 and later in most other developed countries. DDT is still being used in developing countries to control disease or combat agricultural insects.

DDE

Substances that are produced by the metabolism of a chemical are called **metabolites;** thus DDE is a metabolite of DDT. The chemical DDE is also produced slowly in the environment by the degradation of DDT under alkaline conditions, and by DDT-resistant insects which detoxify DDT by this transformation. Unfortunately, in some birds DDE interferes with the enzyme that regulates the distribution of calcium, so contaminated birds produce eggs which have insufficient shell (calcium carbonate) thickness to withstand the weight of the parents who sit on them to make them hatch.

In humans, most ingested DDT is slowly but eventually eliminated. Most of the "DDT" stored in human fat is actually the DDE that was present in the food we have eaten, having previously been converted from DDT originally in the environment. Unfortunately DDE is almost nondegradable biologically and is very fat-soluble, so it remains in our bodies for a long time.

For environmental reasons, DDT is now banned from use in most Western industrialized countries; its use had been declining anyway as resistant insect populations evolved that could metabolize DDT to the noninsecticidal DDE and thus render it inactive. In susceptible insects, DDT kills by severely disrupting the nervous system, as discussed in more detail later.

The toxicity of chemicals is often expressed in terms of a *lethal dose*, **LD.** Experiments on test animals are done to establish the dose, called LD_{50}, of the chemical that is needed to kill half the animals in the study. If the substance is administered orally, the *lethal oral dose* is sometimes quoted as LOD_{50}. The values of the LD_{50} or LOD_{50} parameters are usually expressed as the mass of the chemical per kilogram of the animal's body weight; these values are often similar in different species of animals and so also may be applicable to humans. For example, the LOD_{50} value for DDT in rats is about 110 mg/kg. Humans are known to have survived DDT doses of about 10 mg/kg, that is, 60 kg (132 lb) humans did not die from ingesting 600 mg of DDT. Presumably, then, the LOD_{50} in humans is greater than the 10 mg/kg value, though it is not known whether the 110 mg/kg value for rats is transferable to humans. DDT is not considered to be very acutely toxic to humans, that is, small doses of it do not immediately kill people. Values of LD_{50}, and for certain other properties of interest, are listed in Table 6-2 for some important insecticides.

Of more concern than the acute toxicity of DDT are its potential chronic effects, that is, detrimental health effects, including cancer, that eventually appear as a consequence of long-term exposure to low levels of the substance. Although not traditionally considered to be a human carcinogen, recent studies by Mary Wolff of Mount Sinai Hospital in

TABLE 6-2   SELECTED DATA FOR SOME PESTICIDES

Pesticide	Solubility in H_2O (ppm)	LD_{50} (mg/kg)	$\log K_{ow}$
HCB	0.0062	3,500–10,000	5.3
DDT	0.0034	115	3.9–6.2
Toxaphene	n/a	85	2.9–3.3
Dieldrin	0.20	46	5.1–6.2
Mirex	0.20	700	5.8
Malathion	145	1,375–2,800	2.7
Parathion	24	3.6–13	n/a
Atrazine	30	1,870–3,080	2.3

Note that values reported in the literature for some of these properties vary depending upon the source laboratory and the test animal (usually rats); in such cases, ranges are listed.

New York and her colleagues have found that the higher the DDE concentration in a woman's blood, the more likely she is to have contracted breast cancer. However, a later study has failed to confirm this association between breast cancer and DDE; the issue will only be resolved by further research.

The environmental levels of DDT and DDE have declined in countries where their use has been restricted or banned, but these substances still enter the environment everywhere as a result of long-range air transport from developing countries where DDT is still in use to control malaria and typhus and for some agricultural purposes. The environmental concentrations of DDT and DDE in developed countries dropped substantially in the early and middle years of the 1970s, and have recently become in large measure stabilized at non-zero levels. Consider, for example, the variations in the concentrations of DDT in trout taken from the various Great Lakes (Figure 6-1). The concentrations have fallen from the 1977–1980 values (which themselves were far lower than those in the late 1960s and early 1970s) in the most polluted lake (Michigan), but show some recent increases in the cleaner lakes (Huron, Superior, and Ontario). As a result of the decline in DDE levels, bald eagles have made something of a comeback around Lake Erie and elsewhere. Similarly, the population of arctic peregrine falcons, a bird that was driven to near-extinction due to the effects of DDE, has now recovered to such an extent that it will probably soon be removed from the list of endangered species in the United States.

FIGURE 6-1

Average DDT concentrations in whole lake trout for selected Great Lakes (1977–1988). Note: when data for consecutive years are not available, dashed lines are used to estimate the trend. (Source: "Toxic Chemicals in the Great Lakes and Associated Effects," Volume 1, Part 2. Ottawa, Canada: Minister of Supply and Services, 1991.)

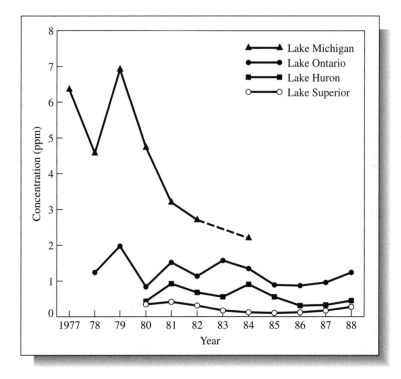

THE ACCUMULATION AND FATE OF ORGANOCHLORINES IN BIOLOGICAL SYSTEMS

Many organochlorine compounds are found in the tissues of fish in concentrations that are orders of magnitude higher than are those in the waters in which they swim. Hydrophobic substances like DDT are particularly liable to exhibit this sort of phenomenon. There are several reasons for this **bioaccumulation** of chemicals in biological systems.

In the first place, many organochlorines are inherently much more soluble in hydrocarbonlike media such as the fatty tissue in fish than they are in water. Thus when water passes through a fish's gills, the compounds selectively diffuse from the water into the fish's fatty flesh and become more concentrated there: this sort of process (which also affects other organisms than fish) is called **bioconcentration.** The **bioconcentration factor** (BCF) represents the equilibrium ratio of the concentration of a specific chemical in a fish relative to that dissolved in the surrounding water if the diffusion mechanism represents the only source of the substance to the fish. BCF values occur over a very wide range, and vary not only from chemical to chemical but also to a certain extent from one type of fish to another. The BCF of a chemical can be predicted, to within about a factor of ten, from a simple laboratory experiment: the chemical is allowed to equilibrate between the liquid layers in a two-phase system made up of water and the alcohol 1-octanol, $CH_3(CH_2)_6CH_2OH$, which has been found experimentally to be an adequate surrogate for the fatty portions of fish. The **partition coefficient,** K_{ow}, for a substance S is defined as

$$K_{ow} = [S]_{octanol} / [S]_{water}$$

where the square brackets denote concentrations in molarity or ppm units. Largely as a matter of convenience, the value of K_{ow} is often reported as its base 10 logarithm since its magnitude is often large— sometimes exceeding one million. (The motivation for doing this is analogous to that prompting the invention of the pH scale for H^+ concentrations in acid solutions.) For example, for DDT (see Table 6-2), K_{ow} is about 100,000 (i.e., 10^5) and so $\log K_{ow} = 5$; experimentally the Bioconcentration Factor for DDT lies in the range of about 20,000 to 400,000, depending upon the type of fish. Thus the K_{ow} value of a compound is a fairly reliable approximation to the BCF values found for fish.

In general, the higher its octanol-water partition coefficient K_{ow}, the more likely a chemical is to be bound to organic matter in soils and sediment and ultimately to migrate to fat tissues of living organisms. However, $\log K_{ow}$ values of 7 or 8 or higher are indicative of chemicals with such strong adsorption to sediments that they are actually unlikely to be mobile enough to enter living tissue. Thus it is chemicals with $\log K_{ow}$ values in the 4–7 range that bioconcentrate to the greatest degree.

PROBLEM 6-2

For HCB, $\log K_{ow} = 5.3$. What would be the predicted concentration of HCB due to bioconcentration in the fat of fish that swim in waters containing 0.000010 ppm of the chemical?

Fish also bioaccumulate organic chemicals from the food they eat. In many such cases, the chemicals are not metabolized by the fish: the substance simply accumulates in the fatty tissue of the fish, where its concentration there increases with time. For example, the concentration of

DDT in trout from Lake Ontario increases almost linearly with the age of the fish, as illustrated in Figure 6-2. The average concentration of many chemicals also increases dramatically as one proceeds up a **food chain,** which is a sequence of species each one of which feeds on the one preceding it in the chain. The **food web,** incorporating interlocking food chains, for the Great Lakes is illustrated in Figure 6-3. Over a lifetime, a fish eats many times its weight in food from the lower levels of the food chain, but retains rather than eliminates most organochlorine chemicals from these meals. A chemical whose concentration increases along a food chain is said to be **biomagnified.** The biomagnification of DDT along some of the Great Lakes food chains is shown in Figure 6-3: note the herring gull's high level of DDT compared with those of the fish below it in the chains leading up to it. Fish at the top of the aquatic

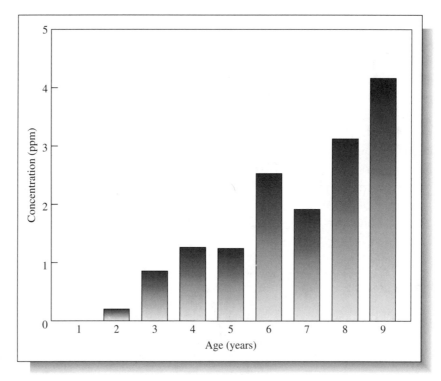

FIGURE 6-2
Variation with age of fish of the average DDT concentration in Lake Ontario trout.
(Source: "Toxic Chemicals in the Great Lakes and Associated Effects," Volume 1, Part 2.
Ottawa, Canada: Minister of Supply and Services, 1991.)

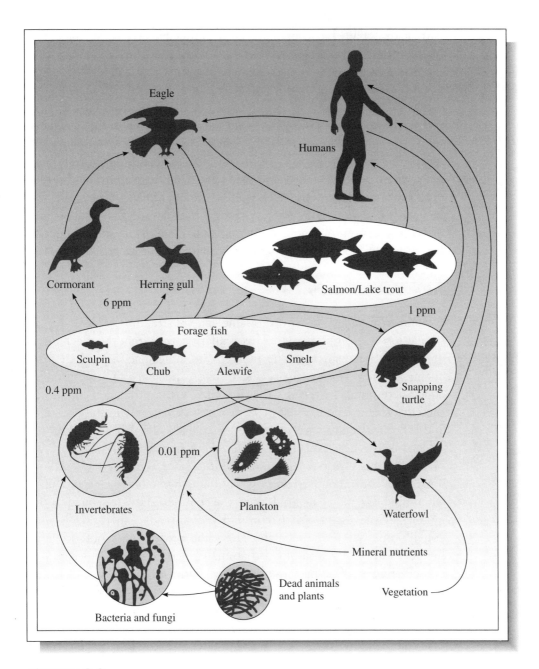

FIGURE 6-3

Simplified food web for the Great Lakes with typical DDT concentrations for some species. (Source: "Toxic Chemicals in the Great Lakes and Associated Effects," Synopsis. Ottawa, Canada: Minister of Supply and Services, 1991.)

part of the chain bioaccumulate DDT rather effectively, so that even higher concentrations are found in the birds of prey that feed on them.

As an example of biomagnification, consider that the DDT concentration in seawater in Long Island Sound and the protected waters of its southern shore is about 0.000003 ppm, but that it reaches 0.04 ppm in the plankton, 0.5 ppm in the fat of minnows and 2 ppm in the needlefish that swim in these waters, and 25 ppm in the fat of the cormorants and osprey that feed on the fish, for a total biomagnification factor of about ten million. It is by such mechanisms that DDE levels in some birds of prey became so great that their ability to reproduce successfully was impaired. The bioaccumulation of organochlorines in fish and other animals is the reason that most of the human daily intake of such chemicals enters via our food supply rather than from the water we drink.

ANALOGS OF DDT

A number of molecules with the same general structure as DDT display similar insecticidal properties. This similarity arises from the mechanism of DDT action, which is due more to its molecular *shape* than from chemical interactions with specific species. The shape of a DDT molecule is determined by the fact that it contains two tetrahedral carbons in the ethane unit, and two flat benzene rings. Apparently in insects, DDT and other molecules with the same general size and 3-D shape become wedged in the nerve channel that leads out from the cell of the nerve. Normally, this channel transmits impulses only as needed via sodium ions. But a continuous series of Na^+-initiated nerve impulses is produced when the DDT molecule holds open the channel. As a consequence, the muscles of the insect twitch constantly, eventually exhausting it with convulsions that lead to death. The same process does not occur in humans and other warm-blooded animals since DDT molecules do not exhibit any such binding action in nerve channels.

Examples of other molecules with DDT-like action include DDD (sometimes called TDE), para-dichlorodiphenyldichloroethane, which is an environmental degradation product of DDT: they differ only in that one chlorine from the $-CCl_3$ group is replaced by a hydrogen. Since the overall shapes and sizes of DDT and DDD are similar, so is their toxicity to insects. Indeed, DDD has itself been sold as an insecticide, but has also been discontinued because it bioaccumulates. Notice that DDE, unlike DDT and DDD, is based upon a *planar* $C=C$ unit rather than a $C-C$ linkage which has tetrahedral groups at each end; thus whereas

DDD is a DDT-like insecticide, DDE is not, since its 3-D shape is very different. DDE is flat rather than propeller-shaped, and so it does not become wedged in the insect's nerve channels.

Scientists have devised analogs to DDT that have its same general size and shape, and that consequently possess the same insecticidal properties, but that are reasonably biodegradeable and thus do not present the bioaccumulation problem associated with DDT. The best known of these analogs is methoxychlor, whose structure is shown below:

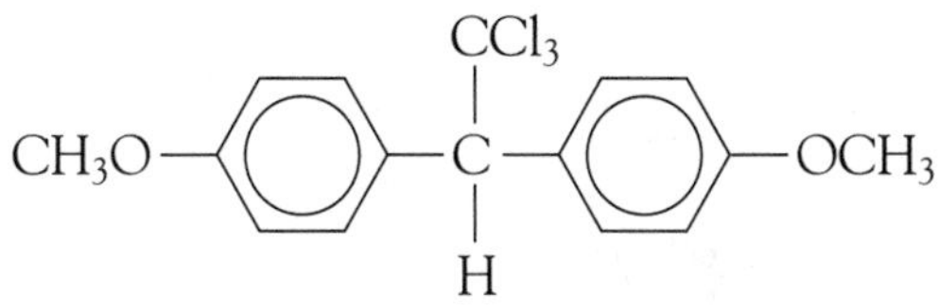

methoxychlor

The para-chlorines of DDT are substituted here by methoxy groups, —OCH$_3$, which are approximately the same size as chlorine but which react much more readily; conveniently enough, these reactions produce water-soluble products which not only degrade in the environment but which organisms excrete rather than accumulate. Methoxychlor is currently used extensively both domestically and agriculturally to control flies and mosquitoes.

PROBLEM 6-3

Draw the molecular structure of "DDD."

PROBLEM 6-4

Methyl groups are approximately the same size as chlorine atoms, but hydrogen atoms are smaller. Would you expect insecticidal properties for DDT molecules in which a. the —CCl$_3$ group is replaced by —C(CH$_3$)$_3$, and in which b. the para-chlorines are replaced by hydrogens?

TOXAPHENES

During the 1970s, after DDT had been banned, the insecticide that replaced it in many agricultural applications such as the growing of cotton and soybeans was toxaphene, which had first been introduced commercially in 1947. It is a mixture of many similar substances, all of which are produced when the naturally occurring hydrocarbon called camphene is partially chlorinated. In the reaction, some rearrangement of atoms within the hydrocarbon structure readily occurs; consequently, most components in the toxaphene mixture are partially chlorinated derivatives of the hydrocarbon shown at the right rather than of camphene itself:

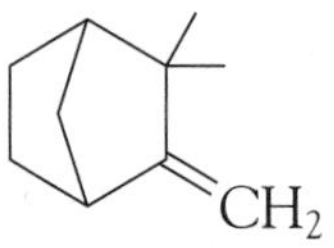

camphene

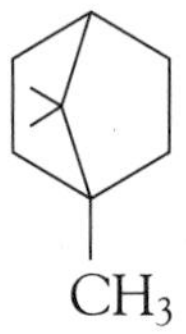

CH_3

Toxaphene became the most heavily used insecticide in the United States before its ban in 1982. It is extremely toxic to fish, and indeed it was used in North America to rid lakes of undesirable fish; however, it was found to be so persistent that the lakes could not be successfully restocked for years thereafter! In addition, toxaphene bioaccumulates in fatty tissues and causes cancer in test rodents. Although it is now banned in developed countries, toxaphene is still being deposited in bodies of water remote from its point of usage because of its long-range transport by air from developing countries that, like Mexico, still make use of it. For example, in relatively clean Lake Superior its concentration in fish exceeds that of any other organochlorine. The current levels of toxaphene in the Great Lakes are sufficiently high that it has been designated as a Priority Pollutant of concern (since it bioaccumulates, persists, is chronically toxic, and is known to harm living matter) by the International Joint Commission for the Great Lakes (IJC). However, the IJC believes that it is unlikely that toxaphene is a risk to human health at current Great Lakes levels provided that fish-consumption advisories are followed.

CHLORINATED CYCLOPENTADIENES

Cyclopentadiene, shown at left below, is an abundant byproduct of petroleum refining. As its name implies, there are two double bonds in each molecule. When fully chlorinated (see diagram below at right) it can be combined with one of several other organic molecules to produce a whole series of insecticidal compounds with properties such as environmental persistence that made them superficially attractive.

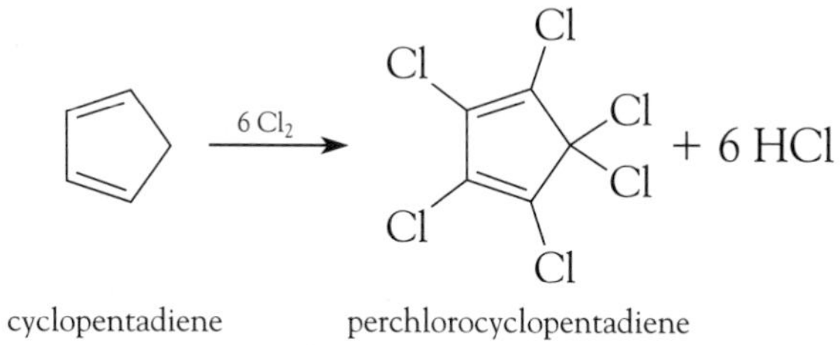

cyclopentadiene   perchlorocyclopentadiene

To form the structurally simplest of these cyclodiene insecticides, perchlorocyclopentadiene is reacted with another organic molecule that has at least one double bond in order to produce a compound with new single bonds joining the two portions of the original molecules; in the process, two of the original three double bonds in the reactants are converted to single bonds:

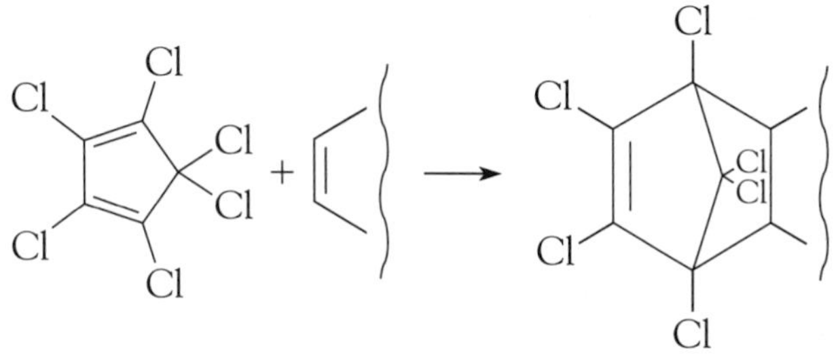

The wavy lines in the formulas indicate that portion of the reactant whose structure need not be specified, as it both can vary from reaction to reaction, and remains unchanged during reaction. Notice that the remaining double bond occurs in the cyclopentadiene ring at the CC link located between the original double bonds.

The most important cyclodiene insecticides in terms of commercial prominence are listed below. They were used to control soil insects, fire ants, cockroaches, termites, grasshoppers, locusts, and other insect pests.

aldrin

dieldrin and its isomer endrin

heptachlor

chlordane (commercial samples contain some heptachlor)

endosulfan

Like most chlorinated organics, these substances have very low solubility in water but are fat soluble.

The cyclodiene pesticides, starting with aldrin and dieldrin, arrived on the market about 1950. Given their persistence, their potential toxicity, their tendency to accumulate in fatty tissues, and the suspicion that—at the least—dieldrin was causing excess mortality of adult bald eagles, the use of almost all of these compounds has by now either been banned or severely restricted in North America and most western European countries. However, some of the compounds are still available elsewhere. Endosulfan is still in use in North America as an insecticide for both domestic and agricultural applications; its bioconcentration and environmental persistence is much lower than those of other cyclodienes.

In the cyclodiene pesticides aldrin and heptachlor, which contain two ring structures, the second ring in the product contains a double bond in which neither $C=$ type carbon is chlorinated but rather is bonded to a hydrogen atom. In the environment, such molecules are converted biologically to the **epoxide:** that is, an oxygen atom is added across the double bond, thereby producing a three-membered ring with single bonds connecting all three atoms:

epoxide

The epoxide compound can also act as an insecticide; indeed the epoxide of aldrin *is* dieldrin and much of the environmental burden of the latter originated as aldrin. Like many other epoxides, dieldrin is a potential carcinogen, but there is no convincing evidence that dieldrin causes cancer in humans. Aldrin and dieldrin become rapidly and strongly adsorbed onto soil particles, and are quite persistent in that medium. Both are acutely toxic to aquatic organisms.

Agricultural uses of dieldrin continued in North America into the early 1970s, but virtually all uses there were prohibited in the mid-1980s. It was used extensively in tropical countries to control the tsetse fly, and is still used in some countries to kill termites. Although banned from most uses in the region for many years, dieldrin is still an IJC Priority Pollutant in Great Lakes waters, since its environmental levels there have not decreased as quickly as have those of the other cyclodiene-based insecticides; it continues to enter water systems from **leachate,** the liquid which percolates from waste disposal sites. The yearly concentration of dieldrin in walleye fish from Lake Erie is illustrated in Figure 6–4; clearly the levels have decreased from those of the late 1970s but have not yet reached zero. The dieldrin concentration in the fat tissue of North American humans has fallen to about one-half its level in the 1970s.

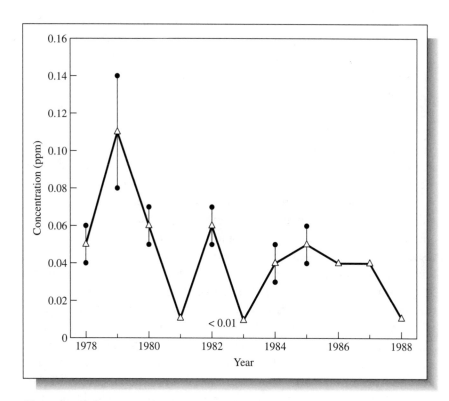

dieldrin

FIGURE 6-4

Variation of average dieldrin concentration in walleye fish taken from Lake Erie (1977–1988). (Source: "Toxic Chemicals in the Great Lakes and Associated Effects," Volume 1, Part 2. Ottawa, Canada: Minister of Supply and Services, 1991.)

If two perchlorocyclopentadiene molecules are chemically combined, the resultant **mirex** molecule also acts as an insecticide, and is particularly effective against the fire ant found in the southeastern United States.

mirex

(All 10 carbon atoms are bonded to chlorine, but for clarity the individual chlorine atoms are not shown; only the total is displayed in the formula.) This compound was also sold under the name "dechlorane" as a flame retardant additive for synthetic and natural materials.

Most mirex and dechlorane use occurred in the 1960s. Although banned since the mid-1970s, mirex is presently an IJC Priority Pollutant in the Great Lakes. Mirex entered Lake Ontario and accumulated there because it was commercially produced along the nearby Niagara River; both accidental spills and its presence in effluent from the manufacturing plant led to deposition of quantities of the chemical in rivers draining into the lake. Because of its great persistence, the mirex will take about a century to be cleared from the area. In the environment in general and in fish in particular, it is found either as the original molecule or as a derivative in which one chlorine atom is substituted by hydrogen; the latter, called *photomirex*, is a degradation product formed by the action of UV light on mirex. Scientists have concerns about its potential as a reproductive toxin for birds and animals that eat fish from Lake Ontario. Human consumption of some types of fish from this lake is also restricted because of mirex levels. As a result of its bioaccumulation in fish or because of transport in air (or for both reasons), mirex has found its way into the other Great Lakes. Some mirex leaves Lake Ontario in the bodies of eels, which travel down the St. Lawrence River and are subsequently eaten by beluga whales that inhabit the estuary of the river. The well-being of these whales is being adversely affected by a number of other pollutants too, besides mirex, as discussed later in this chapter.

For the most part, the chlorinated cyclodiene pesticides are chemical products of the past. Their use has been phased out or at least severely restricted because of environmental and human health considerations.

OTHER TYPES OF MODERN INSECTICIDES

ORGANOPHOSPHATE INSECTICIDES

Organophosphate pesticides as a class are nonpersistent; in this respect, they represent an advance over organochlorines. However, they are generally much more acutely toxic to humans and other mammals than are organochlorines. Many organophosphates represent an acute danger to the health of those who apply them and to others who may come into contact with them. Exposure to these chemicals by inhalation, swallowing, or absorption through the skin can lead to acute health problems. Like chlorinated hydrocarbons, organophosphates concentrate in fatty tissues, but on the other hand, the organophosphates decompose within days or weeks, and thus are seldom found in food chains.

Organophosphate pesticides all contain a central, pentavalent phosphorus atom to which are connected

a. an oxygen or sulfur atom doubly bonded to the P atom

b. two methoxy ($-OCH_3$) or ethoxy ($-OCH_2CH_3$) groups singly bonded to the P atom

c. a longer, more complicated R group singly bonded to phosphorus by an oxygen or sulfur atom

The three subclasses of the organophosphates, each falling under one of these headings, are illustrated in Figure 6-5. All decompose fairly rapidly in the environment since oxygen in air alters $P{=}S$ bonds to $P{=}O$, and water molecules add to and therefore split $P{-}O$ bonds, ultimately yielding nontoxic substances such as phosphoric acid, $O{=}P(OH)_3$, and alcohols.

Dichlorvos is an example of a type A (Figure 6-5) organophosphate, in which no sulfur is incorporated into the molecule. It is a relatively volatile insecticide, and is used as a domestic fumigant released from impregnated "fly strips" hung from ceilings and light fixtures. The chemical slowly evaporates, and its vapor kills flies in the room. Plastic is impregnated with dichlorvos for use in flea collars. It is relatively toxic to mammals; its LOD_{50} is 25 mg/kg in rats.

Parathion is an example of category B, organophosphates in which the doubly bonded oxygen has been replaced by sulfur. It is very toxic (LOD_{50} = 3 mg/kg in rats), and is probably responsible for more deaths of agricultural field workers than any other pesticide. Since it is nonspecific to insects, its use can inadvertently kill birds and other non-targeted organisms. Bees, too, often economically valuable (in honey production and the pollinating of fruit crops, for instance) are indiscriminately destroyed by parathion. It now is banned in some Western industrialized countries, but is still widely used in developing countries. The structure of fenitrothion is very similar to parathion but it has a lower toxicity to mammals (LOD_{50} = 250 mg/kg in rats). It has been used

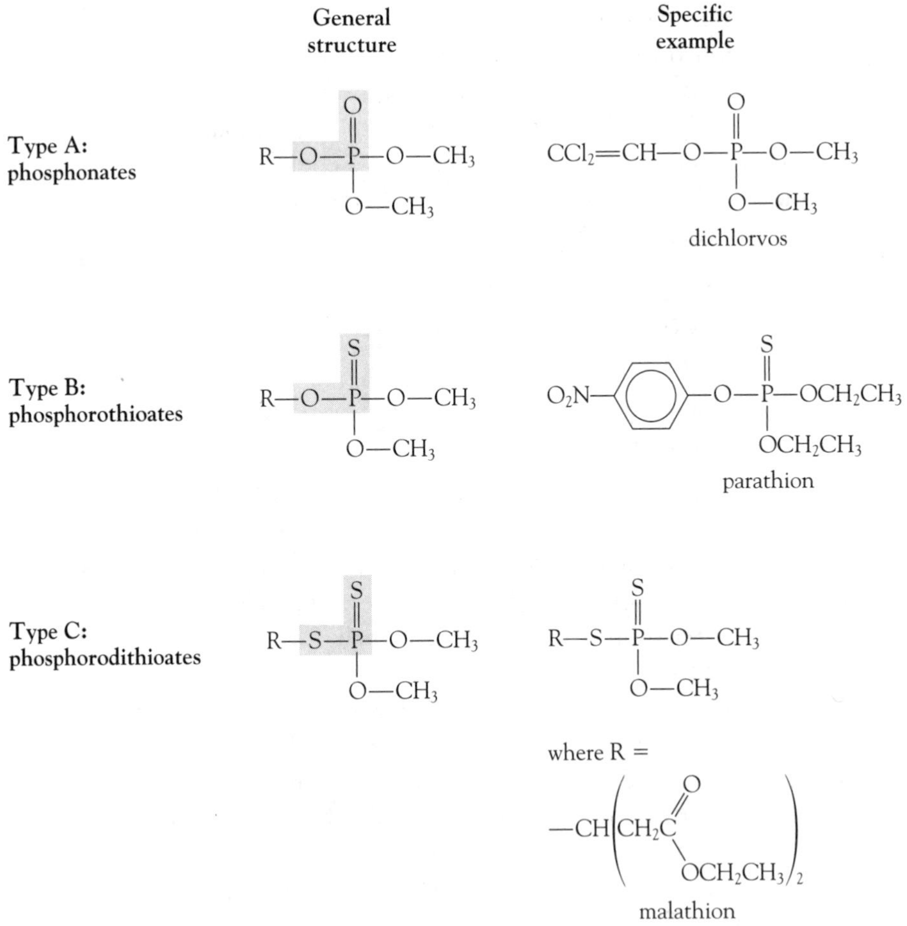

General structure   **Specific example**

Type A: phosphonates

Type B: phosphorothioates

Type C: phosphorodithioates

dichlorvos

parathion

where R =

malathion

FIGURE 6-5

Structural types of organophosphate insecticides. In some molecules (e.g., parathion), the methoxy group, —OCH_3, is replaced by the ethoxy group, —OCH_2CH_3.

extensively as a spray against spruce budworm in evergreen forests in eastern Canada, though not without some controversy. Diazinon, also a member of category B, is commonly used for insect control in homes, gardens, and on pets, and was thought to be relatively safe (LOD_{50} = 300 mg/kg in rats). Because it is toxic to birds, however, its usage now has been restricted.

Malathion is the most important example of category C, in which two oxygens have been replaced by sulfur. Introduced in 1950, malathion is not particularly toxic to mammals (LOD_{50} = 885 mg/kg in rats) but it is nevertheless fatal to many insects since they metabolize it in a different way. It is still used in domestic fly sprays and to protect agricultural crops. In combination with a protein bait, low concentrations of malathion have been sprayed from helicopters over several areas in the United States (California, Florida, and Texas) to combat infestations of the Mediterranean fruit fly, a dangerously destructive pest. Dimethoate (LOD_{50} = 250 mg/kg), commercially known as Cygon®, is another member of this group; it is often used for insect control on food crops, including those grown in backyards.

The organophosphates are toxic to insects because they inhibit enzymes in the nervous system, and thus they function as nerve poisons. In particular, oroganophosphates disrupt the communication that is carried on between cells by the acetylcholine molecule. This cell-to-cell transmission cannot operate properly, however, unless the acetylcholine molecule is destroyed after it has executed its function. Organophosphates block the action of the enzymes whose job it is to destroy the acetylcholine by selectively bonding to them. The presence of the insecticide molecule thus has the effect of suppressing the continued transmission of impulses between nerve cells that is essential to the coordination of the organism's vital processes, and death ensues. (At the atomic level, it is the phosphorus atom of the organophosphate molecule that attaches to the enzyme and stays bound to it for many hours.)

CARBAMATE AND NATURAL INSECTICIDES

The mode of action of **carbamate** insecticides is similar to that of the organophosphates; they differ in that it is a carbon atom rather than a phosphorus that attacks the acetylcholine-destroying enzyme. The **carbamates,** introduced as insecticides in 1951, are derivatives of carbamic acid, H_2NCOOH. One of the hydrogens attached to the nitrogen is replaced by an alkyl group, usually methyl, and the hydrogen attached to the oxygen is replaced by a longer, more complicated organic group symbolized below simply as R:

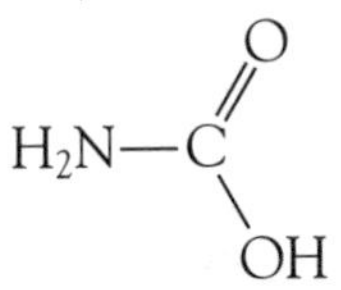

$$H_2N-\underset{\underset{\textstyle OH}{\big\backslash}}{\overset{\overset{\textstyle O}{\big/\big/}}{C}}$$

carbamic acid

$$CH_3-\overset{\overset{\textstyle H}{|}}{N}-\overset{\overset{\textstyle O}{\|}}{C}-O-R$$

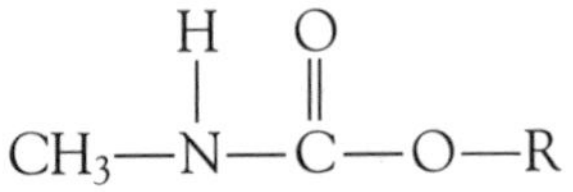

the general formula of a carbamate

Like the organophosphates, the carbamates are short-lived in the environment since they react with water and decompose to simple, nontoxic products. (The reaction with water involves the splitting of one of the single bonds to the central carbon.)

Important examples of the *carb*amate pesticides are *carb*ofuran (LOD_{50} = 8 mg/kg in rats), *carb*aryl (LOD_{50} = 307), and aldi*carb* (LOD_{50} = 0.9): the latter is very toxic indeed to humans. Although carbaryl, a widely used lawn and garden insecticide with the trade name Sevin®, has a low toxicity to mammals, it is particularly toxic to honeybees.

In summary, the organophosphates and carbamates solve the problem of environmental persistence and accumulation associated with organochlorine compounds, but sometimes at the expense of dramatically increased acute toxicity to the humans and animals who encounter them while the chemicals are still in the active form. They are a particular problem in developing countries, where widespread ignorance about their hazards and failure to use protective clothing has led to many deaths among agricultural workers. Indeed, an estimate in a recent World Health Organization publication put the number of persons who have suffered from short-term exposure to pesticides in the millions, and the number of resulting deaths, including suicides, in the hundreds of thousands.

As pointed out earlier, many plants can themselves manufacture certain molecules for their own self-protection that either kill or disable insects. Chemists have isolated some of these compounds so that they can be used to control insects in other contexts. Examples are nicotine, rotenone, the pheromones, and juvenile hormones.

One group of "natural pesticides" that has been used by humans for centuries is the pyrethrins. The original compounds, the general structure for which is illustrated below, were obtained from the flowers of a certain species of chrysanthemum.

In the form of dried, ground-up flower heads, pyrethrins were used in Napoleonic times to control body lice. They are generally considered to be safe to use; like organophosphates, they paralyze insects; but they usually do not kill them. Unfortunately, these compounds are unstable in sunlight. For that reason, synthetic pyrethrinlike insecticides which are stable outdoors have now been developed by chemists.

HERBICIDES

Herbicides are chemicals that destroy plants. They are often employed to kill weeds without causing injury to desirable vegetation, for example, to eliminate broad-leaf weeds from lawns without killing the grass. The agricultural use of herbicides has replaced human and mechanical weeding in developed countries, and has sharply reduced the number of people employed in agriculture. Herbicides also are used to eliminate undesirable plants from roadsides, railway and powerline rights-of-way, and so on, and sometimes to defoliate entire regions.

In Biblical times, armies sometimes used salt or a mixture of brine and ashes to sterilize land that they had conquered, the intent being to make it uninhabitable by future generations of the enemy. In the first half of the twentieth century, several inorganic compounds were used as weed killers—principally sodium arsenite, Na_3AsO_3, sodium chlorate, $NaClO_3$, and copper sulfate, $CuSO_4$. The latter are only two of a large number of salts formerly used as herbicidal sprays whose means of operation is to kill plants by the rather primitive action of extracting the water from them, while at the same time leaving the land treated in this way still capable of supporting agriculture.

Eventually, organic derivatives of arsenic replaced its inorganic compounds as herbicides since they are less toxic to mammals. Inorganic and organometallic herbicides have in general been largely phased out because of their persistence in soil, and completely organic herbicides now dominate the market. Their utility is based partially on the fact that they are much more toxic to certain types of plants than to others, so they can be used to eradicate the former while leaving the latter unharmed.

TRIAZINE HERBICIDES

One modern class of herbicides is the **triazines,** which are based upon the symmetric, aromatic structure shown below, which has alternating carbon and nitrogen atoms in a six-membered ring:

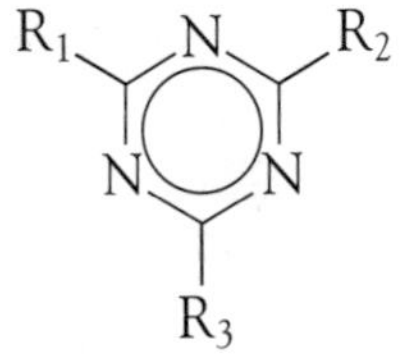

In herbicidal triazines, R_1 = Cl and R_2, R_3 = amino groups

the general formula of the triazines

In triazines that are useful as herbicides, one carbon atom in the ring is bonded to chlorine, and the other two to amino groups, which are nitrogen atoms singly bonded to hydrogens and/or carbon chains.

The best-known member of this group is atrazine, which was introduced in 1958 and is used in huge quantities to destroy weeds in corn fields. In atrazine, R_2 is $-NH-CH_2CH_3$ and R_3 is $-NHCH(CH_3)_2$.

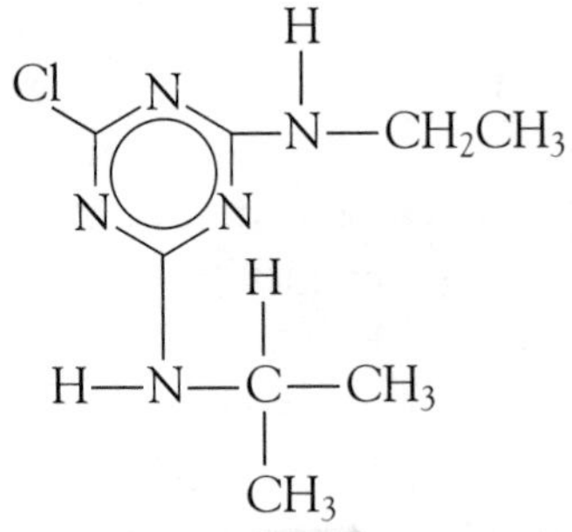

atrazine

It usually is applied to cultivated soils, at the rate of a few kilograms per hectare, in order to kill grassy weeds. In higher concentrations, it has been used to kill off all plant life, for example to create parking lots. Biochemically, it acts as a herbicide by blocking the operation of photosynthesis in the plant.

Atrazine is moderately soluble (30 ppm) in water. In waterways that drain agricultural land on which it is used, its concentration typically is found to be a few parts per billion. Usually atrazine is detectable in well water in such regions. In Canada, the maximum concentration allowable in drinking water is 60 ppb.

Since atrazine's K_{ow} value is about 10^3, some bioconcentration does occur, although it does not represent a significant problem. Atrazine is not considered to be a very toxic compound (its LOD_{50} is 1,870 mg/kg). However, some preliminary surveys on the health of farmers and other individuals exposed to it in high concentrations show disturbing links to higher cancer rates and higher incidents of birth defects. No definitive studies linking atrazine use to human health problems have as yet been reported.

OTHER ORGANIC HERBICIDES

In some regions where soybeans and corn are grown intensively—southwestern Ontario, in particular—atrazine is yielding its status as herbicide of choice to metolachlor, which is a derivative of chloroacetic acid, $ClCH_2COOH$, in which the —OH group is replaced by an amino group:

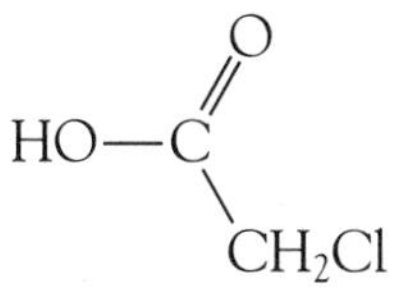

$$
HO-\overset{\displaystyle O}{\underset{\displaystyle CH_2Cl}{C}} \qquad R_1R_2N-\overset{\displaystyle O}{\underset{\displaystyle CH_2Cl}{C}}
$$

chloroacetic acid 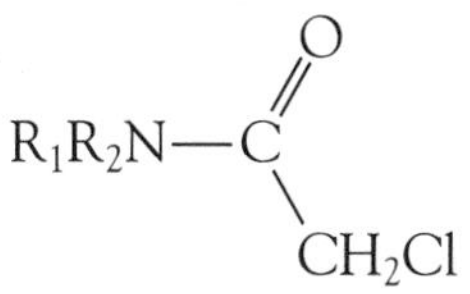   metolachlor

It degrades in the environment by the action of sunlight and of water.

Another well-known herbicide is paraquat, which has acquired a sort of fame owing to its use to destroy crops of marijuana. The -quat ending in its name signifies that it contains quaternary nitrogen atoms,

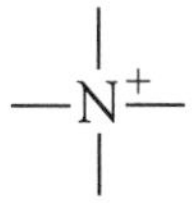

$$
-\overset{\displaystyle |}{\underset{\displaystyle |}{N^{\pm}}}-
$$

in its structure; such ionic nitrogens make the substance water-soluble. Paraquat is a preemergent herbicide: that is, it is used to eliminate weeds from fields before they sprout in the spring. Residual paraquat may be the reason some marijuana smokers suffer severe lung damage. Paraquat poisoning has become a significant health problem in many developing countries, since it has been used there for the treatment of health problems such as louse infestations.

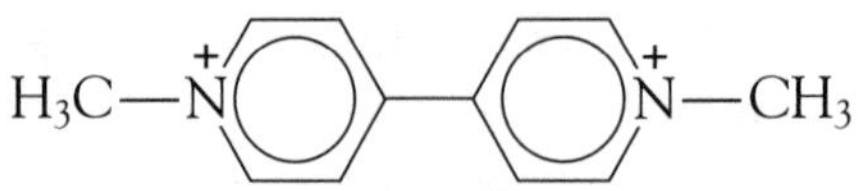

paraquat

PHENOXY HERBICIDES

These weed killers were introduced at the end of the Second World War. Environmentally, the byproducts contained in phenoxy herbicides are often of greater concern than the herbicides themselves. For that reason, we begin by discussing the chemistry of phenol, the fundamental component of these compounds.

Chlorobenzene will react with NaOH at high temperatures to substitute —OH for —Cl, giving the compound called phenol:

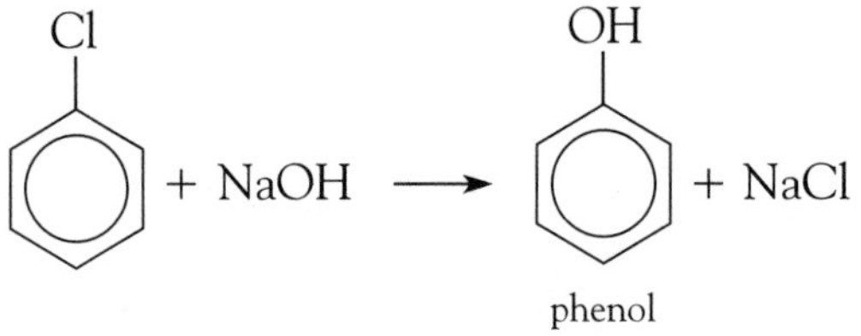

phenol

$$\text{or}\quad C_6H_5Cl + NaOH \longrightarrow C_6H_5OH + NaCl$$

Phenols are mildly acidic; in the presence of concentrated solutions of a strong base like NaOH, the hydrogen of the OH group is lost as H^+ (as with any common acid) and the phenoxide anion, $C_6H_5O^-$, is produced in the form of its soldium salt:

$$C_6H_5OH + NaOH \longrightarrow C_6H_5O^-Na^+ + H_2O$$

The O^-Na^+ group is a reactive one, and this property can be exploited in order to prepare molecules containing the C—O—C linkage. Thus if an R—Cl molecule is heated together with a salt containing the phenoxide ion, NaCl is eliminated and the phenoxy oxygen links the benzene ring to the R group:

$$C_6H_5O^-Na^+ + Cl—R \longrightarrow NaCl + C_6H_5—O—R$$

Such a reaction is the most direct commercial route to the large-scale preparation of the herbicide, introduced in 1944, whose well-known commercial name is 2,4,5-T. Here (see the reaction immediately above), the R group is acetic acid, CH_3COOH, minus one of its methyl group hydrogens, so that R = $-CH_2COOH$, and the Cl—R reactant is Cl—CH_2COOH. Thus according to the reaction format we obtain C_6H_5—O—CH_2COOH, called *phenoxyacetic acid* as an intermediate in the production of the actual herbicides.

In the commercial herbicides, some of the five remaining hydrogen atoms of the benzene ring in phenoxyacetic acid are replaced by chlorine atoms.

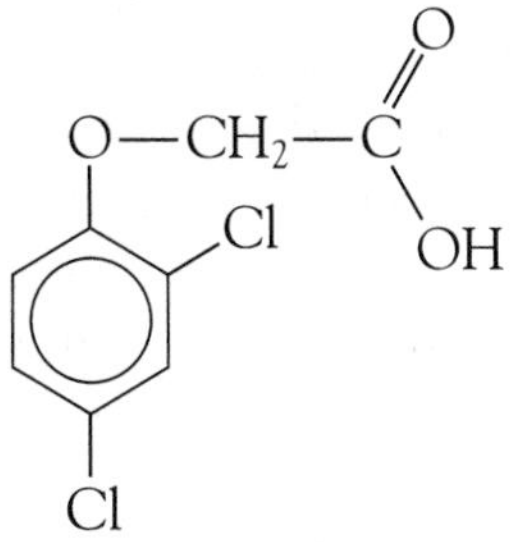

2,4-D	2,4, 5-T
2,4-dichlorophenoxyacetic acid	2,4,5-trichlorophenoxyacetic acid

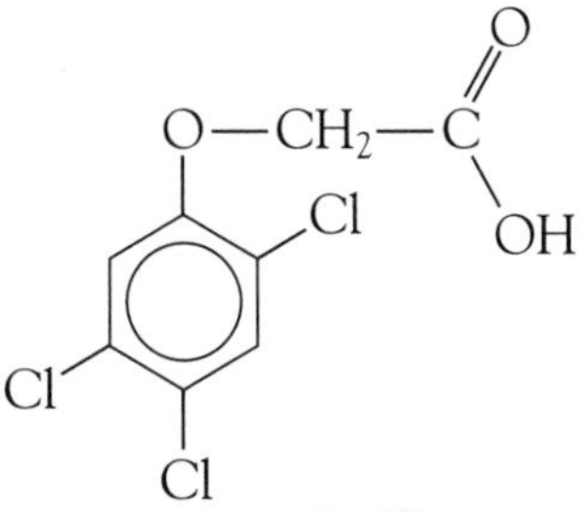

Note that the numbering scheme for the benzene ring begins at the carbon attached to the oxygen.

The 2,4-D compound (which is produced by a strategy that does not involve first forming the corresponding chlorinated phenol) is used to kill broad-leaf weeds in lawns, golf course fairways and greens, and agricultural fields. In contrast, 2,4,5-T is effective in clearing brush, for instance, on roadsides and power line corridors.

The herbicide known as MCPA is 2,4-D with the chlorine in the 2 position replaced by a methyl group, CH_3. The herbicides called dichlorprop, silvex, and mecoprop are identical to 2,4-D, to 2,4,5-T, and to MCPA, respectively, except that their molecules have a methyl group replacing one hydrogen in the —CH_2— group of the acid chain; thus they are phenoxy herbicides that are based upon propionic acid, $CH_3 CH_2COOH$, rather than upon acetic acid.

Huge quantities of 2,4-D are used in developed countries for the control of weeds in both agricultural and domestic settings. In some communities, its continued use on lawns has become a controversial practice because of its suspected effects on human health. In particular, farmers in the midwestern United States who mix and apply large quantities of 2,4-D to their crops are found to have an increased incidence of the cancer known as non-Hodgkin's lymphoma. Indeed, this once relatively rare disease is becoming more prevalent, though this development is apparently due more to its association with AIDS than to exposure to herbicides.

DIOXIN CONTAMINATION OF HERBICIDES AND WOOD PRESERVATIVES

Traditionally, the industrial synthesis of 2,4,5-T started with 2,4,5-trichlorophenol, which was produced by reacting NaOH with the appropriate tetrachlorobenzene. Unfortunately, during the reaction in which the phenol was produced from tetrachlorobenzene, there occurs a side reaction which converts a very small portion of the trichlorophenol product into "dioxin." In this side reaction, *two* trichlorophenoxy anions react with each other, resulting in the elimination of two chloride ions:

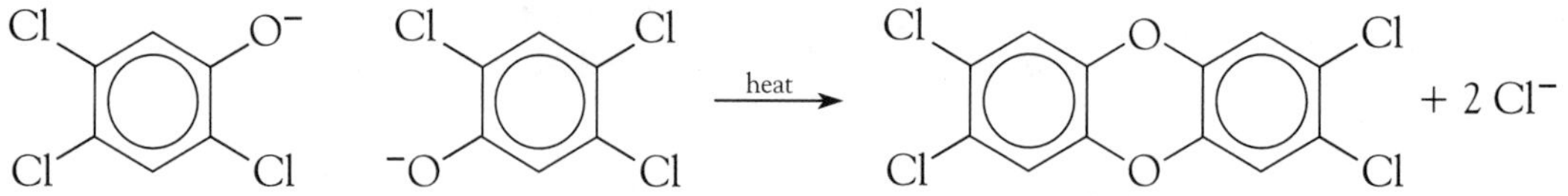

"dioxin"
(tetrachlorodibenzo-*p*-dioxin)

In this process, a new six-membered ring is formed which links the two chlorinated benzene rings. This central ring has two oxygen atoms located *para* to each other, as is found in the simple molecule 1,4-dioxin or *para*-dioxin (*p*-dioxin).

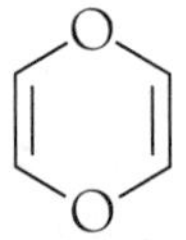

1,4-dioxin

Although the molecule labeled "dioxin" above is correctly known as a tetrachlorodibenzo-*p*-dioxin, it has become popularly known simply as "dioxin," with the understanding that it is the most toxic of a class of related compounds.

The side reaction which produces dioxin is "second order in chlorophenoxide," that is, the reaction rate depends on the square (or second power) of the ion's concentration. Thus the rate of dioxin production increases dramatically as the initial chlorophenoxide ion concentration increases. In addition, the rate of this side reaction increases rapidly with increasing reaction temperature. Thus, the extent to which the trichlorophenol and consequently the herbicide become contaminated with the dioxin byproduct can be minimized by controlling concentration and temperature in the preparation of the original trichlorophenol. Today, the contamination of 2,4,5-T by so-called dioxin can be kept to about 0.1 ppm by keeping both the phenoxide concentration and the temperature low. Nevertheless, its manufacture and use in North America were phased out in the mid-1980s because of concerns about its dioxin content, however small.

A 1:1 mixture of 2,4-D and 2,4,5-T, called *Agent Orange,* was used extensively as a defoliant during the Vietnamese War. Since the mixture contained dioxin levels of about 10 ppm, it is clear that the reaction used to produce the 2,4,5-T was not carefully controlled so as to minimize contamination. (The human health consequences of exposure to dioxins are discussed in a later section.)

Environmental contamination by "dioxin" also occurred as the result of an explosion in a chemical factory in Seveso, Italy, in 1976. The factory produced 2,4,5-trichlorophenol from tetrachlorobenzene by reaction with base, as described above. On one occasion, the reaction was not brought to a complete halt before the workers left for the weekend. The reaction continued unmonitored and the heat subsequently released by the reaction resulted in an explosion. Since the phenol had been heated to a high temperature, considerable dioxin—probably a few kilograms— was produced; the explosion effectively distributed the toxin into the environment. Many wildlife deaths apparently resulted from the contamination. Although a large number of humans, both adults and children, were also exposed to the chemical as a result of this explosion, no serious health effects to humans were found for many years. A recent study, however, established that the rates of a number of cancers has increased in people who lived in the zones most exposed to dioxin from the explosion, although their incidence of breast cancer is lower than in nonexposed residents.

The nomenclature and numbering system used for ring systems like the dioxins is a little unusual. Since the dioxin ring is bounded on either side by benzene rings, the three-ring unit is properly known as dibenzo-*p*-dioxin. The chlorine substitution on the outer rings also should be recognized, so the dioxin shown below is a *tetrachlorodibenzo-p-dioxin*, or TCDD for short.

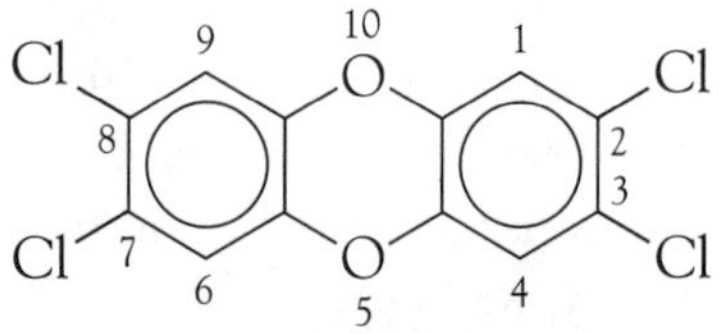

2,3,7,8-tetrachlorodibenzo-*p*-dioxin
(2,3,7,8-TCDD)

The numbering scheme for the ring carbons takes into account the fact that the carbons shared between two rings carry no hydrogen atoms and so need not be numbered; thus C—1 is the carbon next to the one joining the rings and the numbering follows a direct path from there. The oxygen atoms *are* also part of the numbered sequence in this scheme, although their locations need not be made explicit in naming any of this family of compounds. The initial position for the numbering system is chosen so as to give the lowest possible value to the first substituent; if there is a choice after this criterion has been applied, then that which gives the lowest number to the second substituent is used, and so on. Thus the dioxin shown above is 2,3,7,8-TCDD, or to give it its full title 2,3,7,8-tetrachlorodibenzo-*p*-dioxin. No wonder it is simply called "dioxin" in the press! However, there actually are 75 different chlorinated dibenzo-*p*-dioxin compounds, when one includes all the possibilities between one and eight chlorines, and given that a number of isomers exist for most of these eight types. Different members of a chemical family that differ only in the number and position of the same substituent are called **congeners.** All dioxins congeners are planar: all carbon, oxygen, hydrogen, and chlorine atoms lie in the same plane.

PROBLEM 6-5

Are 1,3-, 2,4-, 6,8-, and 7,9-dichlorodibenzo-*p*-dioxins all unique compounds or are they all really the same compound? Are 1,2- and 1,8-dichlorodibenzo-*p*-dioxins unique compounds? Deduce the structures of all unique dichlorodibenzo-*p*-dioxins, keeping in mind that the two rings are equivalent and that the molecule has "top-bottom" symmetry.

OTHER SOURCES OF DIOXINS

In addition to their use as starting materials in the production of herbicides, chlorophenols find use as wood preservatives (fungicides) and as
slimicides. In particular, some trichlorophenol isomers and some tetrachlorophenol isomers are sold as wood preservatives. The most common
preservative, in use since 1936, is pentachlorophenol (PCP, though not
the "angel dust" compound known by the same initials); all the benzene's hydrogens have been substituted in this compound:

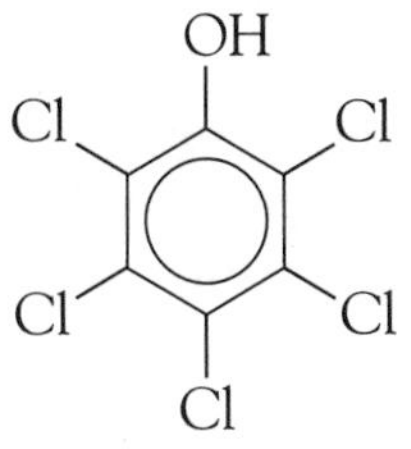

pentachlorophenol

Commercial PCP is not pure pentachlorophenol but is significantly contaminated (about 20%) with 2,3,4,6-tetrachlorophenol. This mixture
has many pesticidal uses: it is used as a herbicide (e.g., as a preharvest
defoliant), an insecticide (termite control), a fungicide (wood preserva-
tion and seed treatment), and a molluscicide (snail control).

Unfortunately, if wood treated with such preservatives is eventually
burned, a small fraction of the chlorophenols can react to eliminate HCl
and thereby produce members of the chlorinated dioxin family.
Thus octachlorodibenzo-*p*-dioxin, OCDD, is produced as an unwanted
byproduct in the low-temperature combustion of pentachlorophenol
products:

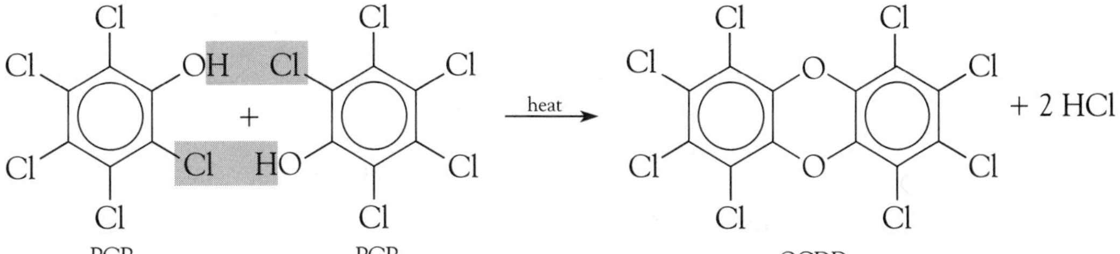

Commercial supplies of chlorinated phenols themselves are known
to be contaminated with various dioxins. Indeed, pentachlorophenols

are one of the largest chemical sources of dioxins to the environment; however the main dioxin involved here, the octachlorodibenzo-*p*-dioxin, is not considered to be particularly toxic, as discussed in a later section.

In general, any two phenol molecules that each have a chlorine *ortho* substituted to the OH group can combine to produce a dibenzo-*p*-dioxin molecule; the two phenols need not be identical, but simply need to make contact when they have been heated sufficiently to facilitate HCl elimination and dioxin formation. Similarly, coupling of phenoxide anions can occur with Cl⁻ elimination, as discussed above in the case of the 2,4,5-T synthesis.

The chlorophenolic source of dioxins found in environmental samples can be deduced by reversing the logic used above. Although the following example may pose a challenge to the reader's powers of consecutive reasoning, efforts spent in pursuing the logic involved will be repaid by the attainment of insight into what amounts to the exercise of common sense in the solution of an important and practical scientific problem.

Consider the congener 1,2,7,8-tetrachlorodibenzo-*p*-dioxin; it could have been formed by elimination of 2 HCl molecules from two chlorophenol molecules in the following two ways:

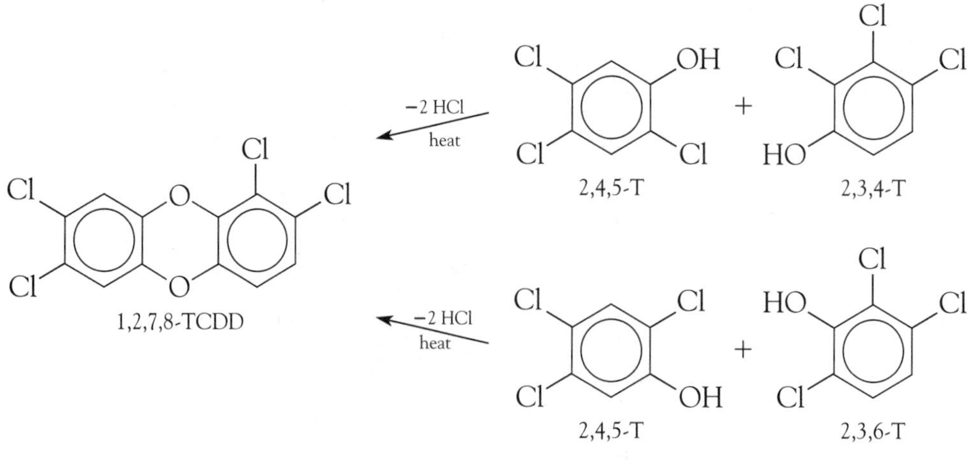

If it is assumed that the oxygen atom at the top of the dioxin congener originates with the chlorophenol congener at the left side of the dioxin molecule, then the bottom oxygen must come from the benzene ring on the right side of the dioxin molecule; with this set of assumptions, the original reactants must have been 2,4,5- and 2,3,4-trichlorophenol.

(Notice in the diagram above that the chlorine atoms eliminated must have arisen from positions adjacent to the oxygen atoms.) The alternative possibility, that the oxygen atom at the top of the dioxin structure came from the benzene ring at the right side of the molecule and the bottom oxygen from the ring on the left, leads to the possibility that the trichlorophenol molecules which combined together were the 2,4,5 and the 2,3,6 congeners. Thus a 1,2,7,8-tetrachlorodibenzo-*p*-dioxin molecule in the environment could have arisen by combination of a 2,4,5-trichlorophenol molecule with either a 2,3,4- or a 2,3,6-substituted congener. Unfortunately, some dioxins undergo rearrangement of substituents during their formation so that such a "retrosynthesis" approach is not an infallible guide to the origin of dioxins discovered in the environment.

In addition to the chlorophenolic sources discussed above, dioxins also enter the environment as byproducts of a number of other processes. Pulp and paper mills that still use chlorine in the bleaching of the pulp are major sources; the dioxins result from the reaction of the chlorine with some of the organic molecules produced from the pulp. Unfortunately, the most common dioxin produced by the pulp and paper bleaching process is the highly toxic 2,3,7,8-TCDD congener. The paper and effluent contain dioxins at parts-per-trillion levels, resulting in total releases, in Canada for example, of several hundred grams of 2,3,7,8-TCDD annually.

Fires of many kinds, including those in incinerators, also release various congeners of the dioxin family into the environment; these chemicals are produced as minor byproducts from the chlorine and organic matter in the fuel. Dioxin production seems unavoidable whenever combustion of organic matter occurs in the presence of chlorine, unless steps are taken to ensure complete combustion by using very high flame temperatures. In many environmental samples of combustion products, several dozen different dioxin congeners are found, all in comparable amounts. Congeners with relatively high numbers of chlorine substituents often are the most common.

As a consequence of their widespread occurrence in the environment and their lipophilicity (tendency to dissolve in fatty matter), dioxins bioaccumulate in the food chain. More than 90% of human exposure to dioxins is attributable to the food we eat, particularly meat, fish, and dairy products. Typically, dioxins and furans (a group of compounds resembling the dioxins in structure) are present in fish and meat at levels of tens or hundreds of picograms (pg, or 10^{-12} gram) per gram of the food; in other words, they occur at levels of a few parts per quadrillion. (Roughly speaking, one would have to consume one million *tons* of food to ingest one gram of dioxin at this rate!) The average concentrations in the United States of 2,3,7,8-TCDD found in common foods are listed in Table 6-3, together with the average daily human intake from each food. Table 6-4 gives a representative analysis of the dioxins present in human fat tissue; their chief source is food. It should be realized, however, that the bulk of dioxins and furans in nature are not present in biological systems: attachment to soil and sediments is their most common sink.

The potential impact on human health of exposure to dioxins is documented later, following a discussion of the properties of PCBs and furans, two types of chemicals with which dioxins share many properties.

PROBLEM 6-6

a. Deduce the structures and the correct numbering for the other two tetrachlorophenol isomers, i.e., in addition to the 2,3,4,6 isomer mentioned in the text. b. For each of these two isomers, deduce the structure and names of the dioxin(s) which would result if two molecules of that isomer were to react together.
c. Deduce which combination(s) of two *different* tetrachlorophenol isomers would produce the following hexachlorodibenzo-*p*-dioxins: i. the 1,2,3,7,8,9 isomer, ii. the 1,2,4,6,8,9 isomer, and iii. the 1,2,3,6,7,9 isomer.

PROBLEM 6-7

Deduce what dioxin(s) would be produced in side reactions if a. 2,4-D is synthesized from 2,4-dichlorophenol, and b. a commercial sample of PCP is burnt at low temperature.

PCBs

The well-known acronym *PCBs* (*polychlorinated biphenyls*) stands for a group of industrial organochlorine chemicals that became a major environmental concern in the 1980s and 1990s. Although not pesticides, they found a wide variety of applications in modern society because of

TABLE 6-3	AVERAGE 2,3,7,8-TCDD ("DIOXIN") CONTENT IN THE AMERICAN FOOD SUPPLY	
Food	Dioxin concentration in picograms per gram (pg/g)	Average dioxin intake (pg/person/day)
Ocean fish	500	8.6
Meat	35	6.6
Cheese	16	0.31
Milk	1.8	0.20
Coffee	0.1	0.04
Ice cream	5.5	0.04
Cream	7.2	0.01
Sour cream	10	0.01
Cottage cheese	2.1	0.01
Orange juice	0.2	0.01
Total		15.9

Data drawn from S. Henry, G. Cramer, M. Bolger, J. Springer, and R. Scheuplein. 1992. *Chemosphere* 25: 235–238.

TABLE 6-4	DIOXIN CONGENERS DETECTED IN HUMAN FAT TISSUE	
Congener	Average concentration (pg/g)	Standard deviation (pg/g)
2,3,7,8-tetrachloro	11	8
1,2,3,7,8-pentachloro	24	12
1,2,3,6,7,8-hexachloro	172	74
1,2,3,7,8,9-hexachloro	22	9
1,2,3,4,6,7,8-heptachloro	232	181
octachloro	1037	712

Data drawn from 1990 autopsies in five cities in Ontario, as reported in G. L. LeBel, D. T. Williams, F. M. Benoit, and M. Goddard, Polychlorinated dibenzodioxins and dibenzofurans in human adipose tissue samples from five Ontario municipalities. *Chemosphere* 21(12): 1465–1475, 1990.

certain other properties they possess. Since the late 1950s, over one million metric tons of PCBs have been produced. Like many other organochlorines, they are very persistent in the environment and they bioaccumulate in living systems. As a result of careless disposal practices, they have become a major environmental contaminant in many areas of the world. Due both to their own toxicity and to that of their furan contaminants, PCBs in the environment have become a cause for concern because of their potential impact on human health, particularly with regard to growth and development. Some recent evidence supporting these worries was discussed in Chapter 1. In the material below, we consider what PCBs are, how they are made, what they are used for, and how they become contaminated and released into the environment. Finally, we discuss what is known about the effects on human health of exposure to PCBs, furans, and dioxins.

THE CHEMICAL STRUCTURES OF PCBS

Although benzene is a very stable compound, heating it to very high temperatures can disrupt the carbon-to-hydrogen bonds. This fact is exploited commercially when benzene is heated to about 750°C, in the presence of lead as a catalyst, to form biphenyl, a molecule in which two benzene rings are linked by a single bond formed between two carbons that have each lost their hydrogen atom:

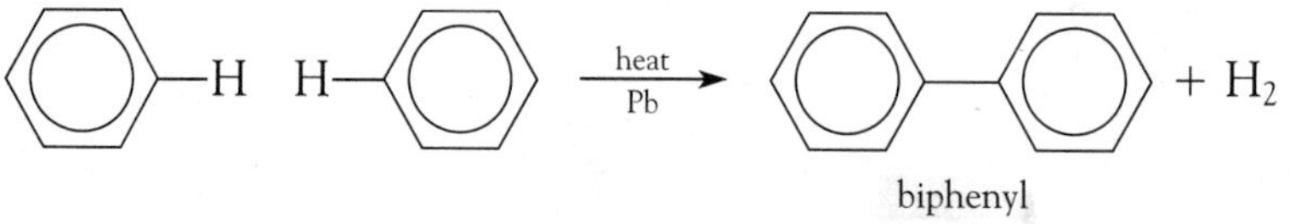

biphenyl

Like benzene, if biphenyl is reacted with Cl_2 in the presence of a ferric chloride catalyst, some of its hydrogen atoms become replaced by chlorine. The more chlorine initially present, and the longer the reaction is allowed to proceed, the greater the extent (on average) of chlorination of the biphenyl molecule. The products are *polychlorinated biphenyls*, or PCBs for short. The reaction of biphenyl with chlorine produces a mixture of many of the 209 congeners of the PCB family; the exact proportions depend upon the ratio of chlorine to biphenyl, and the reaction time and temperature. Although many individual PCB compounds are solids, the mixtures usually are liquids or low–melting–point solids. Commercially, individual PCB compounds were not isolated; rather they were sold as partially separated mixtures, with the average chlorine content in different products ranging from 21% to 68%.

PROBLEM 6-8

The general formula for any PCB congener is $C_{12}H_{10-n}Cl_n$, where n ranges from 1 to 10. Determine the average number of chlorine atoms per PCB molecule in a mixture that is 60% chlorine by mass.

The numbering scheme used for individual PCB congeners begins with the carbon of one ring that is joined to a carbon in the other ring; it is given the number 1, and the other carbons around the ring are numbered sequentially. As shown below, the positions in the second ring are distinguished by primes. The 2′ position in the second ring lies on the same side of the C—C bond joining the rings as does the 2 position in the first ring, and so on.

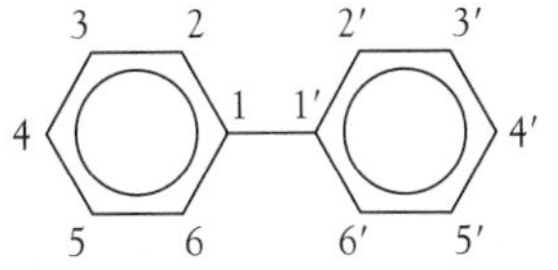

In most instances, the two rings in a chlorinated biphenyl molecule are *not* equivalent since the patterns of substitution differ. The unprimed ring is chosen to be that which would give a substituent with the lowest-numbered carbon. An example of a PCB is

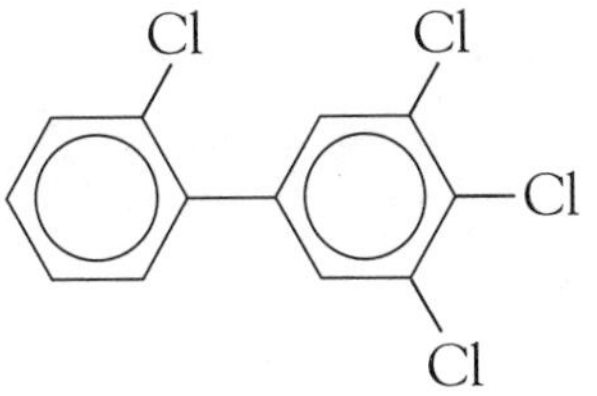

2,3′,4′,5′-tetrachlorobiphenyl

Very rapid rotation occurs around carbon-carbon single bonds in most organic molecules, including the C—C link joining the two rings in biphenyl and in most PCBs. Thus it is not normally possible to isolate compounds corresponding to different orientations of the two rings in a PCB relative to each other. For example, 3,3′- and 3,5′-dichlorobiphenyl are not individually isolatable compounds; one form is constantly being converted into the other and back again by rapid rotation about the C—C bond linking the rings:

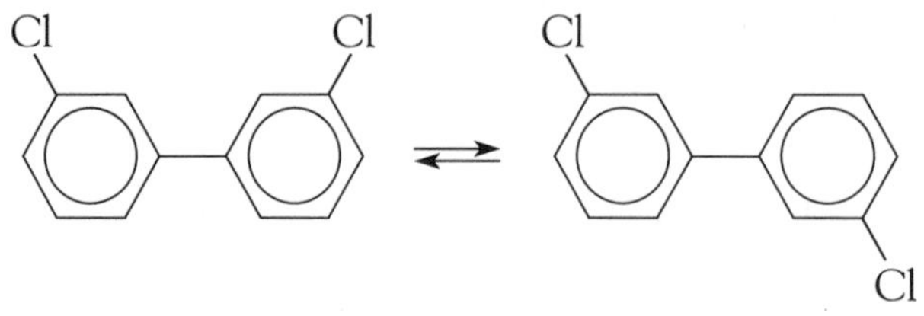

3,3'-dichlorobiphenyl   3,5'-dichlorobiphenyl

The label used for the compound is that which has the lowest number for the second chlorine, so the system shown above is called the 3,3' isomer. Although the rings rotate rapidly with respect to each other, the energetically optimum orientation is the one having the rings coplanar or close to it, except, as we shall see later, when large atoms or groups occupy the 2 and 6 positions.

PROBLEM 6-9

Draw the structures of all unique dichlorobiphenyls, assuming first that free rotation about the bond joining the rings does *not* occur. Then deduce which pairs of structures become identical due to free rotation.

THE PROPERTIES AND USES OF PCBS

All PCBs are practically insoluble in water but are soluble in hydrophobic media, such as fatty or oily substances. Commercially, they were attractive because they are chemically inert liquids and are difficult to burn, have low vapor pressures, and are excellent electrical insulators. As a result of these properties, they were used extensively as the coolant fluids in power transformers and capacitors. Later, they were also employed as **plasticizers,** that is, agents used to keep other materials such as PVC products more flexible; in "carbonless" copy paper; as de-inking solvents for recycling newsprint; as heat transfer fluids in machinery; as waterproofing agents; and even further uses were found for them.

Because of their stability and extensive usage, together with inattentive disposal practices, PCBs became widespread and persistent environmental contaminants. When their accumulation and harmful effects became recognized, **open uses,** that is, those in which their disposal could not be controlled, were terminated. Although North American production of PCBs was halted in 1977, the substances remain in use in many electrical transformers currently in service. The liquid in such transformers is mainly PCBs, but often contains some polychlorinated benzenes as well. As these electrical units are gradually decommissioned, their PCB content usually is stored in order to prevent further contamination of the environment; in some locales, PCBs are destroyed by incineration. Previously, PCB-containing transformers were often just dumped into landfills, and their PCB content was allowed to leak into the ground. In summary, PCBs were released into the environment during their production, their use, their storage, and their disposal.

If released into the environment, PCBs persist for years because they are so resistant to breakdown by chemical or biological agents. Although their solubility in water is very slight, the tiny amounts of PCBs in surface waters are constantly being volatilized and subsequently redeposited on land or in water after traveling in air for a few days. By such mechanisms, PCBs have been transported worldwide; there are measurable background levels of PCBs even in polar regions and at the bottom of oceans. This environmental load of PCBs will continue to be recycled among air, land, and water, including the biosphere for decades to come.

Because of their persistence and their solubility in fatty tissue, PCBs in food chains undergo biomagnification; an example is shown in Figure 6-6: note that the ratio of PCBs in the eggs of herring gulls in the Great Lakes is 50,000 times that in the phytoplankton in the water. Similarly, the levels in the 1980s of PCBs in the eggs of the bird called Forster's tern in Green Bay, Wisconsin was about 180,000 times those in the water. The good news is that the average levels of PCBs in such eggs has been falling with time in many locations in the Great Lakes basin, as the data in Figure 6-7 illustrates for one site in Lake Ontario.

FURAN CONTAMINATION OF PCBS

Strong heating of PCBs in the presence of a source of oxygen can result in the production of small amounts of dibenzofurans. These compounds are structurally similar to dioxins; they differ in that they are missing one oxygen in the central ring. The basic furan ring contains five atoms, one of which is oxygen and the other four of which are carbon atoms that participate in double bonds:

furan

FIGURE 6-6

The bioaccumulation and biomagnification of PCBs in the Great Lakes aquatic food chain. (Source: "The State of Canada's Environment." Ottawa, Canada: Minister of Supply and Services, 1991.)

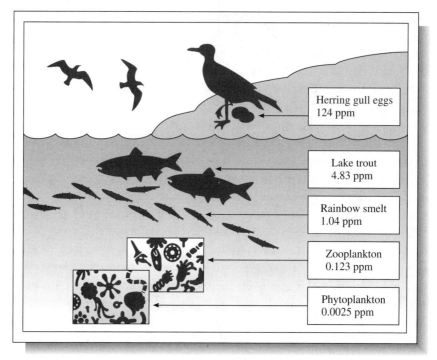

Herring gull eggs
124 ppm

Lake trout
4.83 ppm

Rainbow smelt
1.04 ppm

Zooplankton
0.123 ppm

Phytoplankton
0.0025 ppm

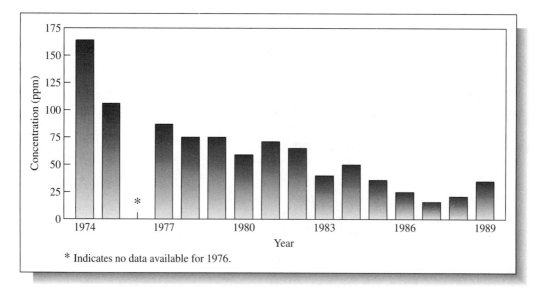

* Indicates no data available for 1976.

FIGURE 6-7

The concentration of PCBs in herring gull eggs at the Toronto shoreline of Lake Ontario (1974–1989). (Source: "The State of Canada's Environment." Ottawa, Canada: Minister of Supply and Services, 1991.)

The *dibenzofurans* (DFs) have a benzene ring fused to opposite sides of the furan ring:

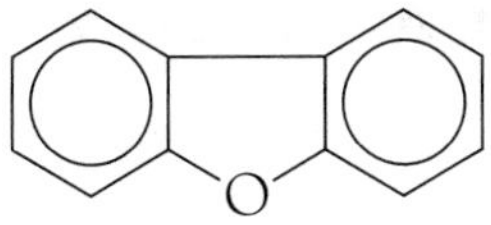

dibenzofuran

As is the case for dioxins, all chlorinated dibenzofuran congeners are planar; that is, all C, O, H, and Cl atoms lie in the same plane. They are formed from PCBs by the elimination of the atoms X and Y (below) bonded to two carbons that are *ortho* in position to those which link the rings and that lie on the same side of the C—C link between the rings:

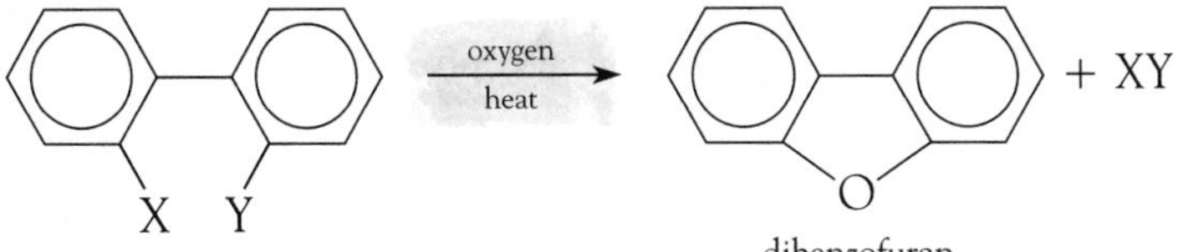

The atoms X, Y can both be chlorine, or one can be hydrogen and the other one chlorine, so the molecule eliminated can be Cl_2 or ClH respectively. A more detailed analysis of the nature of the specific furans that result from particular PCB congeners is given in Box 6-2.

Most of the chlorine in the original PCB molecule is still present in the dibenzofuran; *polychlorinated dibenzofurans* are known commonly as PCDFs. The numbering scheme for substituents is the same as that for dioxins (PCDDs); note however that the numbering starts next to the carbon which forms the single bond *opposite* the oxygen.

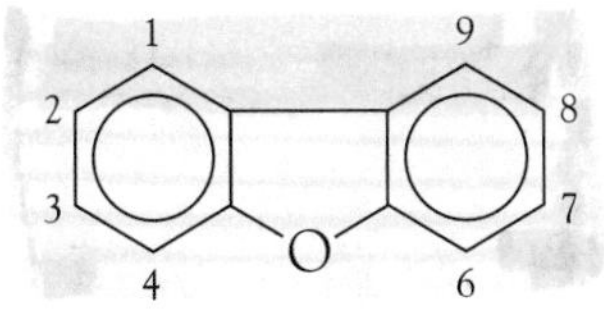

While there exist 75 different chlorine-substituted dibenzo-*p*-dioxins, there are 135 dibenzofuran congeners, since the symmetry of the ring system is lower for furans.

BOX 6-2   PREDICTING THE FURANS THAT WILL FORM FROM A GIVEN PCB

In deducing the nature of the polychlorinated dibenzofuran (PCDF) that would be formed from a particular PCB, it should be remembered that free rotation occurs about the single bond joining the two rings in the original biphenyl in all PCBs at the elevated temperatures of the reaction. Thus HCl elimination in 2,3'-dichlorobiphenyl gives both 4- and 2-chlorodibenzofuran.

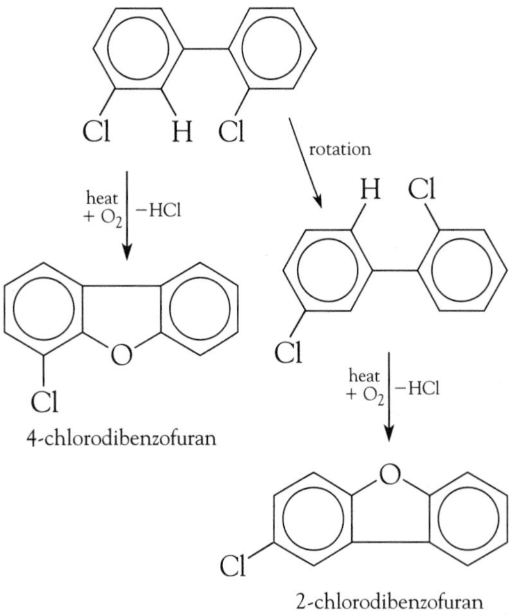

At the high temperatures of this reaction, some interchange of the adjacent substituents in the 2 and 3 positions (*ortho* and *meta*) of any given ring can occur as a prelude to HCl elimination; in particular *chlorine can move from an ortho to a meta position, and hydrogen from meta to ortho, preceding HCl elimination*. For example, when 2,6,2',6'-tetrachlorobiphenyl (shown on the opposite page) is heated in air, some of its molecules lose a pair of ortho chlorines to give a dichlorodibenzofuran, and some first

4-chlorodibenzofuran

2-chlorodibenzofuran

PROBLEM 6-10

For each PCB shown below, deduce which furans would be expected to be produced by Cl_2 or HCl elimination when it is heated in air. Write the correct name for each PCDF.

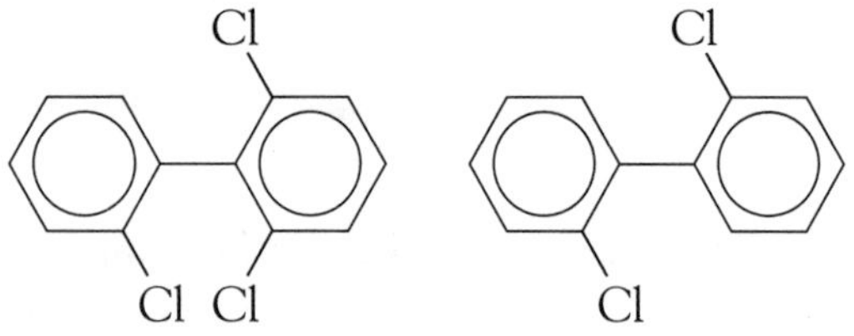

interchange Cl and H in one ring to eliminate HCl and produce a trichlorodibenzofuran.

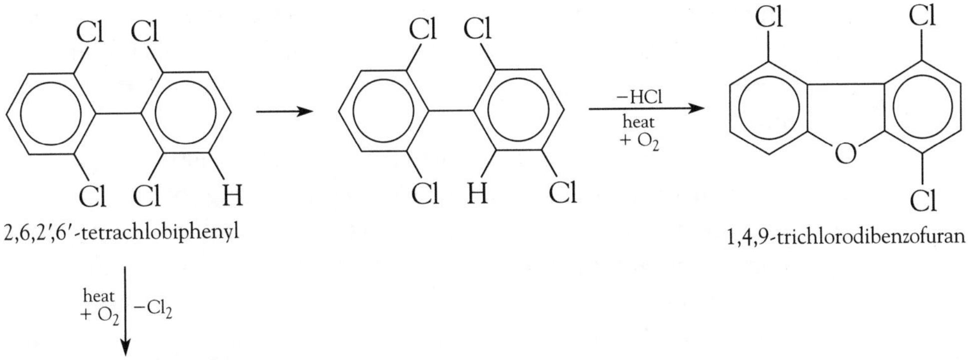
Cl
Cl
H
Cl
Cl
Cl
Cl
Cl
H
Cl
−HCl
heat
+ O2
Cl
Cl
O
Cl
heat
+ O2
−Cl2

2,6,2′,6′-tetrachlobiphenyl

1,4,9-trichlorodibenzofuran

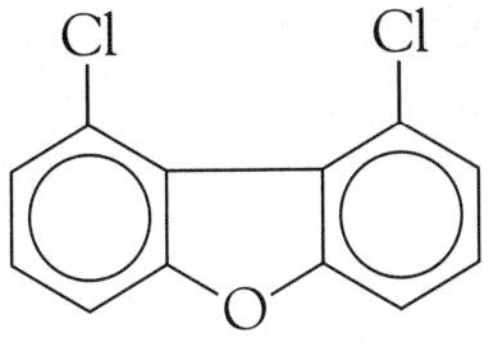
Cl
Cl
O

1,9-dichlorodibenzofuran

Free rotation about the C—C bond does not occur after the interchange, as presumably the elimination occurs immediately.

PROBLEM 6-11

Recently it has been discovered that upon strong heating in air PCBs can also react by elimination of two *ortho* hydrogen atoms (one on each ring) as H_2 . Decide which, if any, additional PCDFs will be produced if the PCBs in the previous problem can eliminate H_2.

PROBLEM 6-12

Consider the PCDF shown below. Decide which PCBs could produce this furan if they are heated in air, given that PCDFs can result from HCl elimination with or without 2, 3 interchange, or from Cl_2 or H_2 elimination.

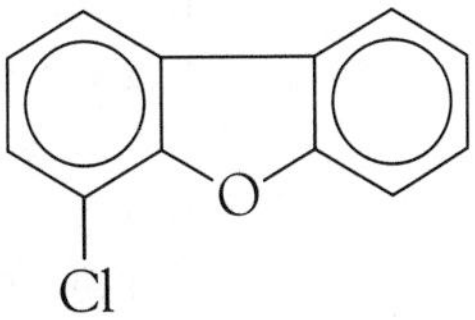
O
Cl

PROBLEM 6-13

Draw the structures of all the 16 unique dichlorodibenzofurans, and give their names.

Almost all commercial PCB samples are contaminated with some PCDFs, but it usually amounts to only a few ppm in the originally manufactured liquids. However, if the PCBs are heated to high temperatures and some oxygen is present, conversion of PCBs to PCDFs increases the level of contamination by orders of magnitude. Furan production also occurs if one attempts to burn PCBs with anything but an unusually hot flame.

Like dioxins, polychlorinated dibenzofurans also are produced in small quantities by a myriad of processes, including the bleaching of pulp and the incineration of garbage. Most furans found in the environment contain an intermediate number, that is, four to six, chlorine atoms, whereas most environmental dioxins are fully chlorinated or almost so. As discussed below, furans with such intermediate amounts of chlorine have toxicities similar to that of 2,3,7,8-TCDD, whereas fully chlorinated dioxin molecules are almost nontoxic. Consequently, the threat to human health from furans in the environment may even exceed that from dioxins.

TOXICOLOGY OF PCBs, DIOXINS, AND FURANS

Over one billion dollars has been spent on research to determine the extent to which dioxins, furans, and PCBs cause toxic reactions in humans. Nevertheless, conclusions about this issue are still tentative and controversial. Evidence about toxicity is derived from two sources: experiments on animals that have been deliberately exposed to the chemicals, and studies of humans who have been accidentally exposed. The test data on animals such as rats and guinea pigs is more "scientific" in the sense that large numbers of experiments at varying levels of concentration can be carried out, but since the immune systems of the animals differ from ours, the results are not necessarily directly applicable to human populations.

It is generally agreed that PCBs are not very *acutely* toxic to humans; in other words their LOD_{50} values are large and so people do not immediately become deathly ill from ingesting them in small quantities. In high doses, PCBs cause cancer in test animals. Most groups of people who have been exposed to relatively high concentrations of PCBs, as a result of their employment in electrical capacitor plants for example, have not experienced a higher overall death rate. (See, however, the comments about human cancer later in this section.) The most common reaction to exposure is **chloracne,** a persistent, disfiguring and painful analog to common acne; it is a biological response by humans to exposure to many types of organochlorine compounds.

PCBs are of concern with respect to their reproductive toxicity in humans and animals, particularly when large quantities of fish have been consumed in which PCBs have accumulated in the fatty tissue. As discussed in Chapter 1, according to one well-documented study, the children of women who consumed large amounts of fish from Lake Michigan suffered some growth retardation and scored significantly lower on certain memory tests; these effects have been blamed upon transfer in utero of PCBs, and persist until the age of four years at least.

However, the most dramatic effects yet observed to human health from exposure to PCB mixtures occurred when two sets of people, one in Japan in 1968 and the other in Taiwan in 1979, unintentionally consumed PCBs that had accidentally been mixed with cooking oil. In the Japanese incident, and probably in the Taiwanese case as well, the PCBs had been used as a heat exchanger fluid in the deodorization process for the oil. Since the PCBs had been heated, the level of PCDF contamination was much greater than occurs in freshly prepared commercial PCBs. The thousands of Japanese and the Taiwanese people who consumed the contaminated oil suffered health effects far worse than did workers at PCB manufacturing and handling plants, even though the resulting PCB levels in their bodies were about the same. From this difference, it has been concluded that the main toxic agents in the poisonings were the PCDFs, and that they were responsible for about 75% of the effects, with the PCBs themselves responsible for the remainder. Indeed, studies on laboratory animals indicate that the furans involved in these incidents are more than 500 times as toxic on a gram-for-gram basis than are pure PCBs.

Test results from studies on animals indicate that the toxicity of dioxins, furans and PCBs depends to an extraordinary degree on the extent and pattern of chlorine substitution. If we refer to the carbon atoms that are bonded to those in the central dioxin ring as "alpha" carbons, and the outlying ones as "beta" carbons (below), then the following generalization can be made: *the very toxic dioxins are those with three or four beta chlorine atoms, and few if any alpha chlorines.*

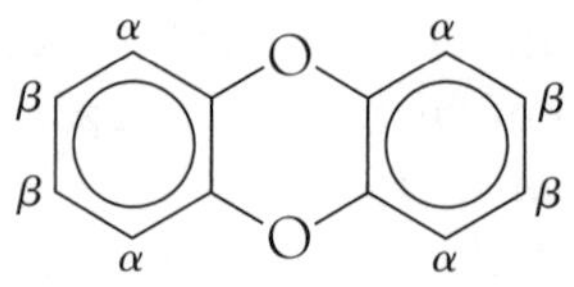

The most toxic is 2,3,7,8-TCDD, which has the maximum number (four) of beta chlorines and no alpha chlorines.

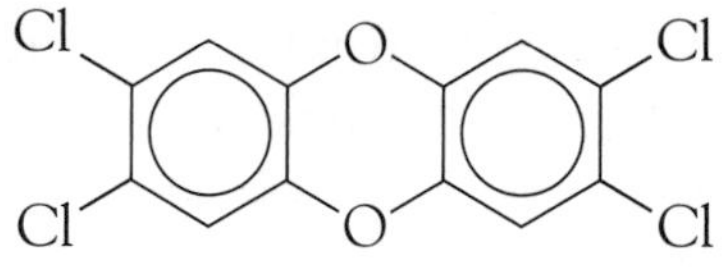

2,3,7,8-TCDD

Recent studies have shown that single doses of 2,3,7,8-TCDD administered to pregnant laboratory animals cause reproductive effects in their offspring; these results have raised some alarm about the potential effects of dioxins on human reproduction. In this connection, many scientists are worried about the dangers posed by environmental chemicals such as dioxins, furans, PCBs, and other organochlorines that can affect sex hormones.

Dioxin congeners that have three beta chlorines, but no (or only one) alpha chlorine, are appreciably toxic, but less so than the 2,3,7,8-system. Fully chlorinated dioxin, that is, octachlorodibenzo-*p*-dioxin (OCDD), has a very low toxicity since all the alpha positions are occupied by chlorine.

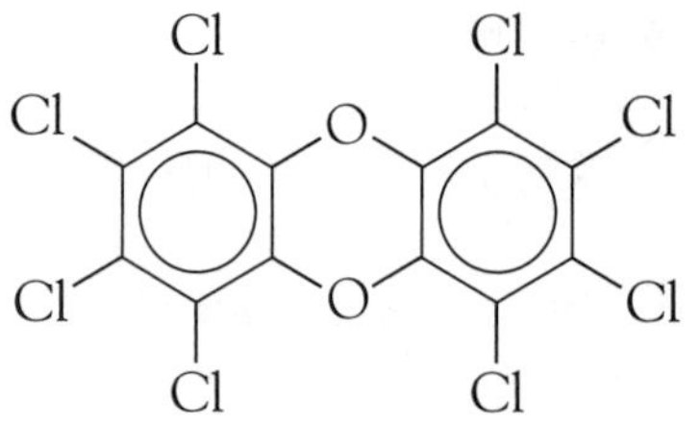

OCDD

Similarly, mono- and di-chloro dioxins are usually not considered highly toxic even if the chlorines are present in beta positions. Both 2,3,7,8-TCDD and 2,3,7,8-TCDF have been named Priority Pollutants by the IJC.

Although they have not been researched as thoroughly as the dioxins, it is generally assumed that the toxicity pattern for furans is similar to that for dioxins in that the most toxic congeners have chlorines in most or all of the beta positions and few chlorines in the alpha positions. In humans, the highly chlorinated furans, dioxins, and PCBs are stored in fatty tissues and are neither readily metabolized nor excreted. This persistence is a consequence of their structure: few of them contain hydrogen atoms on adjacent pairs of carbons at which hydroxyl groups, —OH, can readily be added in the biochemical reactions necessary for their elimination (as discussed later in this chapter); in contrast, those compounds with few chlorines always contain one or more such adjacent pairs of hydrogens and tend to be excreted rather than stored for a long time.

PROBLEM 6-14

Predict the order of relative toxicities of the following three dioxin congeners, given that for systems not too dissimilar to TCDD, the presence of an alpha chlorine reduces the toxicity less than does the absence of a beta chlorine:

2,3,7-trichlorodibenzo-*p*-dioxin

1,2,3-trichlorodibenzo-*p*-dioxin

1,2,3,7,8-pentachlorodibenzo-*p*-dioxin

According to animal tests, the most toxic PCBs are those having no chlorine atoms (or at most one) in the positions that are *ortho* to the carbons that join the rings, that is, in the 2, 2, 6, and 6' carbons. Without *ortho* chlorines, the two benzene rings can easily adopt an almost coplanar configuration, and rotation about the C—C bond joining the rings is rapid. However, because of the large size of chlorine atoms, they get in each other's way if they are present in both *ortho* positions on the same side of the two rings; this interaction forces the rings to twist away from each other, and prevents the rings from adopting the coplanar geometry:

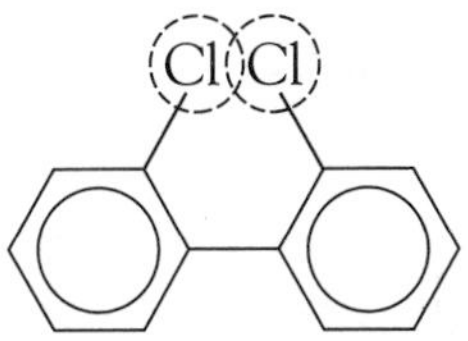

Thus, PCB molecules with chlorines in three or four of the *ortho* positions cannot adopt a coplanar geometry.

If the rings are not kept from coplanarity by interference between chlorine atoms, and if certain *meta* and *para* carbons are substituted by chlorine, then the PCB molecule can readily attain the coplanar geometry which happens to be similar in size and shape to 2,3,7,8-TCDD; such PCB molecules are highly toxic. Apparently 2,3,7,8-TCDD and other molecules of its size and shape readily fit into the same cavity in a specific biological receptor; the complex of the molecule and the receptor can pass through cell membranes and thereby initiate toxic action. By comparing molecular models, it is not difficult to see, for example, that 3,4,4′,5′-tetrachlorobiphenyl is almost the same size and shape as 2,3,7,8-TCDD. Only a very small fraction of commercial PCB mixtures corresponds to coplanar PCBs having no *ortho* chlorines. Although individually less toxic, PCBs with one *ortho* chlorine, and with chlorines in both *para* and at least one *meta* position, contribute more substantially to the overall toxicity of PCB mixtures since they are far more prevalent than those having no *ortho* chlorines. It is now believed that some of the toxic effects observed in the cooking oil incidents were due to coplanar PCB congeners.

Since most organisms, including humans, possess a mixture of many dioxins, furans, and PCBs stored in their body fat, it is useful to have a measure of their net toxicity. To this end, scientists often report concentrations of these organochlorines in terms of the equivalent amount of 2,3,7,8-TCDD which if present alone would produce the same toxic effect. To this end, an **international toxicity equivalency factor,** or TEQ, has been devised which rates each dioxin, furan, and PCB congener's toxicity relative to that of 2,3,7,8-TCDD, which is arbitrarily assigned a value of 1.0.

A summary of the TEQ values for some representative dioxins and furans is given in Table 6-5. As an example, consider an individual who ingests 30 picograms (pg) of 2,3,7,8-TCDD, 60 pg of 1,2,3,7,8-PCDF, and 200 pg of OCDD. Since the TEQ factors for these three substances are, respectively, 1.0, 0.05, and 0.001, the intake is equivalent to

$$30 \text{ pg} \times 1.0 + 60 \text{ pg} \times 0.5 + 200 \text{ pg} \times 0.001 = 33.2 \text{ pg}$$

Thus, even though a total of 290 pg of dioxins and furans were ingested by this person, it is equivalent in its toxicity to an intake of 33.2 pg of 2,3,7,8-TCDD. A discussion of the average TEQ exposure to humans is given in Box 6-3.

TABLE 6-5	TOXICITY EQUIVALENCE FACTORS (TEQ) FOR SOME IMPORTANT DIOXINS AND FURANS	
Dioxin or furan	Toxicity Equivalency Factor	
2,3,7,8-tetrachlorodibenzo-p-dioxin	1	
1,2,3,7,8-pentachlorodibenzo-p-dioxin	0.5	
1,2,3,4,7,8-hexachlorodibenzo-p-dioxin 1,2,3,7,8,9-hexachlorodibenzo-p-dioxin 1,2,3,6,7,8-hexachlorodibenzo-p-dioxin	0.1	
1,2,3,4,6,7,8-heptachlorodibenzo-p-dioxin	0.01	
octachlorodibenzo-p-dioxin	0.001	
2,3,7,8-tetrachlorodibenzofuran	0.1	
2,3,4,7,8-pentachlorodibenzofuran	0.5	
1,2,3,7,8-pentachlorodibenzofuran	0.05	
1,2,3,4,7,8-hexachlorodibenzofuran 1,2,3,7,8,9-hexachlorodibenzofuran 1,2,3,6,7,8-hexachlorodibenzofuran 2,3,4,6,7,8-hexachlorodibenzofuran	0.1	
1,2,3,4,6,7,8-heptachlorodibenzofuran 1,2,3,4,7,8,9-heptachlorodibenzofuran	0.01	
octachlorodibenzofuran	0.001	

TEQ data are drawn from the Canadian Environmental Protection Act Priority Substance List. Assessment Report No. 1 (1990).

PROBLEM 6-15

Using the TEQ values in Table 6-5, calculate the number of equivalent picograms of 2,3,7,8-TCDD that corresponds to an intake of 24 pg of 1,2,3,7,8,9-hexachlorodibenzo-p-dioxin, 52 pg of 2,3,4,7,8-pentachlorodibenzofuran, and 200 pg of octachlorodibenzofuran.

Although there is not too much argument as to the *relative* toxicities of various dioxin and furan congeners, their *absolute* risk to humans is

---

BOX
6-3

AVERAGE EXPOSURE OF HUMANS TO DIOXIN AND FURAN

According to data gathered in the mid-1980s, the average total concentration of all dioxins and furans in the fat tissue of adult Canadians amounts to about 1,130 ppt. However since highly chlorinated and therefore less toxic congeners predominate, the toxic equivalent concentration amounts to only about 32 ppt of 2,3,7,8-TCDD. Presumably, similar values apply also to Americans.

The average North American adult contains about 15 kg of fat, so his/her total body burden of 2,3,7,8-TCDD equivalents amounts to about 0.48 μg. Given that the average residence time T_{avg} of dioxins and furans in the human body is about seven years, and using the relationship from Chapter 4 connecting T_{avg} to the total amount C and the input rate R, i.e.,

$$T_{avg} = C/R$$

then the average human rate of intake of 2,3,7,8-TCDD equivalents is

$$R = C/T_{avg} = 0.48 \ \mu g/7 \ y = 0.07 \ \mu g \ y^{-1}.$$

This value is quite close to the intake estimated by considering the average PCDD and PCDF concentrations in food, air, and water consumed by a nonsmoking adult.

---

very controversial. The amounts of 2,3,7,8-TCDD per kilogram body weight required to kill a guinea pig is extraordinarily small—about one microgram, making it the most toxic synthetic chemical known for that species. However, the LD_{50} required to kill many other types of animals is hundreds or thousands of times this amount, and it appears that 2,3,7,8-TCDD's acute toxicity in humans is not substantial. The lethal doses of 2,3,7,8-TCDD for a number of animal species are listed in Table 6-6.

In contrast to the requirements discussed above for acute toxicity behavior for PCBs, experiments indicate that the PCBs that lead to the neurological effects in children probably are *non*coplanar molecules having at least one chlorine in the *ortho* position.

HUMAN HEALTH EFFECTS OF PCBs, DIOXINS, AND FURANS

Scientists now are more worried about the long-term effects of exposure to dioxins than about their short-term toxicities. A recent study of

TABLE 6-6	ACUTE TOXICITIES OF 2,3,7,8-TCDD IN EXPERIMENTAL ANIMALS	
Species	Route	LD_{50} (micrograms per kilogram)
Guinea pig (male)	Oral	0.6
Guinea pig (female)	Oral	2.1
Rabbit (male, female)	Oral	115
Rabbit (male, female)	Dermal	275
Rabbit (male, female)	Intraperitoneal	252–500
Monkey (female)	Oral	<70
Rat (male)	Oral	22
Rat (female)	Oral	45–500
Mouse (male)	Oral	<150
Mouse (male)	Intraperitoneal	120
Dog (male)	Oral	30–300
Dog (female)	Oral	>100
Frog	Oral	1,000
Hamster (male, female)	Oral	1,157
Hamster (male, female)	Intraperitoneal	3,000

Data are taken from F. H. Tschirley (1986). Dioxin. *Scientific American* 254 (February): 29–35.

American workers who were employed in industries that produced chemicals contaminated with 2,3,7,8-TCDD indicated that exposure to it at relatively high levels may cause cancer. The current theory concerning the action of dioxin predicts that there should be a threshold below which no toxic effects will occur, and the recent studies of workers exposed to 2,3,7,8-TCD support this hypothesis.

The Seveso study discussed previously was the first to show an increased rate of cancer among people exposed accidentally to TCDD. Another widespread exposure of humans to 2,3,7,8-TCDD occurred in the early 1970s in and around Times Beach, Missouri. Waste oil containing PCBs and 2,3,7,8-TCDD from manufacture of 2,4,5-trichlorophenol was used for dust control. Some horses died on account of exposure in the arena where the contamination was particularly high, and some children became ill. A decade later, widespread contamination of the soil in the town was discovered. Although no long-term human health effects from this incident have as yet been found, the town was evacuated.

For the furans, direct evidence of human susceptibility is available from the incidents of PCB-contaminated cooking oil consumption mentioned above. The most common symptoms observed were chloracne and other related skin problems. Unusual pigmentation occurred in the skin of babies born to some of the mothers who had been exposed. The children also often had low birth weight and experienced a rather high infant mortality rate. Children who directly consumed the oil showed retarded growth and abnormal tooth development. Many of the victims also reported aching or numbness in various parts of their bodies, and frequent bronchial problems. Other than chloracne, such symptoms are not observed in workers who are occupationally exposed to PCBs, in which the PCDF concentration is orders of magnitude lower, and whose body burdens of PCBs are comparable to those who consumed the contaminated oil. However, some children of these workers had mild cases of the less serious problems seen in the poisoned group.

There continues to be vigorous debate in scientific, industrial, and medical communities regarding the environmental dangers of dioxins, furans, and PCBs. In one camp are those who feel that the dangers from these chemicals have been wildly overstated in the media and by some special-interest groups; they point to the very low concentrations of these substances that exist in the environment, to the lack of human fatalities that have resulted from them, and to the enormous economic costs associated with instituting effective controls on them. At the other extreme are persons who point to the substantial biomagnification and high toxicity per molecule of these substances, and to their presence in almost all environments; they consider the detrimental effects such as cancer and birth deformities caused by these chemicals in wildlife to be "warning canaries" that signal potential ill effects in humans. Discovering where the "truth" lies between these opposing viewpoints presents a challenge even for environmental science students, to say nothing of the public at large!

In animal studies, a threshold of about 1 nanogram of 2,3,7,8-TCDD equivalent per kilogram of body weight per day is observed with respect to the cancer-causing ability of dioxins and furans. In determining the maximum tolerable human exposure to such compounds, the Canadian government has applied a safety factor of 100, resulting in a guideline for maximum exposure of 0.010 ng/kg/day averaged over a lifetime. (Regulatory agencies in the United States do not use a threshold model for dioxin action.) Currently, the average Canadian is estimated to be exposed to about one-quarter this amount, about 95% of which originates from his/her food. Exposure levels near the guideline limit are expected for persons consuming large amounts of fish that have elevated dioxin and furan levels.

Toxic Organic Waste Disposal

Given their potential for long-term environmental contamination, it would be highly desirable to destroy PCBs rather than to simply store them after they are removed from electrical equipment. Unfortunately, this aim is difficult to achieve because the compounds are so stable with respect to heat and so resistant to other chemicals; indeed it was these characteristics that made them suitable for many of their uses. The same problems arise in the disposal of other organochlorine compounds.

The only practical technique at present for the destruction of bulk PCBs is to incinerate them at very high temperatures (1,200°C) in the presence of oxygen. Under these conditions they undergo combustion to produce CO_2, HCl, H_2O and a small amount of Cl_2. Very high temperatures are used so that the conversion is nearly 100% complete (99.9999% conversion, called "six nine's," is the usual standard), and so that any dibenzofurans formed as intermediates are also destroyed. In order to achieve these high temperatures, special combustion chambers or cement kilns must be employed. Particular attention must be paid to the gases which would be exhausted if the process were to be interrupted, for example by a power failure, since lower combustion temperatures will then occur for a brief period, during which time dibenzofurans will be created and perhaps released. Despite the successes that have been achieved with such methods, public opposition to the siting and use of high-temperature incineration facilities for PCBs and other organochlorines remains strong in some locales.

When an oil or other substance is contaminated at a low level with PCBs, chemical decontamination is an alternative to incineration. A number of different reactive processes have been devised for such purposes. They involve the removal of the chlorine from the PCBs by the action of sodium or potassium metal or a reactive salt of one of these metals, thereby removing the chlorine in the form of harmless NaCl or KCl. The hydrocarbon portion of the PCB is converted into biphenyl itself if the reactive medium contains a hydrogen donor to supply H to the sites which have lost their Cl atoms; otherwise, a complicated polymer of benzene rings joined by single bonds is formed. In a modern variation of this process, concentrated PCBs are hydrogenated to produce methane and HCl. Soils contaminated with PCBs can be treated with a mixture of potassium hydroxide (KOH) and polyethylene glycol. The active ingredient in this mixture is an alkoxide ion, RO^-; this species will displace chloride ion from PCB molecules, thereby producing ether linkages. The polyether that results in less toxic and more water soluble than the original PCB.

Further aspects of the disposal and destruction of toxic waste are discussed in Chapter 7.

PROBLEM 6-16

Construct and balance chemical equations for the destruction of a PCB molecule having the formula $C_{12}H_6Cl_4$

a. by combustion with oxygen to yield CO_2, H_2O, and HCl, and

b. by hydrogenation to yield methane and HCl.

POLYNUCLEAR AROMATIC HYDROCARBONS (PAHS)

THE STRUCTURE OF PAHS

There is a series of benzenelike hydrocarbons that contain several six-membered rings connected together by the sharing of a pair of adjacent carbon atoms between adjoining **fused** rings. The simplest example is naphthalene, $C_{10}H_8$:

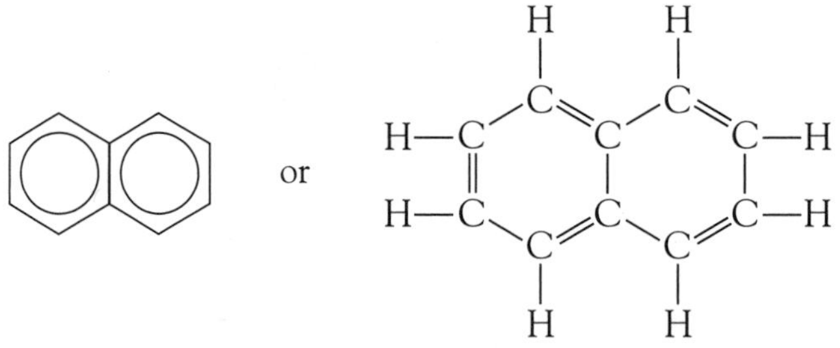

naphthalene

Notice that there are ten, not twelve, carbon atoms in total and that there are only eight hydrogen atoms, since the shared carbons have no attached hydrogens. As a compound, naphthalene is a volatile solid whose vapor is toxic to some insects. It has found use as one form of "moth balls," the other being 1,4-dichlorobenzene.

There are two ways to fuse a third benzene ring to two carbons in naphthalene; one results in a linear arrangement for the centers (the "nuclei") of the rings while the other is a "branched" arrangement:

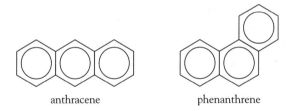

anthracene   phenanthrene

Both anthracene and phenanthrene are pollutants associated with in-complete combustion, especially of wood and coal, and are also released into the environment from the dumpsites of industrial plants that con-vert coal into gaseous fuel, and from petroleum and shale refineries. In rivers and lakes, they are found mainly attached to sediments rather than dissolved in the water; both are found to be subsequently partially incor-porated by freshwater mussels.

PROBLEM 6-17

Draw the full structural diagram for phenanthrene, showing all atoms and bonds explicitly.

PROBLEM 6-18

Show that the molecules below are not additional isomers of $C_{14}H_{10}$:

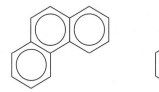

PROBLEM 6-19

Deduce the structural formulas for the five unique isomers of $C_{18}H_{12}$, which contains four fused benzene rings.

In general, hydrocarbons that display benzene-like properties are called aromatic; those which contain fused benzene rings are called **polynuclear** (or polycyclic) **aromatic hydrocarbons,** or PAHs for short. Like benzene itself, many PAHs possess unusually great stability and a planar geometry. Other than naphthalene, they are not manufactured commercially since they have no uses. Some PAHs are found in coal tar derivatives in commerce, however.

PAHs AS POLLUTANTS

PAHs are common air pollutants and are strongly implicated in the degradation of human health in some cities. Typically, the concentration of PAHs in urban outdoor air amounts to a few nanograms per cubic meter, although it can reach ten times this amount in very polluted environments. PAHs are formed when carbon-containing materials are incompletely burnt. PAHs containing four or fewer rings usually remain as gases if they are released into air. Usually after spending less than a day in outside air, they are degraded by a sequence of free radical reactions that begin, as expected from our previous analysis of air chemistry (Chapter 3), by the addition of the OH$^{\bullet}$ radical to a double bond. Elevated PAH concentrations in indoor air are typically due to the smoking of tobacco and the burning of wood and coal.

In contrast to their smaller analogs, PAHs with more than four benzene rings do not exist for long in air as gaseous molecules. Owing to their low vapor pressure, they quickly condense and become adsorbed onto the surfaces of soot and ash particles. Even PAHs with two to four rings adsorb onto particles in the wintertime, since their vapor pressure decreases sharply with lowering of the temperature. Since many soot particles are of respirable size, the PAHs can be transported into the lungs by breathing.

Soot itself is mainly graphitic-like carbon; it consists of collection of tiny crystals ("crystallites"), each of which is composed of stacks of planar layers of carbon atoms, all of which occur in fused benzene rings.

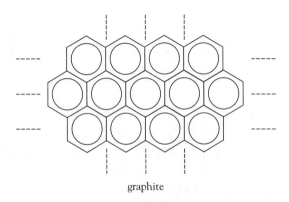

graphite

Graphite is the ultimate PAH: its parallel planes of fused benzene rings each contain a vast number of carbon atoms. There are no hydrogen atoms in graphite except at the periphery of the layers. The surfaces of soot particles are excellent adsorbers of gaseous molecules.

Polycyclic aromatic hydrocarbons also are serious water pollutants. The PAHs are generated in substantial quantity in the production of such coal tar derivatives as creosote, a wood preservative. The leaching of PAHs from the creosote used to preserve the immersed lumber of fishing docks and the like represents a significant source of pollution to crustaceans such as lobsters. Indeed, the lobster fishery in some parts of Atlantic Canada was closed in the early 1980s due to high levels of PAHs from this source.

PAHs also enter the aquatic environment as a result of spills of oil from tankers, refineries, and offshore oil drilling sites. In drinking water, the PAH level typically amounts to a few nanograms per liter, and usually is an unimportant source of these compounds to humans. The larger PAHs have been found to bioaccumulate in the fatty tissues of some marine organisms; they have been linked to the production of liver lesions and tumors in some fish, and are thought to play a role in the devastation of the populations of beluga whales in the St. Lawrence River.

The mechanism of PAH formation during combustion is complex, but apparently is due primarily to the repolymerization of hydrocarbon fragments that are formed during the **cracking,** that is, the splitting into several parts, of larger fuel molecules in the flame. The repolymerization reaction occurs particularly under oxygen-deficient conditions; generally the PAH formation rate increases as the oxygen:fuel ratio decreases. The fragments often lose some hydrogen, which forms water after combining with oxygen during the reaction steps. The carbon-rich fragments combine to form the polynuclear aromatic hydrocarbons, which are the most stable molecules that have a high C/H ratio. Details of these reactions are illustrated in Box 6-4.

PAHs are introduced into the environment from a number of sources: the exhaust of gasoline and diesel combustion engines, the "tar" of cigarette smoke, the surface of charred or burnt food, the smoke from burning wood or coal, and other combustion processes in which the carbon of the fuel is not completely converted to CO or CO_2. Although they constitute only about 0.1% of airborne particulate matter, their existence as air pollutants is of concern since many PAHs are carcinogenic, at least in test animals.

MORE ON THE FORMATION OF PAHs DURING COMBUSTION

Fragments containing two carbon atoms are particularly prevalent after cracking and partial combustion has occurred. In the prototype reaction below, a two-carbon free radical is shown reacting with an acetylene molecule, C_2H_2, to produce a four-carbon radical:

$$H_2C=\overset{\bullet}{C}H + HC\equiv CH$$
$$\longrightarrow\ H_2C=CH-CH=\overset{\bullet}{C}H$$

The resultant radical can subsequently add another acetylene, and cyclize to produce a six-membered ring:

$$H_2C=CH-CH=\overset{\bullet}{C}H + HC\equiv CH$$

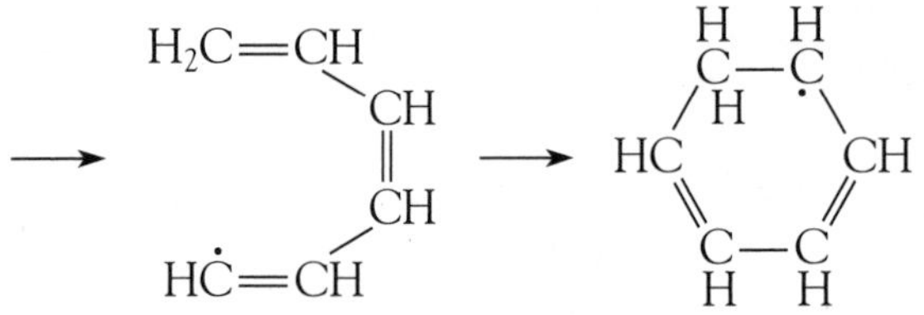

Loss of a hydrogen atom from the CH_2 carbon would yield benzene; alternatively the carbon which is the radical center could add more acetylene molecules and thereby build side chains that could form additional fused benzene rings:

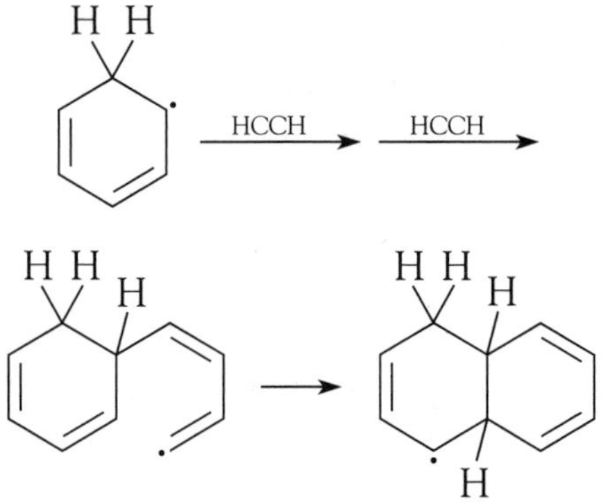

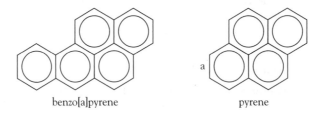

The most notorious and common carcinogenic hydrocarbon of this type is benzo[a]pyrene, BaP, which contains five fused benzene rings:

benzo[a]pyrene   pyrene

The molecule is named as a derivative of pyrene, which has the structure shown at right above; conceptually, if an additional benzene ring is added at the "a" bond of pyrene, the benzo[a]pyrene molecule is obtained.

Benzo[a]pyrene is a common byproduct of the incomplete combustion of fossil fuels, of organic matter (including garbage), and of wood. It is a carcinogen in test animals, and a probable human carcinogen. Its level in sediments at several urban sites in the Great Lakes exceeds the current guidelines, and consequently it is an IJC Priority Pollutant. BaP is worrisome since it bioaccumulates in the food chain: its log K_{ow} value is 6.3, comparable to that of many organochlorine insecticides (see Table 6-2).

A second example of a PAH known to be carcinogenic is the four-ring hydrocarbon benz[a]anthracene, which is anthracene with another benzene ring fused to the "a" bond:

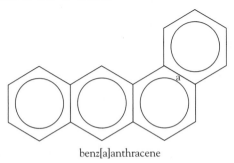

benz[a]anthracene

Some PAHs with certain of their hydrogen atoms replaced by methyl groups are even more potent carcinogens than are the parent hydrocarbons.

Has exposure to PAHs been demonstrated to produce cancer in humans? The answer is both yes and no. For over 200 years, it has been known that prolonged exposure in occupational settings to very high levels of coal tar, the principal toxic ingredient of which is benzo[a]pyrene, leads to cancer in humans. In 1775, the occurrence of scrotal cancer in chimney sweeps was associated with the soot lodged in the crevices of the skin of their genitalia. Modern workers in coke oven and gas production plants likewise experience increased levels of lung and kidney cancer due to this PAH. The evidence for cancer induction in the general public, whose exposure is at levels which are orders of magnitude lower than in these occupational environments, is less clear-cut. The main cause of lung cancer is the inhalation of cigarette smoke, which contains many carcinogenic compounds in addition to PAHs; the deduction from health statistics of the much smaller influences of pollutants such as PAHs from other sources is difficult to accomplish.

The relative positions in space of the fused rings in PAHs play a major role in determining their level of carcinogenic behavior in

BOX 6-5   MORE ON THE MECHANISM OF PAH CARCINOGENESIS

Research has established that the PAH molecules themselves are not carcinogenic agents; rather they must be transformed by several metabolic reactions in the body before the actual cancer-causing species is produced.

The first chemical transformation that occurs in the body is the formation of an epoxide ring across one C=C bond in the PAH. The specific epoxide of interest to the carcinogenic behavior of benzo[a]pyrene is shown below:

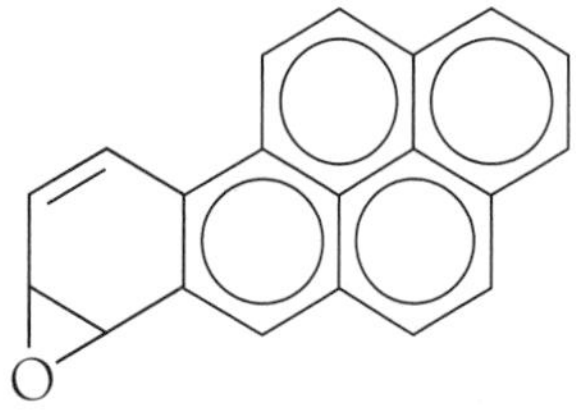

A fraction of these epoxide molecules subsequently add H_2O, to yield two —OH groups on adjacent carbons:

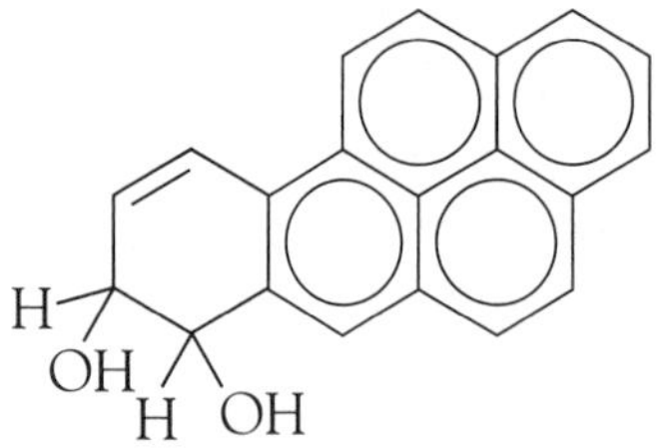

The double bond (shown in the structure above) that remains in the same ring as the two —OH groups subsequently undergoes epoxidation, thereby yielding the molecule that is the active carcinogen:

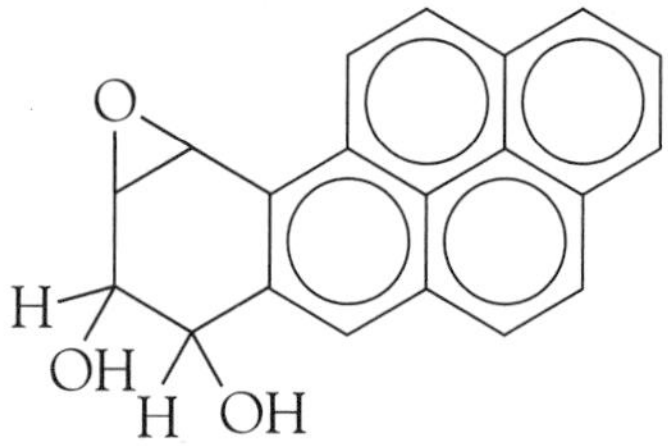

By adding H^+, this molecule can form a particularly stable cation that can bind to molecules such as DNA, thereby inducing mutations and cancer.

The metabolic reactions of epoxide formation and H_2O addition are part of the body's attempt to introduce —OH groups into hydrophobic molecules like PAHs and thereby make them more capable of becoming water-soluble and eliminated. For BaP and other PAHs that possess a bay region, one of the intermediate products in this multistep process can be diverted instead into the formation of a very stable cation that induces cancer.

animals. The PAHs that are the most potent carcinogens each possess a **bay region** formed by the branching in the benzene ring sequence: the organization of carbon atoms as a bay region imparts a high degree of biochemical reactivity to the PAH, as explained in Box 6-5.

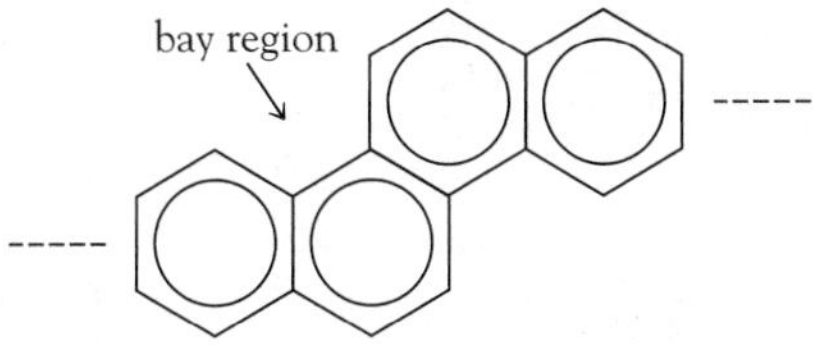

bay region

PROBLEM 6-20

Based upon the bay region theory, would you expect naphthalene to be a carcinogen? How about anthracene or phenanthrene? What about the PAH called benzo[ghi]perylene whose structure is shown below?

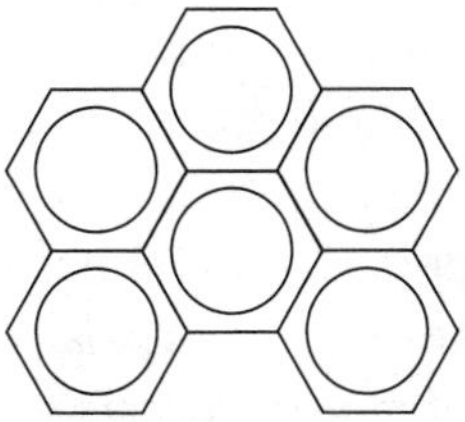

benzo[ghi]perylene

Diesel engine exhaust, which recently has been labeled a "probable human carcinogen," contains not only PAHs but also some of their derivatives containing the nitro group, $-NO_2$, as a substituent; these substances are even more active carcinogens than are the corresponding PAHs. For example, the nitropyrene and dinitropyrene molecules shown below are responsible for much of the mutagenic character of diesel exhaust, that is, its ability to cause mutations that could ultimately produce cancer:

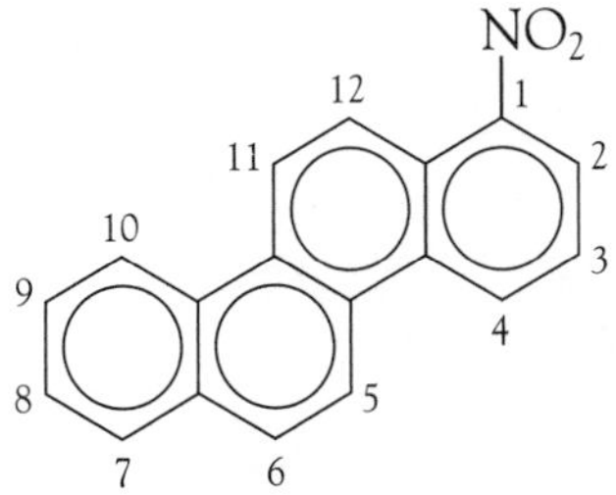

1-nitropyrene

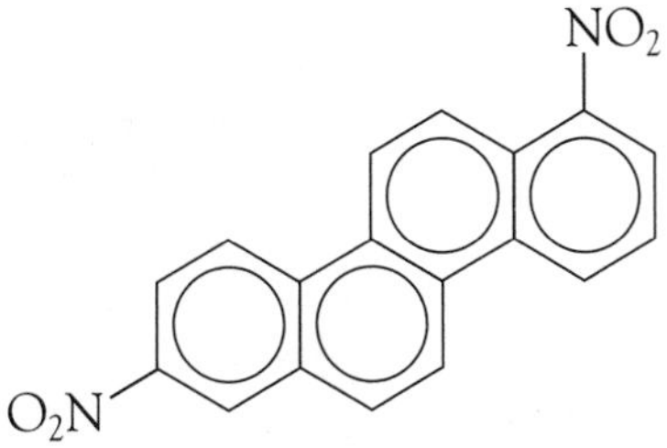

1,8-dinitropyrene

These compounds are formed within the engines by the reaction of pyrene with NO_2^+ and N_2O_4. There is also evidence that PAHs are nitrated by some of the constituents of photochemical smog. The emission of particulates from heavy-duty diesel engines was to be controlled in the United States beginning in the early 1990s by means of "trap" devices in the exhaust stream; these devices temporarily retain the solids, including soot, in the exhaust and oxidize them further. Some scientists have worried that during their residence in the traps, PAHs could undergo reaction to produce even greater quantities of nitrated PAHs. While this apparently does occur, tests indicate that the total mutagenic activity of a given amount of engine exhaust is actually decreased, since most of it becomes oxidized.

Recently, researchers have established tentative links between particulate air pollution and allergies. Apparently, exposure to diesel fumes and other airborne particulates enhances sensitization and allergic responses to pollen and household dust.

Many cities in developing countries have chronic problems with carbon-based particulate air pollution. The particulate matter is traceable not only to smoke from the burning of coal, but also to the exhaust of diesel-fueled vehicles and of motor scooters with two-stroke engines. For example, the air in Bombay, India is of such poor quality that breathing it for a day is said to be equivalent in toxicity to smoking ten cigarettes!

For most nonsmokers in developed countries, by far the greatest exposure to carcinogenic PAHs arises from their diet, rather than directly from pollution of the air, water, or soil to which they are exposed. As expected from their mode of preparation, charcoal broiled and smoked meat and fish contain some of the highest levels of PAHs found in food. However, leafy vegetables such as lettuce and spinach can constitute an even larger source of carcinogenic PAHs due to the deposition of these substances from air onto the leaves of these vegetables while they are growing. Unrefined grains also contribute significantly to the total amount of PAHs ingested from food.

REVIEW QUESTIONS

1. What is meant by the term "hydrophobicity"?

2. What are the three main categories of pesticides?

3. Draw the structure of DDT, and state what the initials stand for.

4. What were the main uses of DDT? Explain why it is no longer used in many developed countries.

5. Explain how DDT functions as an insecticide.

6. Draw the structure of DDE. Is it a pesticide or not? Explain.

7. Explain what is meant by the terms "bioconcentration" and bioconcentration factor (BCF).

8. Explain what is meant by the term "biomagnification," and how it differs from "bioconcentration."

9. Write the defining equation for the partition coefficient K_{ow}. How is it related to a compound's BCF? What is octanol supposed to be a surrogate for in this experiment?

10. Describe one analog of DDT that works in the same fashion but does not bioaccumulate.

11. In general terms, explain what toxaphene is and why it is no longer in use. Why is it still being deposited by airborne transport in the Great Lakes?

12. Draw the structure of perchlorocyclopentadiene. Name at least three insecticides produced from it.

13. What is the general structure of organophosphate insecticides? Explain how organophosphates function as insecticides.

14. In what way are organophosphate insecticides considered superior to organochlorines as pesticides? In what way are they more dangerous?

15. What is the general structure of carbamate insecticides?

16. What is the function of a herbicide? Name a few "old-fashioned" insecticides that contained metals.

17. What is phenol? Draw its structure and that of 2,4-dichlorophenol.

18. Draw the structures, and write out the names, of the two most important phenoxy herbicides.

19. Draw the structure of 2,3,7,8-TCDD. What is the full name for this "dioxin"?

20. Using structural diagrams, write the reaction by which 2,3,7,8-TCDD is produced from 2,4,5-trichlorophenol.

21. What, chemically speaking, was "Agent Orange" and how was it used?

22. Draw the structure of pentachlorophenol. What is its main use as a compound? What is the main dioxin congener that it could produce?

23. Other than the chlorophenols, what are some of the other sources of dioxin in the environment? From what medium—air, food, or water—does most human exposure to dioxin come about?

24. What does *PCB* stand for? Draw the structural diagram of a representative PCB molecule.

25. What were the main uses for PCBs? What is meant by an "open" use?

26. Draw the structure of a representative polychlorinated dibenzofuran congener. What are some of the sources of PCDFs in the environment?

27. What molecules can be eliminated by a PCB molecule when it is heated to moderately high temperatures?

28. Are PCBs acutely toxic to humans? What is the basis for health concerns about them?

29. Are all dioxin congeners equally toxic? If not, what pattern of chlorine substitution leads to the greatest toxicity? What is the most toxic dioxin?

30. What is meant by a "coplanar" PCB? What structural features give rise to non-coplanarity?

31. What are the main methods used at present for the disposal of concentrated PCBs and, on the other hand, of dilute PCBs?

32. What does PAH stand for? Draw the structures of two examples.

33. How are PAHs formed during incomplete combustion processes?

34. Name several sources of PAHs, and their nitrated derivatives, as pollutants of the environment.

35. By means of a structural diagram, show what is meant by the "bay region" in certain PAHs. How is the presence of this region related to the health effects of PAHs?

SUGGESTIONS FOR

FURTHER READING

1. Ware, G. W. 1983. *The Pesticide Book*. San Francisco: W. H. Freeman.

2. Menzie, C. A., B. B. Potocki, and J. Santodonato. 1992. Exposure to carcinogenic PAHs in the environment. *Environmental Science and Technology*. 26: 1278. Reviews PAH concentration data for U.S. air, soil, water, and food, and derives human exposure doses.

3. *Our Planet, Our Health*. 1992. Geneva: World Health Organization.

4. Schwarzenbach, R. P., P. M. Gschwend, and D. M. Imboden. 1993. *Environmental Organic Chemistry*. New York: John Wiley & Sons. Discusses the physical chemistry of organic substances of environmental significance.

5. Ney, R. E., Jr. 1990. *Where Did That Chemical Go?—A Practical Guide to Chemical Fate and Transport in the Environment*. New York: Van Nostrand Reinhold.

6. Mellanby, K. 1992. *The DDT Story*. Farnham, Surrey, U.K.: The British Crop Protection Council.

7. Carson, R. L. *Silent Spring*. 1962. Boston: Houghton Mifflin.

8. Manahan, S. E. 1992. *Toxicological Chemistry*. Boca Raton, FL: Lewis Publishers.

9. Hileman, B. Concerns broaden over chlorine and chlorinated hydrocarbons. *Chemical and Engineering News*. April 19, 1993: 11–20.

10. Anon, 1991. *Toxic chemicals in the Great Lakes and associated effects*. Environment Canada.

11. Tschirley, F. H. 1986. Dioxin. *Scientific American* 254 (February): 29–35.

Interchapter:
Toxic Organics—Dioxin

An Interview with James Worman

James Worman is Visiting Professor in chemistry at Dartmouth College in Hanover, New Hampshire. He has been involved in environmental research as well as community service since 1967. He operates a consulting business that performs environmental assessments and does experimental work on high-energy pyrolysis. He also is heard on a radio show approximately once a month in New Hampshire that gives him the opportunity to interview experts on the environment and respond to calls and questions from listeners.

What is the primary source of dioxins today? Has their incidence increased or decreased in the past ten years?

In order to answer this question or any questions regarding such halogenated compounds fully, one needs to refer to the article "Naturally-Occurring Organohalogen Compounds: A Survey" by Gordon W. Gribble of Dartmouth College [*Journal of Natural Products* 55/10 (1992): 1353–1393] and *Environmental Science Technology* 28/7 (1994).

The primary sources of dioxins today are waste incineration, smelters, and forest fires. Forest fires occur worldwide and are probably the largest source of dioxins. Since forest fires are caused mainly by lightning, this natural source of dioxin cannot be controlled. Other significant sources are cigarette smoke, unleaded gasoline combustion, and the chlorination of paper pulp. Other very minor sources probably exist also.

Because we have become more proficient in techniques and technology that detect dioxins, it appears that their abundance has increased only because now we can actually find them. Ten years ago the technology to identify dioxins was not available. In order to answer the question whether or not they have in fact increased or decreased, scientists will have to follow the research over the next ten years. Then we can determine if any increase is the result of human contributions.

What can be done to reduce the risk of contamination of the environment in terms of decreased production of dioxins or minimization of the release of dioxins into the environment?

To reduce the risk of contamination, there has to be efficient scrubbing of emissions from waste incinerators and more efficient use of chlorine in the paper industry. Both of these actions are already being taken. In the United States, when waste is destroyed according to EPA methods at the present time, PCBs and other organohalogenated compounds are incinerated in the

284

presence of oxygen, and this incineration tends to form dioxins in the emission gases. A change in the method of destruction of hazardous wastes needs to be made. High-temperature pyrolysis methods that work in the absence of oxygen should be employed. These methods are available today.

What groups or agencies have measured the levels of dioxins in our environment? What methods are used to measure them? Are these methods satisfactory?

The Environmental Protection Agency in the United States and its counterpart agencies in many countries have measured the levels of dioxins in the environment. The methods used are gas chromatography coupled with mass spectrometry. These techniques are satisfactory, but, like all measurement techniques, there is a limit to their sensitivity. For example, scientists can detect contaminants present in concentrations in the range of parts per trillion, or picograms per gram (a picogram—10^{-12} grams—is a very small quantity indeed). To get an idea of the sensitivity involved here, you may visualize a parts-per-trillion level as analogous to a single crouton in a five hundred thousand-ton salad! But under normal circumstances we can't do any better than that. Despite the great precision of the methods at our disposal, 100 million molecules still go undetected at the levels we're working with. This means that it's scientifically impossible to satisfy the requirement of the Delaney Clause here in the United States that mandates a zero level of carcinogenic chemicals on anything that humans ingest. The average person thinks that scientific measurement means little more than that scientists carry out a test and then declare "we detected this or that chemical," and the case is closed. But science cannot detect in a typical test sample the 100 million background molecules that are now present in the environment and have been there since the beginning of time.

Considering the effects of bioaccumulation, where should measurements of dioxins be taken in order to predict an effect on human health?

This questions cannot be answered accurately because science just doesn't have enough data at this time. However, everyone knows that dioxin occurs in nature and has been around since the existence of the first forest fire. A natural background level exists due to natural sources. Nobody knows the cutoff between natural and nonnatural levels.

Since not all environmental groups and agencies are in agreement as to the hazards posed by dioxins, how can a citizen evaluate conflicting data that are supplied by different groups?

First the average person has to do some personal research. That is, he or she must go to a public library and read to some extent about the issue. After that, the person should go to a nearby university or college library and ask the librarian there for assistance in finding easily understandable reading material. In addition, the individual should consult experts who live in the community. If the scientists at local colleges and universities cannot be of help, they will direct the person to people who can. People have to put all the information together before making an educated choice. The number one rule is as follows: Do not believe what is printed or announced in the media without examining the information closely. Average people believe what they see and hear in the news media without bringing much critical judgment to bear upon it. However, the interpretation of scientific articles by the news media often takes information out of context and changes the scientific meaning of the work. People must be aware of that.

What lesson can be learned from the Times Beach, Missouri incident?

The entire town was relocated at taxpayers' expense due to the presence of small amounts of dioxin in oil used to control dust on gravel roads. The Centers for Disease Control admits that knowing what we know now the evacuation was unnecessary. A nonpolitical group or agency must be available that will not buckle under pressure from alarmists, politicians, industrialists, and public interest groups. Scientists must be brought into the decision-making process along with politicians. Moreover, politicians should be encouraged to seek advice from "real scientists" who are working in the field.

NATURAL WATERS: CONTAMINATION AND PURIFICATION

All life forms on Earth depend upon water. Each human being needs to consume several liters of fresh water every day to sustain his or her life. Unfortunately, in many areas of the world it is not always possible to obtain a ready source of sufficiently pure drinking water. Through the ages, the contamination of drinking water by biological sources such as sewage and animal wastes has probably been responsible for more human deaths from environmental origins than has any other cause.

WATER SUPPLIES AND THEIR CONTAMINATION

Ground water, that is, water present underground and often found in large reservoirs called *aquifers*, is accessed by wells and is used as a supply of drinking water by almost half the North American population and one-third that of Great Britain. It has been tradition-ally considered to be a pure form of water: because of its filtration through soil and its long residence time underground, it contains much less natural organic matter and many fewer disease-causing microorganisms than does water in lakes and rivers.

However, in many locations, the ground water has been contaminated by chemicals for many decades, though this form of pollution was not recognized as a serious environmental problem until the 1980s. The inorganic contaminant of greatest concern is the nitrate ion, NO_3^-, most of which originates from farms (in a way which we will examine later in this chapter). In rural areas, the contamination of shallow aquifers by organic pesticides such as atrazine leached from the surface has become a concern. Liquid that contains dissolved matter that drains from a terrestrial source is called a *leachate*. However, the typical organic contaminants in most major ground water supplies are two chlorinated solvents mentioned in Chapters 5 and 6, namely tri-chloroethlyene, C_2HCl_3, and tetrachloroethlyene (perchloroethlyene), C_2Cl_4, and the hydrocarbons benzene, C_6H_6, and its methylated deriva-tives toluene $C_6H_5CH_3$ and the isomers of xylene, $C_6H_4(CH_3)_2$. The sources of these organic substances include leaking chemical waste dumps, leaking underground gasoline storage tanks, leaking municipal landfills, and accidental spills of chemicals on land. In the United States, organic leachates at most of the "Superfund" toxic waste sites (see Box 7-1) have contaminated the ground water that lies below them. Within a contaminated aquifer, most of the organic chemicals are con-tained within oily blobs that exist below their original points of entry into the ground. Over the years, the oily liquid gradually dissolves, pro-viding a continuous supply of contaminants to the ground water as it flows by. In view of this contamination, many wells have had to be closed.

Surface fresh waters, by which we mean water that resides on the Earth's surface in rivers, streams, and lakes, are important not only as a major source of drinking water but also as habitats for the plant and ani-mal life that they contain, and for the recreation and transportation that they provide.

One of the world's most famous cases of water pollution involves Lake Erie, which in the 1960s was said to be dying. Indeed, the author of this book can recall visiting a once-popular beach on its north shore in the early 1970s and being repulsed by the sight and smell of dead, rotting fish on the shoreline. Lake Erie's problems stemmed primarily from an excess input of phosphate ion, PO_4^{3-}, in the waters of its tributaries. The phosphate sources were the polyphosphates in detergents (as explained in detail later), raw sewage, and the runoff from farms that used phosphate fertilizers. Since there is commonly an excess of other dissolved nutrients in lakes, phosphate ion usually functions as the limit-ing (or controlling) nutrient for algal growth: the larger the supply of the ion, the more abundant the growth of algae, and its growth can be quite abundant indeed. When the vast mass of excess algae eventually dies

BOX 7-1 THE SUPERFUND PROGRAM

In 1980, the federal government of the United States established a program now known as Superfund to clean up abandoned and illegal toxic waste dumps, since dangerous chemicals from many such sites were polluting ground water. The cleanup costs are shared by chemical companies, the current and past owners of the sites, and the government. Many billions of dollars have already been spent on remediation, and many billions more will eventually be required. Progress in the cleanups has been rather slow on account of the litigation involved and the huge amounts of money at stake. Many decades will pass before even the highest priority sites are all cleaned up.

The Superfund program is administered by the Environmental Protection Agency, which has identified nearly 1,300 waste sites having such serious potential to cause harm to humans and the environment that they have been placed on a National Priorities List. New Jersey, Pennsylvania, and California have the greatest number of such priority dumps.

By 1992, the EPA had finished cleanup work at 40 sites, begun work at more than 300 others, and conducted emergency removal of materials at more than 3000 more. In all, over 30,000 sites have been identified as potentially in need of cleanup.

The most common contaminants at the Superfund sites are the heavy metals lead, cadmium, and mercury, and the organic compounds benzene, toluene, ethylbenzene, and trichloroethylene.

and starts to decompose by oxidation, the water becomes depleted of dissolved oxygen, with the result that fish life is adversely affected. The lake water also becomes foul-tasting, green, and slimy, and masses of dead fish and aquatic weeds rot on the beaches.

To correct the problem, the United States and Canada in 1972 signed the Great Lakes Water Quality Agreement. Since that time, over eight billion dollars have been spent in building sewage treatment plants to remove phosphates from wastewater before it reaches the tributaries and the lake itself. In addition, the levels of phosphates in laundry detergents were restricted in Ontario (to a maximum of 2.2%) and in many of the states (to a maximum of 0.5%) that border the Great Lakes. The total amount of phosphorus entering Lake Erie has now decreased by more than two-thirds. As a result, Lake Erie has sprung back to life: its once-fouled beaches are regaining popularity with tourists and its commercial fisheries have been revived.

As the examples discussed above indicate, it is important to understand the sorts of chemical activity that prevail in natural waters and how the science and application of chemistry can be employed to purify water intended for drinking purposes. As a framework for our study of these questions, it will be convenient to divide our considerations of water chemistry into the two most common reaction categories: acid-base reactions and oxidation-reduction (redox) reactions. Acid-base and solubility phenomena control the concentrations of dissolved inorganic ions such as phosphate and carbonate in waters, whereas the organic content of water is dominated by redox reactions. The pH and principal ion concentrations in most natural water systems are dominated by the dissolution of atmospheric carbon dioxide and soil-bound carbonate ions; such reactions are considered in detail in Chapter 8.

PHOSPHATE IN NATURAL WATERS

As we have pointed out, the presence of excess phosphate ion in natural waters can have a devastating effect on an aquatic ecology because it overfertilizes plant life. Formerly, one of the largest sources of phosphate as a pollutant was detergents, and in the material that follows, the role of such phosphates is discussed.

The reaction of synthetic detergents with calcium and magnesium ions, to form complex ions, diminishes the cleansing potential of the detergent. Polyphosphate ions, which are anions containing several phosphate units linked by shared oxygens, are added to detergents as "builders" that preferentially form soluble complexes with these metals and thereby allow the molecules of the detergent to operate as cleansing agents rather than being complexed with Ca^{2+} and Mg^{2+}. Another role of the builder is to make the washwater somewhat alkaline, which is necessary to help remove the dirt from certain fabrics.

In particular, great quantities of sodium tripolyphosphate (STP), $Na_5P_3O_{10}$, were formerly added as the builder in most synthetic detergent formulations. As shown in Figure 7-1a, STP contains a chain of alternating phosphorus and oxygen atoms, with one or two additional oxygens attached to each phosphorus. In solution, one tripolyphosphate ion can form a complex with one calcium ion by forming interactions between three of its oxygen atoms and the metal ion (Figure 7-1b). Substances like STP that have more than one site of attachment to the metal ion, and thereby produce ring structures that each incorporate the metal, are called *chelating* agents (from the Greek word for *claw*); because

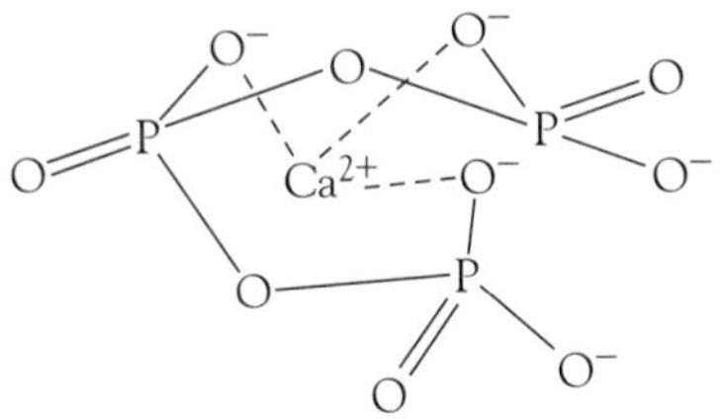

FIGURE 7-1
Structure of the tripolyphosphate ion, $P_3O_{10}^{5-}$: a. uncomplexed; b. complexed with Ca^{2+}. (Redrawn after T. G. Spiro and W. M. Stigliani. 1980. *Environmental Issues in Chemical Perspective*. Albany, NY: State University of New York Press.

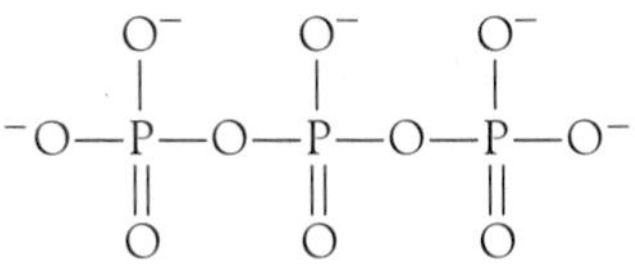

several bonds are formed, the resulting chelates are very stable and do not normally release their metal ions back into a free form.

Tripolyphosphate ion, like phosphate ion itself, is a weak base in aqueous solution and thus provides the alkaline environment that is required for effective cleaning:

$$\underset{\text{tripolyphosphate ion}}{P_3O_{10}^{5-}} + H_2O \longrightarrow P_3O_{10}H^{4-} + OH^-$$

Unfortunately, when washwater containing STP is discarded, the excess polyphosphate enters waterways where it slowly reacts with water and is transformed into phosphate ion, PO_4^{3-}, (sometimes called *orthophosphate*):

$$\underset{\text{tripolyphosphate ion}}{P_3O_{10}^{5-}} + 2\,H_2O \longrightarrow \underset{\text{phosphate ion}}{3\,PO_4^{3-}} + 4\,H^+$$

Note that when the system decomposes, STP behaves as an acid rather than a base (since H^+ is formed in the reaction).

Because of environmental concerns, phosphates are now used only sparingly as builders in detergents in many areas of the world. In Canada and parts of Europe, STP was replaced largely by sodium nitrilotriacetate (NTA) (see Figure 7-2a). The anion of NTA acts in a similar fashion to that of STP, chelating calcium and magnesium ions using three of its oxygen atoms and the nitrogen atom (Figure 7-2b). NTA is not used as a builder in the United States because of concerns that its slow rate of degradation in solution might present a health hazard in drinking water.

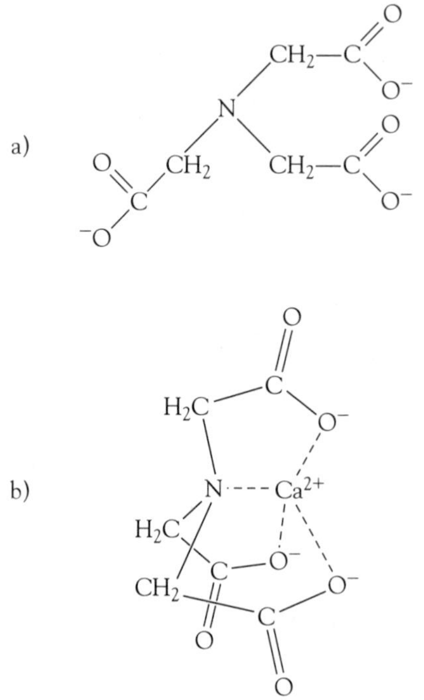

FIGURE 7-2
Structure of the
nitrilotriacetate ion
(NTA): a. uncomplexed;
b. complexed with Ca^{2+}.
(Redrawn after T. G.
Spiro and W. M. Stigliani.
1980. *Environmental Issues
in Chemical Perspective*.
Albany, NY: State
University of New York
Press.)

Other builders now used include sodium citrate, sodium carbonate ("washing soda"), and sodium silicate. Currently, substances called *zeolites* are also employed as detergent builders. Zeolites are aluminosilicate minerals (see Chapter 9) consisting of sodium, aluminum, silicon, and oxygen; in the presence of calcium ion, they exchange their sodium ions for Ca^{2+}. Further aspects of the chemistry of soaps and detergents are discussed in Box 7-2.

Phosphate ion can be removed from municipal and industrial waste-water by the addition of sufficient calcium as the hydroxide, $Ca(OH)_2$, so that insoluble calcium phosphates such as $Ca_3(PO_4)_2$ and $Ca_5(PO_4)_3OH$ are formed as precipitates that can then be readily removed. As mentioned at the end of this chapter, phosphate removal should be a standard practice in the treatment of wastewater, but it is not yet practiced in all cities.

Geographically, phosphate ion enters waterways from both point and nonpoint sources. Point sources are specific sites such as towns and cities that discharge a large quantity of a pollutant—in this case phosphates as a component of human wastes in the municipality's untreated sewage. Nonpoint sources are entities such as farms; each provides a much smaller amount of pollution, in this case phosphates from animal wastes

BOX **7-2** # SOAPS AND DETERGENTS

Solid soaps are the sodium salts of unbranched, long-chain carboxylic "fatty" acids such as $CH_3(CH_2)_{16}COOH$; the soaps are usually derived from natural sources such as animal fat by the addition of sodium hydroxide ("caustic soda"). Indeed, it is the fatty-acid anion that is the active cleansing ingredient of the soap. Liquid and semisolid soaps contain the potassium salts of such acids. Both sodium and potassium soaps are soluble in water; otherwise the fatty-acid anion would be unavailable to do its work.

So-called *hard water* contains metallic ions (usually calcium and/or magnesium) that react with the large, fatty-acid anion of soaps to form insoluble salts apparent as a "scum" in the washwater. Not only is the cleansing power of the soap canceled by the formation of such precipitates, but the insoluble matter clings to the clothes being "cleaned" and leaves residues in them even after rinsing.

For these reasons, synthetic detergents were developed and since World War II have largely displaced soap in laundry and dishwashing applications. They are sodium salts of an organic sulfonic acid, $R—SO_3H$, rather than of a carboxylic acid. Synthetic detergents are much more soluble, and therefore more effective, in water than are soaps, and they do not form insoluble precipitates with calcium or magnesium ions.

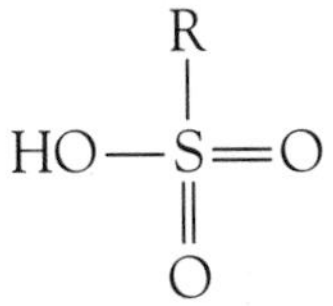

The first synthetic detergents contained a benzene ring bonded to an alkyl carbon chain that was highly branched. This is an example:

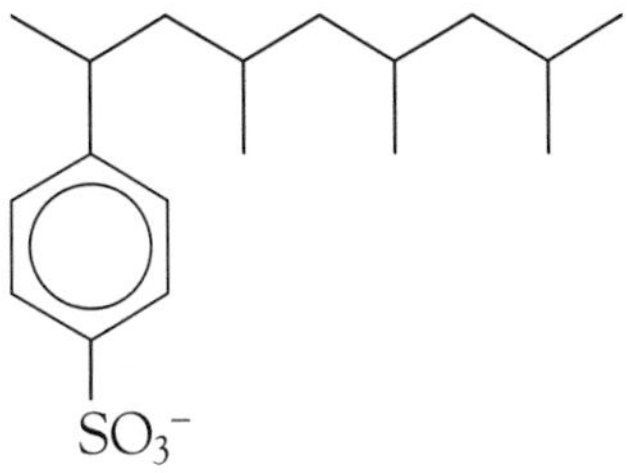

Because of the branched nature of the chain, such detergents are only slowly biodegraded; this property gave rise to the frothing notoriously observable in many rivers and lakes and even in drinking water in the 1950s. Beginning in the 1960s, nonbranched rather than branched chains of carbon atoms have been used to bond to the benzene ring attached to the SO_3 unit, since this change helps detergents to biodegrade much more quickly. For instance:

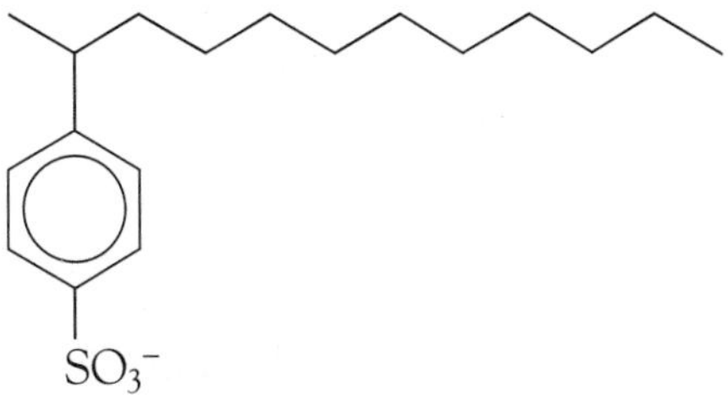

an organic sulfonic acid

and fertilizers in runoff from the property. However, on account of the large number of farms involved, nonpoint sources can generate larger total quantities than do point sources. Indeed, now that sewage treatment plants and detergent controls have been instituted around the Great Lakes, much of the remaining phosphate arises from nonpoint sources, mainly farms.

DISSOLVED OXYGEN IN NATURAL WATERS

The only significant oxidizing agent in natural waters is dissolved molecular oxygen, O_2. Upon reaction, each of its oxygen atoms is reduced from the zero oxidation state to the -2 state in H_2O or OH^-. The half-reaction that occurs in acidic solution is

$$O_2 + 4\,H^+ + 4e^- \longrightarrow 2H_2O$$

whereas that which occurs in basic aqueous solution is

$$O_2 + 2\,H_2O + 4e^- \longrightarrow 4\,OH^-$$

The concentration of dissolved oxygen in water is small and therefore precarious from the ecological point of view. For the reaction

$$O_2(g) \; \rightleftharpoons \; O_2(aq)$$

the appropriate equilibrium constant is the Henry's Law constant K_H (see Chapter 3), which for oxygen at 25°C has the value 1.3×10^{-3} mol L^{-1} atm^{-1}:

$$K_H = [O_2(aq)] / P = 1.3 \times 10^{-3} \text{ mol L}^{-1} \text{ atm}^{-1}$$

Since in dry air, the partial pressure P of oxygen is 0.21 atm, it follows that the solubility of O_2 is 8.7 milligrams per liter of water (see Problem 7-1). This value can also be stated as 8.7 ppm since, as discussed in Chapter 6, for condensed phases, ppm concentrations are based on mass rather than moles.

PROBLEM 7-1

Confirm by calculation the value of 8.7 mg/L for the solubility of oxygen in water.

Because the solubilities of gases increase with decreasing temperature, the amount of O_2 that dissolves at 0°C (14.7 ppm) is greater than is the amount that dissolves at 35°C (7.0 ppm). The median concentration of oxygen found in natural, unpolluted surface waters in the United States is about 10 ppm.

River or lake water that has been artificially warmed can be considered to have undergone thermal pollution in the sense that it will contain less oxygen than colder water because of the decrease in gas solubility with increasing temperature. To sustain their lives, fish require water containing at least 5 ppm of dissolved oxygen; their survival in warmed water therefore can be problematic. Thermal pollution often occurs due to the operation of electric power plants, since they draw cold water from a river or lake, use it for cooling purposes, and then return the warmed water to its source.

OXYGEN DEMAND IN NATURAL WATERS

The most common substance oxidized by dissolved oxygen in water is organic matter having a biological origin, such as dead plant matter and animal wastes. If the organic matter—by way of simplifying the case—is considered to be polymerized carbohydrate (plant fiber would be an example), with an approximate empirical formula of CH_2O, the oxidation reaction would be

$$CH_2O(aq) + O_2(aq) \longrightarrow CO_2(g) + H_2O(aq)$$
carbohydrate

Similarly, dissolved oxygen in water is consumed by the oxidation of dissolved ammonia (NH_3) and ammonium ion (NH_4^+) to nitrate ion (NO_3^-)(see Problem 7-3).

PROBLEM 7-2

Show that 1 liter of water saturated with oxygen at 25°C is capable of oxidizing 8.2 milligrams of polymeric CH_2O.

PROBLEM 7-3

Determine the balanced redox reaction for the oxidation of ammonia by O_2 in alkaline solution.

Water that is aerated by flowing in shallow streams and rivers is constantly replenished with oxygen. However, stagnant water or that near the bottom of a deep lake is usually almost completely depleted of oxygen because of its reaction with organic matter and the lack of any mechanism to replenish it.

The capacity of the organic matter in a sample of natural water to consume oxygen is called its **biological** (or biochemical) **oxygen demand** (BOD). It is evaluated experimentally by determining the concentration of dissolved O_2 at the beginning and at the end of a 5-day period in which a sealed water sample is maintained at 25°C. The BOD equals the amount of oxygen consumed in this period as a result of the oxidation of dissolved organic matter in the sample. The oxidation reactions are accelerated in the lab by the addition of microorganisms known to be catalysts for the process. If it is suspected that the sample will have a high BOD, it first is diluted with pure water so that sufficient oxygen will be available overall to oxidize all the organic matter; the results are corrected for this dilution. The median BOD for unpolluted surface water in the United States is about 0.7 milligrams of O_2 per liter, which is considerably less than the maximum solubility of O_2 in water (of 8.7 mg/L at 25°C).

A faster determination of oxygen demand can be made by evaluating the **chemical oxygen demand** (COD) of a water sample. Dichromate ion, $Cr_2O_7^{2-}$, can be dissolved as one of its salts, such as $Na_2Cr_2O_7$, in sulfuric acid: the result is a powerful oxidizing agent. It is this preparation, rather than O_2, that is used to ascertain COD values. The reduction half-reaction for dichromate when it oxidizes the organic matter is

$$Cr_2O_7^{2-} + 14\,H^+ + 6\,e^- \longrightarrow 2\,Cr^{3+} + 7\,H_2O$$

dichromate ion   chromium (III) ion

The number of moles of O_2 that the sample would have consumed equals $6/4 = 1.5$ times the number of moles of dichromate, since the latter accepts 6 electrons per ion whereas O_2 accepts only 4:

$$O_2 + 4\,H^+ + 4e^- \longrightarrow 2\,H_2O$$

(In practice, excess dichromate is added to the sample and the resulting solution is back-titrated with Fe^{2+} to the end point.)

PROBLEM 7-4

A 25 mL sample of river water was titrated with 0.0010 M $Na_2Cr_2O_7$ and required 8.3 mL to reach the end point. What is the chemical oxygen demand, in milligrams of O_2 per liter, of the sample?

The difficulty with the COD index as a measure of oxygen demand is that acidified dichromate is such a strong oxidant that it oxidizes substances that are very slow to consume oxygen in natural waters and that therefore pose no real threat to their oxygen content. In other words, dichromate oxidizes substances that would not be oxidized by O_2 in the determination of the BOD. Because of this excess oxidation, namely, of stable organic matter to CO_2, and of Cl^- to Cl_2, the COD value for a water sample is as a rule greater than its BOD value. Neither method of analysis oxidizes aromatic hydrocarbons or many alkanes, which resist degradation in natural waters.

It is not uncommon for water polluted by organic substances associated with animal or food waste or sewage to have an oxygen demand which exceeds the maximum equilibrium solubility of dissolved oxygen. Under such circumstances, unless the water is continuously aerated, it will soon be depleted of its oxygen, and fish living in the water will die. The treatment of wastewater to reduce its BOD is discussed at the end of this chapter.

ANAEROBIC DECOMPOSITION OF ORGANIC MATTER IN NATURAL WATERS

Dissolved organic matter will decompose in water under anaerobic (oxygen-free) conditions if appropriate bacteria are present. Anaerobic conditions occur naturally in stagnant water such as swamps and at the bottom of deep lakes. The bacteria operate on carbon so as to *disproportionate* it: in other words, some carbon is oxidized (to CO_2) and the rest is reduced (to CH_4):

$$2 \, CH_2O \longrightarrow CH_4 + CO_2$$

organic matter   methane   carbon dioxide

This is sometimes called a *fermentation* reaction, which is defined as one in which both the oxidizing and the reducing agents are organic materials. Since the methane produced in this process is almost insoluble in water, it forms bubbles that can be seen rising to the surface in swamps; indeed, methane was originally called "marsh or swamp gas." The same chemical reaction occurs in "digestor" units used by rural inhabitants in semitropical developing countries (India, for instance) to convert animal wastes into methane gas that can be used as a fuel.

Since anaerobic conditions are reducing conditions in the chemical sense, insoluble Fe^{3+} compounds that are present in sediments at the bottom of lakes are converted into soluble Fe^{2+} compounds which then dissolve:

$$Fe^{3+} + e^- \longrightarrow Fe^{2+}$$

It is not uncommon to find aerobic and anaerobic conditions in different parts of the same lake at the same time, particularly in the summertime when a stable stratification of distinct layers often occurs (see Figure 7–3). Water at the top of the lake is warmed by the absorption of sunshine by biological materials; that below the level of penetration of sunlight remains cold. Since warm water is less dense than cold, the upper layer "floats" on the lower, and little transfer between them occurs. The top layer usually contains near-saturation levels of dissolved oxygen, due both to its contact with air and to the O_2 produced in photosynthesis by algae. Thus conditions in the top layer are aerobic, and

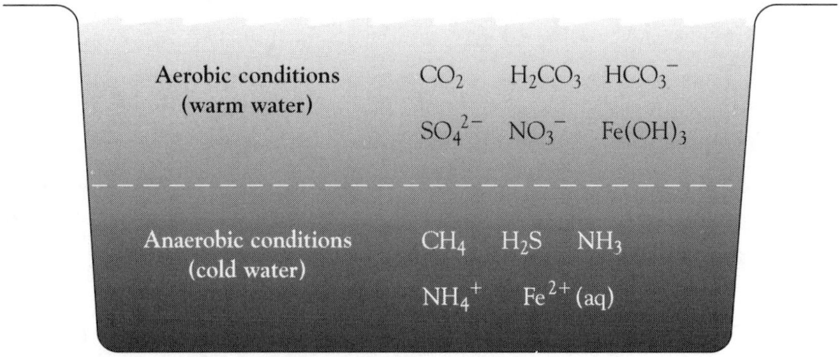

FIGURE 7-3
The stratification of a lake in summer, showing the typical forms of the elements it contains.

consequently elements exist in their most oxidized forms: carbon as CO_2 or H_2CO_3 or HCO_3^-, sulfur as SO_4^{2-}, nitrogen as NO_3^-, and iron as insoluble $Fe(OH)_3$. Near the bottom, the water is oxygen-depleted since there is no contact with air and since O_2 is consumed when biological material decomposes. Under such anaerobic conditions, elements exist in their most reduced forms: carbon as CH_4, sulfur as H_2S, nitrogen as NH_3 and NH_4^+, and iron as soluble Fe^{2+}. Anaerobic conditions usually do not last indefinitely. In the fall and winter, the top layer of water is cooled by cold air passing above it, so that eventually the oxygen-rich water at the top becomes more dense than that below it and gravity induces mixing between the layers. Thus in the winter and early spring, the environment near the bottom of a lake usually is aerobic.

The common inorganic oxidation states in which sulfur is encountered in the environment are illustrated in Table 7-1; they range from the highly reduced -2 state that is found in hydrogen sulfide gas, H_2S, and insoluble minerals containing the sulfide ion, S^{2-}, to the highly oxidized $+6$ state that is encountered in sulfuric acid, H_2SO_4 and in salts containing the sulfate ion, SO_4^{2-}. In organic and bioorganic molecules such as amino acids, intermediate levels of sulfur oxidation are present. When such substances decompose anaerobically, hydrogen sulfide and other gases such as methanethiol, CH_3SH, and dimethyl sulfide, CH_3SCH_3, containing sulfur in highly reduced forms are released, thereby giving swamps their unpleasant odor. The occurrence of such gases as air pollutants was discussed in Chapter 3.

TABLE 7-1   COMMON OXIDATION STATES FOR SULFUR

	Increasing levels of sulfur oxidation				
Oxidation state of S	-2	-1	0	$+4$	$+6$
Aqueous solution and salts	H_2S			H_2SO_3	H_2SO_4
	HS^-			HSO_3^-	HSO_4^-
	S^{2-}	S_2^{2-}		SO_3^{2-}	SO_4^{2-}
Gas phase	H_2S			SO_2	SO_3
Molecular solids			S_8		

As would be expected on the basis of the principles of air chemistry discussed in Chapter 3, hydrogen sulfide is oxidized in air first to sulfur dioxide, SO_2, and then fully to sulfuric acid or a salt containing the sulfate ion. Similarly, hydrogen sulfide dissolved in water can be oxidized by certain bacteria to elemental sulfur or more completely to sulfate. Overall the complete oxidation reactions correspond to

$$H_2S + 2\,O_2 \longrightarrow H_2SO_4$$
$$\text{hydrogen sulfide} \qquad\qquad \text{sulfuric acid}$$

Some anaerobic bacteria are able to use sulfate ion (rather than the conventional O_2) as the oxidizing agent to convert organic matter such as polymeric CH_2O to carbon dioxide; the SO_4^{2-} ions are reduced in the process to elemental sulfur or even to hydrogen sulfide:

$$2\,SO_4^{2-} + 3\,CH_2O + 4\,H^+ \longrightarrow 2\,S + 3\,CO_2 + 5\,H_2O$$
$$\text{sulfate} \qquad\qquad\qquad\qquad \text{sulfur}$$

PROBLEM 7-5

Balance the reduction half-reaction that converts SO_4^{2-} to H_2S under acidic conditions, and then obtain a balanced overall equation for the reduction by CH_2O of sulfate ion to hydrogen sulfide gas.

One characteristic reaction of ground water, which by definition is not well-aerated since it has spent much time far from air, is that when it reaches the surface and O_2 has an opportunity to dissolve in it, its rather high level of soluble Fe^{2+} is converted to insoluble Fe^{3+}, and an orange-brown deposit of $Fe(OH)_3$ is formed:

$$4\,Fe^{2+} + O_2 + 2\,H_2O \longrightarrow 4\,Fe^{3+} + 4\,OH^-$$
$$4\,[\,Fe^{3+} + 3\,OH^- \longrightarrow Fe(OH)_3\,(s)\,]$$

The overall reaction is

$$4\,Fe^{2+} + O_2 + 2\,H_2O + 8\,OH^- \longrightarrow 4\,Fe(OH)_3\,(s)$$
$$\text{soluble iron (II)} \qquad\qquad\qquad\qquad \text{iron (III) hydroxide}$$

ACID MINE DRAINAGE

An analogous reaction occurs in some underground mines. Normally FeS_2, called iron pyrites or "fool's gold," is a stable, insoluble component of underground rocks as long as it does not come into contact with air. However, as a result of the mining of coal and other substances, some of it is exposed to oxygen and becomes partially solubilized. The disulfide ion, S_2^{2-}, which contains sulfur in the -1 oxidation state, is oxidized to sulfate ion, SO_4^{2-}, which contains sulfur in the $+6$ state:

$$S_2^{2-} + 8\,H_2O \longrightarrow 2\,SO_4^{2-} + 16\,H^+ + 14\,e^-$$

disulfide ion $\qquad\qquad$ sulfate ion

The oxidation is accomplished in the main by O_2:

$$7\,[\,O_2 + 4\,H^+ + 4\,e^- \longrightarrow 2\,H_2O\,]$$

When this half-reaction is added to twice the oxidation half-reaction, the net redox reaction is obtained:

$$2\,S_2^{2-} + 7\,O_2 + 2\,H_2O \longrightarrow 4\,SO_4^{2-} + 4\,H^+$$

Since the sulfate of the ferrous ion, Fe^{2+}, is soluble in water, the iron pyrites is solubilized by the reaction. More importantly, the reaction produces a large amount of concentrated acid (note the H^+ product), only some of which is consumed by the air oxidation of Fe^{2+} to Fe^{3+} which accompanies the process:

$$4\,Fe^{2+} + O_2 + 4\,H^+ \longrightarrow 4\,Fe^{3+} + 2\,H_2O$$

This reaction is catalyzed by bacteria. Combining the last two reactions in the correct ratio, that is, 2:1, we obtain the overall reaction for the oxidation of both the iron and the sulfur:

$$4\,FeS_2 + 15\,O_2 + 2\,H_2O \longrightarrow 4\,Fe^{3+} + 8\,SO_4^{2-} + 4\,H^+$$
$$\text{(i.e., } 2\,Fe_2(SO_4)_3 + 2\,H_2SO_4)$$

In other words, the oxidation of the fool's gold produces soluble iron (III) sulfate—$Fe_2(SO_4)_3$—and concentrated sulfuric acid, H_2SO_4. The Fe^{3+} ion is soluble in the highly acidic water that is first produced (see

Problem 7-6); however once the drainage from the highly acidic mine water becomes diluted and its pH rises, a yellowish-brown precipitate of $Fe(OH)_3$ forms from Fe^{3+}, discoloring the water and waterway. Thus the pollution associated with acid mine drainage is characterized in the first instance by the seeping from the mine of copious amounts of both acidified water and a rust-colored solid. Unfortunately, the concentrated acid can also liberate toxic heavy metals from their ores in the mine, further adding to the pollution. The phenomenon of acid drainage currently is of particular importance in the many abandoned mines in the mountains of Colorado.

Interestingly, the oxidation of disulfide ion to sulfate ion in the above process is accomplished to some extent by the action of Fe^{3+} as an oxidizing agent, rather than by O_2 :

$$S_2^{2-} + 14\,Fe^{3+} + 8\,H_2O \longrightarrow 2\,SO_4^{2-} + 14\,Fe^{2+} + 16\,H^+$$

PROBLEM 7-6

The K_{sp} values for $Fe(OH)_2$ and $Fe(OH)_3$ are 7.9×10^{-15} and 6.3×10^{-38} respectively. Calculate the solubilities of Fe^{2+} and Fe^{3+} at a pH of 8, assuming they are controlled by their hydroxides. Also calculate the pH value at which the ion solubilities reach 100 ppm.

THE PURIFICATION OF DRINKING WATER

The quality of "raw" (untreated) water intended eventually for drinking varies widely, from almost pristine to highly polluted. Because both the type and quantity of pollutants in raw water vary, the processes used in purification also vary from place to place. The most commonly used procedures are shown in schematic form in Figure 7-4.

Aeration is commonly used in the improvement of water quality. Municipalities aerate drinking water that is drawn from aquifers in order to remove dissolved gases such as the foul-smelling H_2S and volatile organic compounds, some of which may have a detectable odor. Aeration of drinking water also results in reactions that produce CO_2 from the most easily oxidized organic material; if necessary, most of the remaining organics can be removed by subsequently passing the water over activated charcoal, although this process is relatively expensive. Another advantage to aeration is that the increased oxygen content of water oxidizes water-soluble Fe^{2+} to Fe^{3+}, which then forms insoluble hydroxides (and related species) that can be removed as solids.

$$Fe^{3+} + 3\,OH^- \longrightarrow Fe(OH)_3\,(s)$$

Most municipalities allow raw water to settle, since this permits large particles to settle out or to be readily filtered. However, much of the insoluble matter will not precipitate spontaneously since it is suspended in water in the form of *colloidal particles*. These are particles that have diameters ranging from 0.001 to 1 μm, and consist of groups of molecules or ions that are weakly bound together and that dissolve as a unit rather than break up and dissolve as individual ions or molecules. In many cases, the individual units within a colloidal particle are spatially organized such that the surface of the particles contain ionic groups. The ionic charges on the surface of one particle repel those on neighboring particles, preventing their aggregation and subsequent precipitation.

To capture the colloidal particles, either iron (III) sulfate—$Fe_2(SO_4)_3$—or aluminum sulfate, $Al_2(SO_4)_3$ ("alum"), is deliberately added to the water; at neutral or alkaline pH values (7 and up), both Fe^{3+} and Al^{3+} form gelatinous hydroxides $Fe(OH)_3$ and $Al(OH)_3$ which physically incorporate the colloidal particles and form a removable precipitate. The water is greatly clarified once this precipitate has been removed. The formation of these two substances also consumes hydroxide ion and results in a decrease in the pH of the water, helping to neutralize alkaline waters.

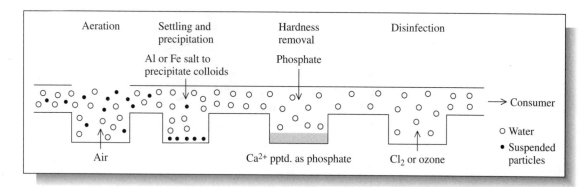

FIGURE 7-4

The stages of purification of drinking water.

If the water comes from wells in areas having limestone bedrock, it will contain significant levels of Ca^{2+} and Mg^{2+} ions which are usually removed during the processing. Calcium can be removed from water by addition of phosphate ion in a process analogous to that previously discussed (see page 292) for phosphate removal; here, however, phosphate is added to precipitate the calium ion. More commonly, calcium ion is removed by precipitation and filtering of the insoluble salt $CaCO_3$; the carbonate ion is either added as sodium carbonate, Na_2CO_3, or if sufficient HCO_3^- is naturally present in the water, hydroxide ion can be added in order to convert bicarbonate ion to carbonate:

$$OH^- + HCO_3^- \longrightarrow CO_3^{2-} + H_2O$$
$$Ca^{2+} + CO_3^{2-} \longrightarrow CaCO_3(s)$$

Magnesium ion precipitates as magnesium hydroxide, $Mg(OH)_2$, when the water is made sufficiently alkaline, that is, when the OH^- ion content is increased. After removal of the solid $CaCO_3$ and $Mg(OH)_2$, the pH of the water is readjusted to near-neutrality by bubbling carbon dioxide into it.

PROBLEM 7-7

Ironically, calcium ion is removed from water often by adding hydroxide ion in the form of $Ca(OH)_2$. Deduce a balanced chemical equation for the reaction of calcium hydroxide with dissolved calcium bicarbonate to produce calcium carbonate. What molar ratio of $Ca(OH)_2$ to dissolved calcium should be added to ensure that almost all the calcium is precipitated?

WATER DISINFECTION

To rid drinking water of harmful bacteria and viruses, purification with an oxidizing agent more powerful than O_2 is usually necessary. In some localities, particularly in Europe but also in some North American cities—Montreal is an example—ozone is used for this purpose. Since O_3 cannot be stored or shipped because of its very short lifetime, it must be generated on-site by a relatively expensive process involving electrical discharge in dry air. The resulting ozone-laden air is bubbled through the water; about 10 minutes of contact is sufficient. Since the lifetime of ozone molecules is short, there is no residual protection in the purified water to protect it from future contamination.

Similarly, chlorine dioxide gas, $ClO_2^•$, is used in more than 300 North American and in several thousand European communities to disinfect water. The $ClO_2^•$ molecules are peroxy free radicals, similar to the $HOO^•$ and $ROO^•$ species previously encountered in air chemistry (Chapters 2 and 3), and operate to oxidize organic molecules by extracting electrons from them:

$$ClO_2^• + 4\,H^+ + 5\,e^- \longrightarrow Cl^- + 2\,H_2O$$
$$\text{chlorine dioxide} \qquad\qquad \text{chloride ion}$$

The organic cations created in the accompanying oxidation half-reaction subsequently react further and eventually become more fully oxidized. Since chlorine dioxide is *not* a chlorinating agent—it does not generally introduce chlorine atoms into the substances with which it reacts—and since it oxidizes the dissolved organic matter, much lesser amounts of toxic organic chemical byproducts are formed than if molecular chlorine is used (see below). As is the case with ozone, $ClO_2^•$ cannot be stored since it is explosive in the high concentrations that its practical use calls for, and so it must be generated on-site. This is accomplished by oxidizing its reduced form, ClO_2^-, from the salt sodium chlorite:

$$ClO_2^- \longrightarrow ClO_2^• + e^-$$
$$\text{chlorite ion} \quad \text{chloride dioxide}$$

By a seeming irony, this process is sometimes accomplished using Cl_2 as the oxidizing agent. (However, no excess chlorine remains when the reaction is complete that might give rise later to undesired chlorinated byproducts.)

The most common water purification agent used in North America is hypochlorous acid, HOCl. This neutral, covalent compound kills microorganisms, readily passing through their cell membranes. Like ozone, HOCl is not stable in concentrated form and so cannot be stored. For large-scale installations, for instance, municipal water treatment plants, it is generated by dissolving molecular chlorine gas, Cl_2, in water; at moderate pH values, the equilibrium in the reaction of chlorine with water lies far to the right and is achieved in a few seconds:

$$Cl_2(g) + H_2O(aq) \rightleftharpoons HOCl(aq) + H^+ + Cl^-$$
$$\text{hypochlorous acid}$$

Thus a dilute aqueous solution of chlorine in water contains very little aqueous Cl_2 itself. If the pH of the reaction water should be allowed to become too high, the result would be the ionization of the weak acid HOCl to the hypochlorite ion, OCl^-, which is less able to penetrate bacteria on account of its electrical charge.

In small-scale applications of chlorination, as in swimming pools, the handling of cylinders of Cl_2 is inconvenient and dangerous, so hypochlorous acid instead is generated from the salt calcium hypochlorite, $Ca(OCl)_2$, or is supplied as an aqueous solution of sodium hypochlorite, NaOCl. In water, an acid-base reaction occurs to convert most OCl^- to HOCl:

$$OCl^- + H_2O \rightleftharpoons HOCl + OH^-$$
hypochlorite ion   hypochlorous acid

Close control of the pH in an environment like a swimming pool is necessary to avoid a shift to the left in the position of equilibrium for this reaction by permitting a very alkaline condition to prevail. On the other hand, corrosion of pool construction materials can occur in acidic water, so the pH usually is maintained above 7 to prevent such deterioration from occurring; maintenance of an alkaline pH also prevents the conversion of dissolved ammonia, NH_3, to the chloramines NH_2Cl, $NHCl_2$, and especially NCl_3, which is a powerful eye irritant.

$$NH_3 + 3\,HOCl \longrightarrow NCl_3 + 3\,H_2O$$
ammonia   hypochlorous   nitrogen
acid   trichloride

It is desirable to adjust the equilibrium point in the $OCl^- \longrightarrow HCl$ reaction so as to favor the predominance of the disinfectant molecular species, HOCl. Since the equilibrium between HOCl and OCl^- shifts rapidly in favor of the ion between pH values of 7 and 9, however, the acidity level must be meticulously controlled. Swimming pool acidity can be adjusted by the addition of acid (in the form of sodium bisulfate, $NaHSO_4$, which contains the acid HSO_4^-) or a base (Na_2CO_3) or a buffer ($NaHCO_3$, which contains the amphoteric anion HCO_3^-). Chlorine must be constantly replenished in outdoor pools since UV-B light in sunshine is absorbed by and decomposes the hypochlorite ion:

$$2\,ClO^- \xrightarrow{\text{UV-B}} 2\,Cl^- + O_2$$
chlorite ion   chloride ion

PROBLEM 7-8

Given that for HOCl, $K_a = 2.7 \times 10^{-8}$, deduce the fraction of a sample of the acid in water that exists in the molecular form at pH values (predetermined by the presence of other species) of 7.0, 7.5, 8.0, and 8.5. (Hint: Derive an expression that relates the fraction of HOCl that is ionized to the $[H^+]$.) Would it be a good idea to allow the pool water's pH to rise to 8.5?

An important drawback to the use of chlorination in disinfecting water is the concomitant production of chlorinated organic substances, some of which are toxic. If the water contains phenol (also called *hydroxybenzene*) or a derivative thereof, chlorine quite readily substitutes for the hydrogen atoms on the ring to give rise to chlorinated phenols: these have an offensive odor and taste and are toxic. Some communities switch from chlorine to chlorine dioxide when their supply of raw water is temporarily contaminated with phenols.

A more general problem with chlorination of water lies in the production of trihalomethanes (THMs), whose general formula is CHX_3, where X is chlorine or bromine. The compound of principal concern is chloroform, $CHCl_3$, which is produced when hypochlorous acid reacts with organic matter that is dissolved in the water (see Box 7-3). Chloroform is a suspected liver carcinogen in humans, and its presence, even at very low levels of approximately 30 ppb, raises the specter that chlorinated drinking water may pose a health hazard. Currently, the limit of THMs in drinking water in the United States has been set at 100 ppb by the Environmental Protection Agency. Several other mutagenic chlorinated organic compounds formed during chlorination have also been detected in water. Recently, an analysis has been reported of all the recent epidemiological studies relating the chlorination of water to cancer rates in various communities in the United States. The conclusion of the analysis was that the risk of bladder cancer in humans increased by 21%, and that of rectal cancer by 38%, for Americans who drank chlorinated surface water in the past. The same risks do not apply to chlorinated well water, since its organochlorine content is much less (only 0.8 ppb on average, versus 51 ppb for surface water) because it contains much smaller amounts of organic matter in the first place. It should be realized that brands of carbonated water which use municipal drinking water as their source also contain dissolved chloroform.

BOX 7-3 THE MECHANISM OF CHLOROFORM PRODUCTION IN DRINKING WATER

Humic acids, with which HOCl reacts to form chloroform, are water-soluble, nonbiodegradable components of decayed plant matter. Of particular importance are humic acids which contain 1,3-dihydroxybenzene rings. The carbon atom (#2) located between those carrying the —OH groups is readily chlorinated by HOCl, as in this elementary case:

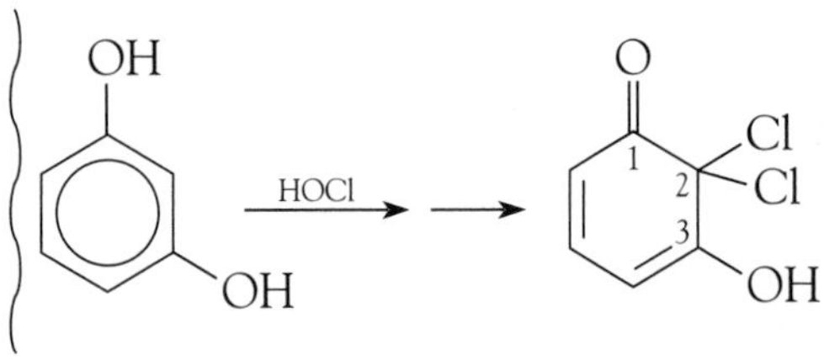

Subsequently the ring cleaves between C-2 and C-3 to yield a chain

$$R{-}\underset{\underset{O}{\|}}{C}{-}CHCl_2$$

In the presence of the HOCl, the terminal carbon becomes trichlorinated, and the $-CCl_3$ group is readily displaced by the OH^- in water to yield chloroform:

$$R{-}\underset{\underset{O}{\|}}{C}{-}CHCl_2 \xrightarrow{HOCl} R{-}\underset{\underset{O}{\|}}{C}{-}CCl_3$$

$$\xrightarrow[H^+]{OH^-} R{-}\underset{\underset{O}{\|}}{C}{-}OH + CHCl_3$$

Analogous sequences of reactions produce bromoform, $CHBr_3$, and mixed chlorine-bromine trihalomethanes from the action on humic materials of hypobromous acid, HOBr, which is formed when bromide ion in water displaces chlorine from HOCl:

$$HOCl + Br^- \rightleftharpoons HOBr + Cl^-$$

Given that slightly more than half the population of the United States drinks surface water, the effects of chlorination is to have increased the bladder cancer incidence by about 4,200 cases per year and the rectal cancer incidence by about 6,500 cases annually. Because of these risks, some communities are considering a switch, or have already switched, to water disinfection by ozone or chlorine dioxide, since these agents produce little or no chloroform. The extent of chlorination has already been reduced in most American communities relative to the levels used when most of the cancer increase noted above would have been initiated. Under no circumstances should disinfection of water be eliminated, however, since it is responsible for the virtual extirpation of fatal waterborne diseases such as cholera and typhoid fever in developed countries. In contrast, about 20 million people, most of them infants, die from waterborne diseases annually worldwide in underdeveloped countries, where water purification is often erratic or even nonexistent.

Ultraviolet light can also be used to purify water; upon absorption of the UV light by organic molecules in water, some of their bonds are broken and free radicals are formed. In a manner reminiscent of the oxidation of stable compounds in air, the dissolved organic matter is oxidized by free radical reactions, and the water thereby is purified. For example, if hydroxyl free radicals, $OH^{\cdot}$, are produced in the process, they initiate the oxidation of hydrogen-containing organic molecules by hydrogen-atom abstraction:

$$OH^{\cdot} + H-\overset{\displaystyle |}{\underset{\displaystyle |}{C}}- \longrightarrow H_2O + C\text{-based radical} \longrightarrow \longrightarrow CO_2$$

An advantage chlorination has over disinfection by ozone or by UV is that some chlorine remains dissolved in water once it has left the purification plant so that the water is protected from subsequent bacterial contamination before it is consumed. Indeed, some chlorine is usually added to water purified by ozone in order to provide this protection. There is very little danger of chloroform production in the purified water since its organic content has been virtually eliminated before the chlorine is introduced. The residual chlorine in water often exists in the form of the chloramines NH_2Cl, $NHCl_2$, and NCl_3, that are produced from reaction with dissolved ammonia gas. Although not as effective as $HOCl$ in disinfecting water, such "combined chlorine" is longer-lived and provides longer residual protection.

Recently, a membrane system has been developed in France that purifies water without the use of chemicals. Water is pumped under pressure through fine membranes that have pores which are only about 1 nanometer wide, and which therefore remove bacteria, viruses, and any organic matter that would nourish the regrowth of bacteria. These "nanofilters" allow water molecules, which are only a few tenths of a nanometer in size, to pass through the filter, but prevent the passage of bioorganic and other large organic molecules, which are much wider than a nanometer. A similar membrane system is used to desalinate sea water in California and the Middle East.

NITROGEN COMPOUNDS IN NATURAL WATERS

In some natural waters, nitrogen occurs in small concentrations in inorganic and organic forms that are nevertheless of concern with respect to human health. As discussed in Chapter 4, there are several environmentally important forms of nitrogen which differ in the extent of oxidation of the nitrogen atom. The most reduced forms are ammonia, NH_3, and its conjugate acid the ammonium ion NH_4^+. The most oxidized form is the nitrate ion, NO_3^-, which exists in salts, aqueous solutions, and in nitric acid, HNO_3. In solution, the most important species between these extremes are the nitrite ion, NO_2^- and molecular nitrogen, N_2. The common oxidation states of nitrogen, along with the most important examples for each, are illustrated in Table 7-2.

Recall from Chapter 4 that in the microorganism-catalyzed process of nitrification, ammonia and ammonium ion are oxidized to nitrate, whereas in the corresponding denitrification process, nitrate and nitrite are reduced to molecular nitrogen, N_2. (Nitrous oxide, N_2O, is a minor byproduct in both cases.) Both processes are important in soils and in natural waters. In aerobic environments such as the surface of lakes, nitrogen exists as the fully oxidized nitrate, NO_3^-, whereas in anaerobic environments such as the bottom of stratified lakes, nitrogen exists as the fully reduced forms ammonia and ammonium ion (Figure 7-3). Nitrite ion occurs in anaerobic environments such as waterlogged soils that are not quite so reducing as to convert the nitrogen all the way to ammonia. Most plants can absorb nitrogen only in the form of nitrate ion, so any ammonia or ammonium ion used as fertilizer must first be oxidized via microorganisms before it is useful to plant life.

TABLE 7-2   COMMON OXIDATION STATES FOR NITROGEN

Oxidation state of N	Increasing levels of nitrogen oxidation						
	−3	0	+1	+2	+3	+4	+5
Aqueous solution and salts	NH_4^+, NH_3				NO_2^-		NO_3^-
Gas phase	NH_3	N_2	N_2O	NO		NO_2	

NITRATES AND NITRITES IN FOOD AND WATER

Concern has been expressed recently about the increasing levels of nitrate ion in drinking water, particularly in well water in rural locations; the main source of this NO_3^- is runoff from agricultural lands into rivers and streams. Initially, oxidized animal wastes (manure) unabsorbed ammonium nitrate and other nitrogen fertilizers were thought to be the main culprits. It now appears that intensive cultivation of land, even without the application of fertilizer or manure, facilitates the oxidation of reduced nitrogen to nitrate in decomposed organic matter in the soil by providing aeration and moisture.

Excess nitrate ion in wastewater flowing into sea water, for example, the Baltic Sea, has resulted in algal blooms that pollute the water after they die. Nitrate ion normally does not cause this effect in bodies of fresh water, where phosphorus rather than nitrogen is usually the limiting nutrient; increasing the nitrate concentration there without an increase in phosphate levels does not lead to an increased amount of plant growth. There are, however, instances where nitrogen rather than phosphorus temporarily becomes the limiting nutrient even in fresh waters.

Excess nitrate ion in drinking water is a potential health hazard since it can result in methemoglobinemia in newborn infants, as well as in adults with a particular enzyme deficiency. The pathological process, in brief, runs as follows.

Bacteria, for example in unsterilized milk-feeding bottles or in the baby's stomach, reduce some of the nitrate to nitrite:

$$NO_3^- + 2\,H^+ + 2\,e^- \longrightarrow NO_2^- + H_2O$$

nitrate ion   nitrite ion

The nitrite combines with and oxidizes the hemoglobin in blood, and thereby prevents the proper absorption and transfer to cells of oxygen. The baby turns blue and suffers respiratory failure. In almost all adults, the oxidized hemoglobin is readily reduced back to its oxygen-carrying form, and the nitrite is readily oxidized back to nitrate. This occurrence of methemoglobinemia, or "blue-baby syndrome," is now rare in industrialized countries: for example, the last reported case in Great Britain was in 1972. However it is still of concern in some developing countries.

As discussed in the next section, excess nitrate ion in drinking water is also of concern because of its potential link with stomach cancer.

Recent epidemiological investigations have, however, failed to establish any positive, statistically significant relationship between nitrate levels in drinking water and the incidence of stomach cancer. As a result, the expenditure of public money on nitrate level reductions in drinking water has become a controversial subject. In Great Britain, in particular, hundreds of millions of dollars have been spent on achieving the 50 ppm maximum level of nitrate ion set by the European Economic Community.

NITROSAMINES IN FOOD AND WATER

Some scientists have warned that excess nitrate ion in drinking water and foods could lead to an increase in the incidence of stomach cancer in humans since some of it is converted in the stomach to nitrite ion; the problem is that the nitrites could subsequently react with amines to produce *N*-nitrosamines, compounds which are known to be carcinogenic in animals. *N*-nitrosamines are amines in which two organic groups and an —N=O unit are bonded to the central nitrogen:

$$
R_2NNO \quad \text{or} \quad
\begin{array}{c}
R \\
\diagdown \\
N\!-\!N\!=\!O \\
\diagup \\
R
\end{array}
$$

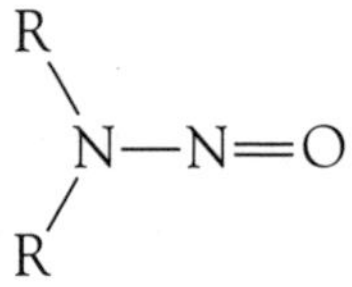

N-nitrosamines

Of concern not only with respect to its production in the stomach and its occurrence in foods and beverages (e.g., cheeses, fried bacon, smoked and/or cured meat and fish, and beer), but also as an environmental pollutant in drinking water, is the compound in which R is the methyl group, CH_3; it is called *N*- nitroso*dimethyl*amine, or NDMA for short:

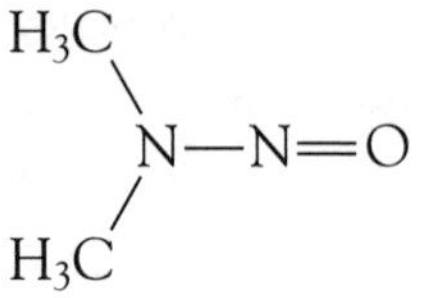

$$
\begin{array}{c}
H_3C \\
\diagdown \\
N\!-\!N\!=\!O \\
\diagup \\
H_3C
\end{array}
$$

NDMA

This organic liquid is somewhat soluble in water (about 4 grams per liter) and somewhat soluble in organic liquids. It is a probable human carcinogen, and probably a potent one if extrapolation from animal studies is a reliable guide. It can transfer a methyl group to a nitrogen or oxygen of a DNA base, and thereby alters the instructional code for protein synthesis in the cell; see Box 7-4 for further details.

In the early 1980s, it was found that NDMA was present in beer to the extent of about 3,000 ppt. Since that time, commercial brewers have modified the drying of malt so that the current levels of NDMA in American and Canadian beer is only about 70 ppt, which is about 3% of the average level found in 1981.

Large quantities of nitrate are used to "cure" pork products such as bacon and hot dogs. In these foods, some of the nitrate ion is biochemically reduced to nitrite ion, which prevents the growth of the organism

---

BOX 7-4 THE ACTION OF NDMA

The methylation of DNA is accomplished not by NDMA itself, but by a metabolite. In the first stage of the sequence, one hydrogen of one methyl group is replaced by —OH. Next, the newly created CH_2OH group is eliminated as formaldehyde (CH_2O), leaving one hydrogen of the CH_2OH group attached to the nitrogen:

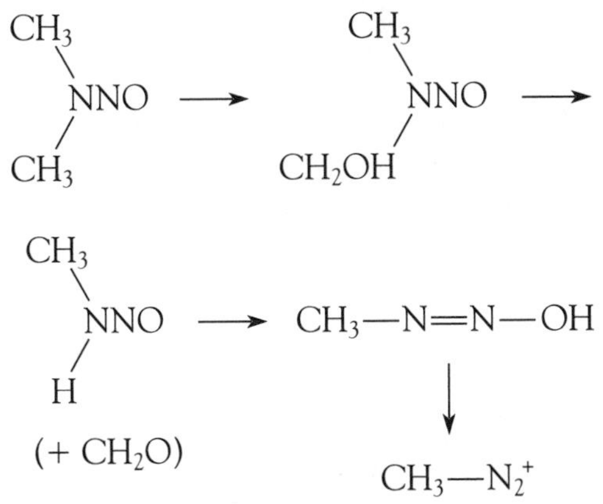

$$CH_3 \backslash NNO / CH_3 \longrightarrow CH_3 \backslash NNO / CH_2OH \longrightarrow$$

$$CH_3 \backslash NNO / H \longrightarrow CH_3{-}N{=}N{-}OH$$

$$(+ CH_2O)$$

$$\downarrow$$

$$CH_3{-}N_2^+$$

The hydrogen readily migrates to the oxygen of the NNO group, yielding $CH_3{-}N{=}N{-}OH$, which readily loses hydroxide ion to form $CH_3{-}N_2^+$, which in turn easily methylates (i.e., donates CH_3^+ to) molecules—including DNA—by loss of molecular nitrogen.

N-nitrosodialkylamines are formed in acidic aqueous media (particularly near a pH of 3) both in the environment and in the body by the combination of an amine of the type R_2NH with nitrite ion or nitrous acid:

$$R_2NH + HO{-}NO \longrightarrow R_2NNO + H_2O$$
nitrous acid

(The mechanism is not as simple as implied by this equation, and involves prior formation of N_2O_3, the anhydride of nitrous acid.)

responsible for botulism. Nitrite ion also gives these meats their charac-teristic taste and color by combining with hemoproteins in blood. Nitrosamines are produced from excess nitrite during frying (e.g., of bacon) and in the stomach, as discussed above. Government agencies have instituted programs to decrease the residual nitrite levels in cured meats. Some manufacturers of these foods now add vitamins C or E to the meat in order to block the formation of nitrosamines. Based upon average levels of NDMA in various foods and the average daily intake for each of them, most of us now ingest more NDMA from consumption of cheese (which is often treated with nitrates) than from any other source.

Even though the commercial production of NDMA has been phased out, it can be formed as a byproduct due to the use of amines in indus-trial processes such as rubber tire manufacturing, leather tanning, and pesticide production.

The levels of NDMA in drinking water drawn from ground water is of concern in some localities that have industrial point sources of the compound. For example, following the discovery that the water supply of one town had been contaminated by up to 100 ppt NDMA from a tire factory, the province of Ontario, Canada recently adopted a guideline maximum of 9 ppt of NDMA in drinking water, which corresponds to a lifetime cancer risk of 1 in 100,000. By contrast, the guideline for water in the United States is set at 0.68 ppt, which corresponds to a cancer risk of 1 in a million, but which lies considerably below the detection limit (about 5 ppt) for the compound.

PROBLEM 7-9

Assuming that equilibrium can be established, which ion—ammonium or nitrate—would predominate under anaerobic conditions at the bottom of a lake? Would your answer be the same for the aerated surface waters?

PROBLEM 7-10

Write balanced redox half-reactions (assuming acidic conditions) for the conversion of NH_4^+ to NO_3^-, and of NO_2^- to N_2.

THE TREATMENT OF WASTEWATER

Most municipalities now treat the raw sewage collected from homes, buildings and industries (including food processing plants) through a "sanitary sewer" system before it is deposited into a nearby source of natural waters, whether a river, lake, or ocean. Since the rainwater and melted snow that drains from streets and other paved surfaces is usually not highly contaminated, it is often collected separately by "storm sewers" and deposited directly into the natural waters.

The main component of sewage is organic matter of biological origin. It occurs mainly as particles—ranging from those of macroscopic size large enough to be trapped (together with such "objets d'art" as facial tissues, socks, tree branches, condoms, and so forth) by mesh screens to those which are microscopic in size and which are suspended as colloids.

In the primary treatment stage of wastewater (see the schematic diagram in Figure 7-5), the larger particles are removed by allowing the water to flow across screens and slowly along a lagoon. A sludge of insoluble particles forms at the bottom of the lagoon, while liquid grease forms a lighter-than-water layer at the top and is skimmed off. About 30% of the Biological Oxygen Demand of the wastewater is removed by the primary treatment process, even though this stage of the process is entirely mechanical in nature. Although the sludge is in fact mainly

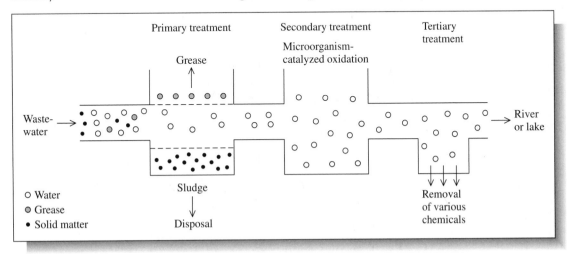

FIGURE 7-5
The stages of treatment of wastewater.

water, the remainder is principally organic matter. In some cases the sludge is incinerated, whereas in other municipalities it is used as landfill or spread on fields as low-grade fertilizer. Unfortunately, sludge may contain heavy metals and other toxic substances.

After passing through conventional primary treatment, the sewage water has been much clarified but still has a very high BOD—typically several hundred milligrams per liter—and would be detrimental to fish life if released at this stage. The high BOD is due mainly to organic colloidal particles. In the secondary treatment stage, most of this organic matter is biologically oxidized by microorganisms to carbon dioxide and water, or converted to additional sludge which can readily be removed from the water. Either the water is sprinkled onto a bed of sand and gravel covered with the microorganisms or it is agitated in a reactor in order to effect the microorganism-driven reaction. Oxidation reduces the BOD to less than 100 mg/L, which is about 10% of the original concentration in the untreated sewage. Upon dilution of the treated water with a greater amount of natural water, aquatic life can be supported. In summary, the secondary treatment of wastewater involves biochemical reactions that oxidize much of the organic material that was not removed in the first stage.

Some municipalities now employ tertiary wastewater treatment as well as primary and secondary. In the tertiary phase, specific chemicals are removed from the partially purified water. Depending upon locale, tertiary treatment can include all the following chemical processes:

further reduction in BOD by removal of most remaining colloidal material using an aluminum salt in a process which forms $Al(OH)_3$ and operates in the same manner as described previously for the purification of drinking water

removal of dissolved organic compounds by their adsorption onto activated charcoal, over which the water is allowed to flow

phosphate removal, usually by its precipitation as the calcium salt, $Ca_5(PO_4)_3OH$, upon the addition of lime, $Ca(OH)_2$

heavy metal removal by the addition of sulfide ion, to form the insoluble metal sulfides

nitrogen removal, by the conversion of NH_4^+ and/or NO_3^- and/or CN^- (cyanide) to N_2 or NH_3, followed in the case of the ammonia product by "air stripping" the gas from water: the water is vigorously aerated, and the ammonia passes from the solution to the gas state

In some areas, river water that has been polluted by effluent from sewage treatment plants is used by municipalities downstream as drinking water. Recent research in Japan has shown that chlorination of the effluent before its release also produces mutagenic compounds, presumably by interaction of chlorine-containing substances with the organic matter that remains in the water.

THE MANAGEMENT OF TOXIC WASTE

The chief danger posed to the well-being of humans and other living things by hazardous waste—even that stored in a nonaqueous environment like a landfill or buried underground in containers—lies in its potential to contaminate natural waters by migrating from its point of deposition by, say, leaching. As discussed in Box 7-1, the accumulation of toxic wastes in past decades has resulted at many sites in North America in widespread contamination of land that is only now being addressed.

Currently, in developed countries hazardous waste is disposed of in a number of different ways by storing it in special landfills; by destroying it by combustion; by injecting it into deep underground wells, caverns and old salt mines; by temporarily storing liquid wastes in lagoons; and by simply placing it in sanitary landfills or discharging it into sewers and natural waters or even onto land. Many warehouses contain stockpiles of toxic chemicals such as PCBs stored in metal drums, presumably awaiting eventual disposal. There have been instances of widespread environmental contamination as the result of fires in such warehouses. The eventual safe disposal of toxic wastes represents a huge financial burden for most industrialized countries—the deferred cost of not regulating waste-disposal practices for most of the twentieth century.

In order to prevent further increases in the accumulated amounts of hazardous waste in the future, management strategies have been devised and partly implemented. Much of the intent of these approaches can be summed up as the "four Rs" of hazardous waste management: **reduction, recycling/reuse,** and **recovery.**

In source reduction, an attempt is made to reduce the total quantities of hazardous materials by conserving them or by using benign substances in their place. For example, it is sometimes feasible to switch industrial operations and products from organic solvents to aqueous ones or sometimes to no solvent at all; thus the quantities of toxic waste solvents is reduced. In recycling/reuse, waste materials from one process, or substances extracted from the wastes, can become the raw material that is needed for another process. Some industrial operations are now adopting a "closed loop" philosophy and designing industrial systems that prevent any pollution from leaving a plant site. With recovery methods, valuable materials in waste are extracted and sold to other industries. An example is the electrolytic recovery of heavy metals from wastewater of industrial operations that use large quantities of such metals. Indeed, solutions containing low concentrations of heavy metals are usually the largest component by volume or mass of hazardous wastes that are generated by polluting industries, followed by other aqueous solutions, and then by organic solvents and residues.

Combustible waste (i.e., organic compounds) can be converted to energy by burning the materials as a supplementary fuel at high temperatures in cement kilns or other types of furnaces. In some cases, toxic wastes are burnt in special incinerators designed to provide very complete combustion (99.99%), with negligible emissions to the environment. A typical toxic waste incinerator is shown schematically in Figure 7-6. "Rich" wastes are those that burn readily. Their combustion in the incinerator generates much heat and the high temperatures that are required to treat "lean" waste. The latter do not burn easily on their own, usually on account of their high water content. In some instances, additional air is forced into the mixture during its combustion to ensure its complete oxidation. The products of combustion are sprayed with an

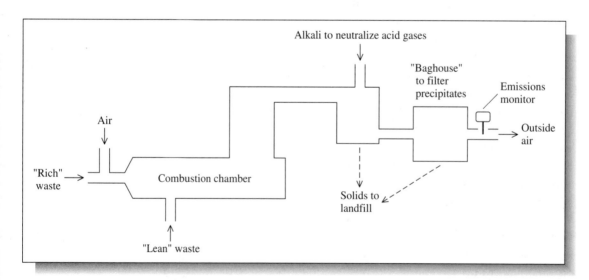

FIGURE 7-6
A toxic waste incinerator.

alkaline substance to neutralize acidic gases such as HCl; this converts such gases to solids that are then separated and sent to a landfill. The combustion products are then passed through a *baghouse* where particulates are separated from gases. Finally the composition of the combustion gases (ideally just CO_2, H_2O, and N_2) is monitored before they are released into the air. Overall, a much larger fraction of hazardous waste is combusted in Europe than in North America, though relatively less combustible waste is produced in Europe in the first place.

Solid hazardous waste itself, or the combustion products obtained when it is burnt, is often stored in a **secure landfill.** If the waste is liquid, it sometimes can be solidified by reacting it with neutralizing or precipitating agents, and/or by immobilizing the substance by mixing it with fly ash or other absorbent matter. The cell of a secure landfill is ideally constructed of clay, since this material is an efficient barrier to the flow of water. Once full, the cell is capped with clay and then covered with topsoil. In principle at least, water does not permeate the clay and there is no movement of materials from the landfill into the external environment. In less secure landfills, water travels through the waste and dissolves some of the chemicals that it contains. In some hazardous waste landfills, there is a drainage system in place that collects any leachate from the cell. Contamination of ground water can occur if leachate is not collected. In the past, landfills in North America and Europe were often left uncovered; this action gave rise not only to pollution of the water beneath the sites by toxic compounds but also of the air above them as volatile compounds gradually evaporated from the site.

REVIEW QUESTIONS

1. Explain the difference between ground water and surface water.

2. Name three organic contaminants commonly found in ground water.

3. What polyphosphate is used in detergents, and why does its use lead to environmental problems? What are the other main sources of phosphate ion in natural waters? What other "builders" are used in detergents? By what chemical reaction can excess phosphate ion be removed from wastewater?

4. Write the half-reaction involving O_2 that occurs in acidic waters when it oxidizes organic matter.

5. How does temperature affect the solubility of O_2 in water? Explain what is meant by "thermal pollution."

6. Define BOD and COD, and explain why their values for the same water sample can differ slightly. Explain why natural waters can have a high BOD.

7. Write the half-reaction, used in the COD titration, which converts dichromate ion to Cr^{3+} ion, and balance it.

8. Write the balanced chemical reaction by which organic carbon, represented as CH_2O, is disproportionated by bacteria under anaerobic conditions.

9. Draw a labeled diagram classifying the top and bottom layers of a lake in summer as either oxidizing or reducing in character, and showing the stable forms of carbon, sulfur, nitrogen and iron in the two layers.

10. What are some examples of highly reduced and of highly oxidized sulfur in environmentally important compounds? Write the balanced reaction by which sulfate can oxidize organic matter.

11. Explain the phenomenon of acid mine drainage, writing balanced chemical equations as appropriate.

12. Explain the various steps, other than disinfection, used in water purification.

13. Explain the chemistry underlying the disinfection of water by molecular chlorine. What is the active agent in destruction of the pathogens? Aside from elemental Cl_2, what other sources of the active ingredient are used?

14. Explain why pH control is important in the water of swimming pools. What compounds are produced when chlorinated water reacts with dissolved ammonia?

15. Discuss the advantages and disadvantages of chlorination in the purification of water, including the nature of the THM compounds. What other chemicals and processes can be used for disinfection?

16. Construct a table which shows the common oxidation states for nitrogen. Deduce in which column the following environmentally important compounds belong: HNO_2, NO, NH_3, N_2O, N_2, HNO_3, NO_3^-. Which of the species become prevalent in aerobic conditions? under anaerobic conditions? What is the oxidation state of nitrogen in NH_2OH?

17. Explain why excess nitrate in drinking water or food products can be a health hazard; include the relevant balanced chemical reaction showing how nitrate becomes reduced.

18. What is an *N*-nitrosamine? Write the structure and the full name for NDMA.

19. What processes are involved in primary wastewater treatment? in secondary treatment? What is the effect on the water's BOD of these processes?

20. List five possible water purification processes that are associated with the tertiary treatment of sewage.

21. Name some of the current practices used to dispose of hazardous waste.

22. What are the "four Rs" of hazardous waste management strategies? Give an example of each.

23. Describe the operation of a modern toxic waste incinerator.

24. What is meant by a *secure landfill*? What material is used to prevent leakage from it?

25. What is meant by the term *leachate*?

SUGGESTIONS FOR

FURTHER
READING

1. Selinger, B. 1989. *Chemistry in the Marketplace*. 4th ed. Sydney, Australia: Harcourt Brace Jovanovich.

2. Keating, M. 1986. *To the Last Drop: Canada and the World's Water Crisis*. Toronto: Macmillan of Canada.

3. Spiro, T. G., and W. M. Stigliani. 1980. *Environmental Issues in Chemical Perspective*. Albany, NY: State University of New York Press.

4. Enger, E. D., J. R. Kormelink, B. F. Smith, and R. J. Smith. 1989. *Environmental Science: The Study of Interrelationships*. 3rd ed. Dubuque, IA: Wm. C. Brown Publishers.

5. Cunningham, W. P., and B. W. Saigo. 1990. *Environmental Science: A Global Concern*. Dubuque, IA: Wm. C. Brown Publishers.

6. Manahan, S. E. 1991. *Environmental Chemistry*. 5th ed. Chelsea, MI: Lewis.

INTERCHAPTER:
WATER POLLUTION

ESSAY BY MICHAEL KEATING

Michael Keating is an international environment writer and consultant, with a special knowledge of water issues. Mr. Keating is author of five books, including, *To the Last Drop, Canada and the World's Water Crisis, The Earth Summit's Agenda for Change*, a report of the Rio conference on environment and development, and *Covering the Environment,* a book on environmental journalism. He is associated with several universities, is a director of the Canadian Global Change Program of the Royal Society of Canada, and is a member of the Institute for Risk Research, the editorial board of the international journal, ECODECISION, and the World Conservation Union's Commission on Environmental Strategy and Planning. He has received awards from a number of organizations, including the United Nations Environment Programme.

> *Water, water, everywhere,*
> *Nor any drop to drink.*
> —Rime of the Ancient Mariner

Humans have been plagued with water pollution problems ever since we first gathered into villages, exposing ourselves to diseases caused by human and animal wastes in drinking water. "As a reservoir and transmission medium of human disease, water has been the leading problem in environmental health throughout history," states the American Academy of Family Physicians.

People long suspected a link between pollution and disease, but it was only in the 1860s that French chemist Louis Pasteur laid the scientific foundations of our understanding of bacterial infection and thus of the dangers posed by polluted drinking water. By the early part of this century, municipalities in North America and Europe began large-scale sterilization of drinking water, particularly with chlorine, to stop the scourge of typhoid, cholera, and similar illnesses. It was several more decades before sewage treatment systems began to be installed, and the task is still not complete. The beaches around many cities are frequently posted as unfit for swimming because of bacterial contamination.

While people in industrialized nations take drinkable tap water for granted, nearly one-quarter of the world's 5.5 billion people do not enjoy a similar security, and more than half the world lacks adequate sanitation to prevent water pollution. As a result, waterborne diseases such as cholera, typhoid, diarrhea, dysentery, malaria, and intestinal worms kill about nine million people a year and afflict hundreds of millions more with serious illnesses.

Another form of pollution causes the excessive fertilization of aquatic plant life. Lake Erie was said to be dying in the 1960s because of the impact of the huge amounts of phosphorus in human sewage and detergents that were discharged into the relatively shallow waters. This chemical fertilized the growth of algae that choked the waters in many areas. When the algae died, their decomposition caused oxygen short-ages that created anoxic "dead" zones in parts of the lake bottom. The same eutrophication process affects lakes and rivers in a number of areas where there are high levels of phosphorus or nitrogen.

As nations industrialized, new forms of water pollution were created. In some cases, so much oil was leaked or dumped into waterways that there have been periodic fires on the surface of rivers. Ducks landing on oil slicks have been coated with the oil and killed. This gross pollution has largely been eliminated in recent years in North America and other developed countries, but traces of highly toxic chemicals remain.

Since the 1960s, scientists have been able to detect ever smaller lev-els of toxic chemicals in the environment, leading to the closure of a number of drinking water supplies and fisheries, and boosting the sales of bottled water and home water filters. Some of the chemicals, such as dioxins and furans, are waste by-products of industrial processes or in-cineration. Others, such as PCBs, are industrial chemicals that have escaped into the environment. Still others, such as DDT, were pesticides that were deliberately sprayed into the environment, where they con-centrated in the food chain to levels that cause biological harm to more species than the pests alone. Many governments have ordered reductions in direct chemical discharges from industries, but found that pollution levels in the environment remained higher than expected.

Scientists discovered that once discharged, some persistent and bioaccumulative chemicals that do not break down easily remain in the food chain, passing from one organism to another. Even when large industrial sources are controlled, pollution still flows into the waterways from a large number of small business that discharge their wastes into municipal sewer systems that are not equipped to treat industrial chemi-cals. Pollution is also traceable to runoff from such diffuse sources as city lawns and farm fields that have been sprayed with weed killers, to waste oil dumped down municipal sewers by do-it-yourself mechanics, and to chemical seepage from dumps and underground storage tanks. Still other chemicals blow in from far away in the form of long-range fallout that can be carried from other countries, even other continents. This is why high levels of chemicals have been found in polar bears and penguins.

The result of pollution by multiple sources can be a chemical soup. Scientists have found 362 contaminants in the Great Lakes ecosystem, including the water, sediments, fish, animals, waterfowl, and humans. There are 32 metals, 68 pesticides, and 262 other organic chemicals, mainly industrial substances and waste by-products. The list includes

126 that can have acute or chronic toxic effects on life. As a result of food chain contamination, a number of fish species in the Great Lakes and other regions are listed by governments as unfit for eating.

Even the chlorination process that disinfects our drinking water poses its own hazards. Low levels of chloroform and other harmful compounds are formed when chlorine reacts with organic material in the water.

For generations we used our waterways as sewers, assuming that they had an infinite capacity to assimilate and neutralize our pollution. The operative rule was, "the solution to pollution is dilution." To some degree this principle of dilution worked when the quantity of pollution discharged was not too great for the ecosystem to assimilate without damage, and the pollutants could be broken down by nature into harmless substances.

Since Rachel Carson's 1962 book, *Silent Spring,* first aroused public concern about chemicals in the environment, a number of laws and regulations have been enacted to control pollution. The result has been tens of billions of dollars in pollution control equipment installed in municipal sewer systems and industrial plants. A number of industrial groups have been undertaking pollution prevention programs to reduce the risk of wastes escaping into the environment.

Critics of the regulatory process typical of North America and many other nations say that the plethora of water pollution laws is still inadequate because it is based mainly on diluting toxic wastes to what governments calculate to be acceptable levels. Furthermore, pollutants and polluters are usually regulated one at a time, in a shortsighted regulatory approach that generally does not reckon with the combined effects of many chemicals and many dischargers who add their pollutants to the same ecosystem.

In recent years, governments have been pushed to take much stronger actions against whole classes of chemicals. For example, the Canada–United States International Joint Commission on boundary waters has recommended the elimination of discharges of persistent, bioaccumulative toxic substances into the Great Lakes.

The challenge ahead is to come to an understanding of what constitutes unacceptable levels of pollution in our waters, even if it appears to be dilute. This will require considerable scientific work to identify pollutants, determine their environmental impacts, and come up with control strategies and safer alternative substances.

NATURAL WATERS: ACID-BASE CHEMISTRY OF THE CARBONATE SYSTEM

Natural waters, even when "pure," contain significant quantities of dissolved carbon dioxide and of the anions it produces, as well as cations of calcium and magnesium. In addition, the pH of such natural water is rarely equal to exactly 7.0, the value expected for pure water. In this chapter, the natural processes that produce these substances in natural waters are analyzed. Since the reactions involved are usually at equilibrium, a mathematical analysis of the chemical systems in natural waters can be made.

THE CO₂/CARBONATE SYSTEM

The acid-base chemistry of many natural water systems, including both rivers and lakes, is dominated by the interaction of the carbonate ion, CO_3^{2-}, which is a moderately strong base, with the weak acid H_2CO_3, carbonic acid. This carbonic acid results from the dissolution of atmospheric carbon dioxide gas in water and from the decomposition of organic matter in the water; there is usually an equilibrium between the gas and the aqueous acid:

$$CO_2(g) + H_2O(aq) \rightleftharpoons H_2CO_3(aq) \qquad (1)$$
$$\text{carbonic acid}$$

The carbonic acid is also in equilibrium in the aqueous medium with bicarbonate ion, HCO_3^- (also called hydrogen carbonate ion), and hydrogen ion:

$$\underset{\text{carbonic acid}}{H_2CO_3} \rightleftharpoons H^+ + \underset{\text{bicarbonate ion}}{HCO_3^-} \qquad (2)$$

The preponderant source of the carbonate ion is limestone rocks, which are largely made up of calcium carbonate, $CaCO_3$. Although this salt is almost insoluble, a small amount of it dissolves when waters pass over it:

$$\underset{\text{calcium carbonate}}{CaCO_3(s)} \overset{H_2O}{\rightleftharpoons} Ca^{2+} + \underset{\text{carbonate ion}}{CO_3^{2-}} \qquad (3)$$

Natural waters that are exposed to limestone are called **calcareous** waters. The dissolved carbonate ion acts as a base, producing bicarbonate ion and hydroxide ion in the water:

$$\underset{\text{carbonate ion}}{CO_3^{2-} + H_2O} \rightleftharpoons \underset{\text{bicarbonate ion}}{HCO_3^- + OH^-} \qquad (4)$$

These reactions that occur in the natural three-phase (air, water, rock) system are summarized in Figure 8-1.

In the discussions that follow, the effects on the composition of a body of water of the simultaneous presence of both carbonic acid and calcium carbonate are analyzed. However, to obtain a qualitative understanding of this rather complicated system, the effect of the carbonate ion alone is first considered. (The chemistry of weak acids and bases is reviewed in Appendix III.)

WATER IN EQUILIBRIUM
WITH SOLID CALCIUM CARBONATE

For simplicity, we first consider a (hypothetical) body of water that is in equilibrium with excess solid calcium carbonate and in which all other reactions are of negligible importance. The only process of interest is reaction 3. The appropriate equilibrium constant for processes that involve the dissolution of slightly soluble salts in water is the "solubility

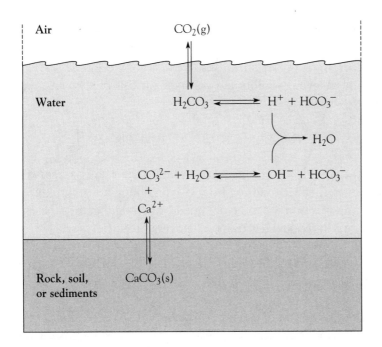

FIGURE 8-1
Reactions among three phases (air, water, rock) of the CO_2/HCO_3^-/CO_3^{2-} system.

product," K_{sp}. For reaction 3, it is related to the equilibrium concentration of the ions by the equation

$$K_{sp} = [Ca^{2+}][CO_3^{2-}]$$

For $CaCO_3$ at 25°C, the numerical value for K_{sp} is 4.6×10^{-9}, where [ ] refers to molar concentrations. It follows from the stoichiometry of reaction 3 that as many calcium ions are produced as carbonate ions, and that both ion concentrations are equal to S, the solubility of the salt:

$$S = \text{solubility of } CaCO_3 = [Ca^{2+}] = [CO_3^{2-}]$$

After substitution for the ion concentrations, we obtain

$$S^2 = 4.6 \times 10^{-9}$$

so that, taking the square root of each side of the equation, a value for S can be extracted:

$$S = 6.8 \times 10^{-5} \text{ M}$$

Thus the solubility of calcium carbonate is estimated to be 6.8×10^{-5} moles per liter of water, assuming that other reactions are negligible.

According to reaction 4, dissolved carbonate ion acts as a base in water; the relevant equilibrium constant for this process (see Appendix III) is K_b, where

$$K_b(CO_3^{2-}) = [HCO_3^-][OH^-]/[CO_3^{2-}]$$

Since the equilibrium in this reaction lies to the right, an approximation of the overall effect resulting from the simultaneous occurrence of reaction 3 and reaction 4 can be obtained by adding together the equations for the two individual reactions. The overall reaction is

$$CaCO_3(s) + H_2O(aq) \rightleftharpoons Ca^{2+} + HCO_3^- + OH^- \qquad (5)$$

calcium carbonate   bicarbonate ion

Thus the dissolution of calcium carbonate in water results essentially in the production of calcium ion, bicarbonate ion, and hydroxide ion.

It is a principle of equilibrium that if several reactions are added together, the equilibrium constant K for the *combined* reaction is the *product* of the equilibrium constants for the individual processes. Thus, since reaction 5 is the sum of reactions 3 and 4, its equilibrium constant K must equal $K_{sp}K_b$, the product of the equilibrium constants for reactions 3 and 4. Since K_a for $HCO_3^- = 4.7 \times 10^{-11}$, and since (Appendix III) for any acid-base conjugate pair such as HCO_3^- and CO_3^{2-}

$$K_aK_b = K_w = 1.0 \times 10^{-14} \text{ (at } 25°C)$$

it follows that for the conjugate base CO_3^{2-}

$$K_b = 1.0 \times 10^{-14}/4.7 \times 10^{-11}$$
$$= 2.1 \times 10^{-4}$$

Thus since K for the overall reaction 5 is $K_{sp}K_b$, its value is

$$4.6 \times 10^{-9} \times 2.1 \times 10^{-4} = 9.7 \times 10^{-13}.$$

Application of the Law of Mass Action to reaction 5 yields the expression

$$K = [Ca^{2+}][HCO_3^-][OH^-]$$

If we make the approximation that reaction 5 is the only process of relevance in the system, then from its stoichiometry we have a new expression for the solubility of CaCO$_3$, namely

$$S = [Ca^{2+}] = [HCO_3^-] = [OH^-]$$

and upon substitution of S for the concentrations we obtain

$$S^3 = 9.7 \times 10^{-13}$$

Taking the cube root of both sides of this equation, we find

$$S = 9.9 \times 10^{-5} \, M$$

Thus the estimated solubility for CaCO$_3$ is 9.9×10^{-5} M, in contrast to the lesser value of 6.8×10^{-5} M obtained when the reaction of carbonate ion was ignored. The actual CaCO$_3$ solubility is greater than estimated from reaction 3 alone since so much of the carbonate ion it produces subsequently disappears by reacting with water molecules. In other words, the equilibrium in reaction 3 is shifted to the right since a large fraction of its product reacts further (reaction 4).

PROBLEM 8-1

Repeat the calculation of the solubility of calcium carbonate by the approximate single equilibrium method using a realistic wintertime water temperature of 5°C; at that temperature, $K_{sp} = 8.1 \times 10^{-9}$ for CaCO$_3$, and $K_a = 2.8 \times 10^{-11}$ for HCO$_3^-$, and $K_w = 0.2 \times 10^{-14}$. Does the solubility of calcium carbonate increase or decrease with increasing temperature?

PROBLEM 8-2

Calculate the solubility of lead (II) carbonate, PbCO$_3$ ($K_{sp} = 1.5 \times 10^{-13}$) in water, given that most of the carbonate ion it produces subsequently reacts with water to form bicarbonate ion. Recalculate the solubility assuming that none of the carbonate ion reacts to form bicarbonate; is your result significantly different from that calculated assuming complete reaction of carbonate with water?

In the calculations above, we have used a single reaction (5) to represent the two-reaction system. We could improve upon our estimates of concentrations now by considering the reaction of the bicarbonate ion with water. Alternatively, we could arrive at the same end result by applying an iterative procedure to the solution of the algebraic equations involved in the original reactions 3 and 4 themselves; the relevant details of the calculation are given in Box 8-1. The solubility of $CaCO_3$, found by the more sophisticated procedure, is 1.24×10^{-4} M

BOX 8-1 EXACT SOLUTION FOR THE COUPLED EQUILIBRIA OF REACTIONS 3 AND 4

Instead of adding together reactions 3 and 4 and obtaining an approximate result, an exact solution for the coupled reactions can be obtained.

For convenience, we define y as the bicarbonate ion concentration; according to the stoichiometry for reaction 4 it is also equal to the hydroxide ion's concentration:

$$y = [HCO_3^-] = [OH^-]$$

Consequently, the equilibrium value for the concentration of the carbonate ion is

$$[CO_3^{2-}] = S - y$$

Upon substitution for the ion concentrations in the expression for K_b, we obtain

$$K_b = y^2 / (S - y)$$

With our initial estimate for S, this expression becomes

$$2.1 \times 10^{-4} = y^2 / (6.8 \times 10^{-5} - y)$$

The usual approximate solution would not be valid here since the concentration and the equilibrium constant are almost equal, so we solve the complete quadratic and obtain $y = 5.4 \times 10^{-5}$ mol^{-1}, which corresponds to 80% conversion of the carbonate ion to bicarbonate ion. Clearly, the original estimate for calcium carbonate's solubility must be invalid since in its calculation, the implicit assumption was made that no significant fraction of the CO_3^{2-} ion would react further.

In this system in which calcium carbonate dissolves in water, then, there are two strongly coupled equilibria, and two equilibrium constant expressions both involving two unknowns, S and y. One way to obtain a simultaneous solution for the two equations is to use the current estimate of y to obtain a revised value for S from the K_{sp} equation, which after substitution of the new expressions for equilibrium concentrations gives the expression

$$S (S - y) = 4.6 \times 10^{-9}$$

(compared to the value of 9.9×10^{-5} M obtained above), the concentrations of HCO_3^- and of OH^- are 8.7×10^{-5} M (compared to 9.9×10^{-5} M), and the CO_3^{2-} concentration is 3.7×10^{-5} M.

From the results above, it is clear that the saturated aqueous solution of calcium carbonate is moderately alkaline; its pH can be obtained by standard procedures (Appendix III):

$$pH = 14 - pOH = 14 - \log_{10}[OH^-] = 14 - \log_{10}(8.7 \times 10^{-5}) = 9.9$$

Then, using this new estimate for S, we obtain a revised value for y from the K_b equation, that is,

$$y^2 / (S - y) = 2.1 \times 10^{-4}$$

If the iteration cycle is repeated a few times, the following self-consistent set of S and y values is eventually obtained:

$$S = [Ca^{2+}] = \text{solubility of } CaCO_3$$
$$= 1.24 \times 10^{-4} \text{ mol L}^{-1}$$

$$y = [HCO_3^-] = [OH^-]$$
$$= 8.7 \times 10^{-5} \text{ mol L}^{-1}$$

The revised solubility of $CaCO_3$ is 82% greater than the first estimate, on account of the secondary reaction of most of the carbonate ion to bicarbonate ion. The equilibrium concentration of carbonate ion, that is, the value of $S - y$, is 3.7×10^{-5} mol L^{-1}, which is only 30% of the calcium ion concentration. Thus 70% of the CO_3^{2-} originally produced from dissolving of the salt reacts to form bicarbonate ion.

PROBLEM 8-3

Consider a body of water in equilibrium with solid calcium sulfate, $CaSO_4$, for which $K_{sp} = 3.0 \times 10^{-5}$ mol^2 L^{-2} at 25°C. Calculate the solubility of calcium sulfate in water assuming other reactions are negligible. From your answer, calculate the fraction of the sulfate ion that reacts with water to form bisulfate ("hydrogen sulfate") ion:

$$SO_4^{2-} + H_2O \rightleftharpoons HSO_4^- + OH^-$$
$$\text{sulfate ion} \qquad\qquad \text{bisulfate ion}$$

For HSO_4^-, $K_a = 1.2 \times 10^{-2}$ mol L^{-1}. Note that an iterative procedure should not be necessary in this case. Does the basicity of the sulfate ion substantially increase the solubility of the salt, as it did for calcium carbonate?

That the solution is alkaline is not surprising, given that the carbonate ion, as weak bases go, is a moderately strong one.

WATER IN EQUILIBRIUM WITH BOTH CaCO₃ AND ATMOSPHERIC CO₂

The system discussed above is somewhat unrealistic since it fails to consider the other important inorganic carbon species in water, namely carbon dioxide and carbonic acid, and the reactions (1 and 2) that involve them. Recall from Chapter 3 that pure water in equilibrium with atmospheric carbon dioxide is slightly acidic due to the operation of these two processes (Problem 3-22). These reactions now will be considered in the context of a body of water that is also in equilibrium with solid calcium carbonate, that is, the three-phase system illustrated in Figure 8-1.

At first sight, it might seem that since reaction 2 provides another source of bicarbonate ion, then by Le Chatelier's Principle the production of bicarbonate from the reaction (4) of carbonate with water should be suppressed. However, a more important consideration is that reaction 2 produces hydrogen ion, which combines with the hydroxide ion that is produced in reaction 4 by the interaction of carbonate ion with water:

$$H^+ + OH^- \rightleftharpoons H_2O \tag{6}$$

Consequently, the equilibrium positions of both reactions that produce bicarbonate ion are shifted to the right due to the disappearance of one of their products by the above reaction.

If reactions 1 through 4, plus reaction 6, are all added together to deduce the net process, then after canceling common terms the result is

$$CaCO_3(s) + CO_2(g) + H_2O(aq) \rightleftharpoons Ca^{2+} + 2\,HCO_3^- \tag{7}$$

calcium carbonate   carbon dioxide   bicarbonate ion

In other words, combining equimolar amounts of solid calcium carbonate and atmospheric carbon dioxide yields aqueous calcium bicarbonate, without any apparent production or consumption of acidity or alkalinity. Natural waters in which this overall process occurs can be viewed as the site of a giant titration of an acid which originates with CO_2 from air with a base which originates with carbonate ion from rocks.

It should be noted that each of the individual reactions added together is itself an equilibrium that does not lie entirely to the right. Since the reactions differ in their extent of completion, it is an approximation to state that the overall reaction shown above is the *only* resulting reaction. Nevertheless, it is the dominant process, and it is mathematically convenient to first consider this process alone in estimating the extent to which CaCO$_3$ and CO$_2$ dissolve in water when both are present.

Since reaction 7 equals the sum of reactions 1 to 4 (plus 6), its equilibrium constant K is the product of their equilibrium constants:

$$K = K_{sp}K_bK_HK_a/K_w$$

Here K_a is the dissociation constant of 4.5×10^{-7} for carbonic acid (i.e., for reaction 2). K_H is the Henry's Law constant (see Chapter 3) for reaction 1. K_w is the ion product for water, and consequently the equilibrium constant for reaction 6 is $1/K_w$. The other constants in the equation above for K have been defined previously. Thus at 25°C, for the overall reaction 7 it follows that

$$K = 1.5 \times 10^{-6} \text{ mol}^3 \text{ L}^{-3} \text{ atm}^{-1}$$

From the balanced equation for the reaction and the Law of Mass Action, it follows that the expression for K is

$$K = [\text{Ca}^{2+}] [\text{HCO}_3^-]^2/P$$

Here P is the partial pressure in the atmosphere of carbon dioxide; currently the value of P is 0.00035 atm since the atmospheric concentration of CO$_2$ is 350 ppm. If the calcium concentration again is called S, then from the stoichiometry of reaction 7 the bicarbonate concentration must be twice as large, or equal to $2S$; after substitution for the concentrations into the equation for K, we obtain

$$S(2S)^2/0.00035 = 1.5 \times 10^{-6}$$

or

$$S^3 = 1.3 \times 10^{-10}$$

Taking the cube root of both sides, we can evaluate S:

$$S = 5.1 \times 10^{-4} \text{ M}$$

The amount of CO_2 dissolved is also equal to S, and is 35 times that which dissolves without the presence of calcium carbonate (Problem 3-22). Furthermore, the calculated calcium concentration is five times that calculated without the involvement of carbon dioxide. Thus the acid reaction of dissolved CO_2 and the base reaction of dissolved carbonate have a synergistic effect on each other which increases the solubilities of both the gas and the solid.

PROBLEM 8-4

Repeat the above calculation for the solubility of $CaCO_3$ in water that is also in equilibrium with atmospheric CO_2 for a water temperature of 5°C. At this temperature, $K_H = 0.065$ M atm^{-1} for CO_2 and K_a for H_2CO_3 is 3.0×10^{-7}; see problem 8-2 for other necessary data.

PROBLEM 8-5

In many rocks of marine origin, the limestone is of the "dolmitic" type and has the formula $CaMg(CO_3)_2$, for which $K_{sp} = 5.1 \times 10^{-7}$ and is equal to $[Ca^{2+}]^{0.5} [Mg^{2+}]^{0.5} [CO_3^{2-}]$ for the reaction

$$0.5 \, CaMg(CO_3)_2 \rightleftharpoons 0.5 \, Ca^{2+} + 0.5 \, Mg^{2+} + CO_3^{2-}$$

Calculate the equilibrium concentrations of Ca^{2+}, Mg^{2+}, and HCO_3^- ions in equilibrium with solid dolomite and atmospheric carbon dioxide at 25°C.

MEASURED ION CONCENTRATIONS IN NATURAL WATERS

The most abundant ions found in samples of unpolluted fresh calcareous water usually are calcium and bicarbonate, as expected from our analysis above. Commonly, such water also contains magnesium ion, Mg^{2+}, principally from the dissolution of $MgCO_3$, plus some sulfate ion, SO_4^{2-}, and smaller amounts of chloride ion, Cl^-, and sodium ion, Na^+. The overall reaction (7) of carbon dioxide and calcium carbonate implies that the ratio of bicarbonate ion to calcium ion should be 2:1, and this is indeed a rule that is closely obeyed on average in river water in North

America and Europe. The calculated calcium ion concentration, 5.1×10^{-4} M, agrees well with the North American river water average value of 5.3×10^{-4} M, and similarly for the bicarbonate ion data. The close agreement between the calculated and the experimental results is somewhat fortuitous since river water temperatures on average lie below 25°C—which results in a higher CO$_2$ solubility than has been assumed—and because several minor factors have been oversimplified in the calculation. In fact, even calcareous river water is usually unsaturated with respect to CaCO$_3$ rather than saturated as has been implicitly assumed above.

Water in rivers and lakes that is not in contact with carbonate salts contains substantially fewer dissolved ions than are present in calcareous waters. The concentration of sodium and potassium ions may be as high as those of calcium, magnesium, and bicarbonate ions in these fresh waters. Even in areas with no limestone in the soil, the waters contain some bicarbonate ion due to the weathering of aluminosilicates in submerged soil and rock in the presence of atmospheric carbon dioxide. The weathering reaction can be written in general terms as

$$M^+(\text{Al-silicate}^-)(s) + CO_2(g) + H_2O \longrightarrow$$
$$\text{aluminosilicate}$$

$$M^+ + HCO_3^- + H_4SiO_4 + \text{new aluminosilicate}$$

Here M is a metal such as potassium, and the anion is one of the many aluminosilicate ions found in rocks (see Chapter 9). The weathering of potassium feldspar is an example of one of the most important sources of potassium ion in natural waters:

$$3 \, KAlSi_3O_8 + 2 \, CO_2 + 14 \, H_2O \longrightarrow$$
$$\text{potassium feldspar}$$

$$2 \, K^+ + 2 \, HCO_3^- + 6 \, H_4SiO_4 + KAl_3Si_3O_{10}(OH)_2$$

Thus bicarbonate normally is the predominant anion in both calcareous and noncalcareous waters since it is produced by the dissolution of limestone and aluminosilicates respectively.

The composition of average river water in the United States is given in Table 8-1. As discussed above, the values for the calcium and magnesium ion concentrations vary significantly from place to place, depending upon whether or not the underlying soil is calcareous. The level of

TABLE
8-1
CONCENTRATIONS OF IONS IN AVERAGE RIVER WATER
IN THE UNITED STATES

Ion	Molar concentration (average for U.S.)	Concentration in ppm		
		Average	Max. recommended concentration (U.S.)	Max. recommended concentration (Canada)
HCO_3^-	9.6×10^{-4}	60		
Ca^{2+}	3.8×10^{-4}	15		
Mg^{2+}	3.4×10^{-4}	8		
Na^+	2.7×10^{-4}	6		200
Cl^-	2.2×10^{-4}	8	250	250
SO_4^{2-}	1.2×10^{-4}	12	250	500
K^+	5.9×10^{-5}	2		
F^-	5.3×10^{-6}	0.1	0.8–2.4	1.5

Note: The value for bicarbonate is actually the total alkalinity.

fluoride ion, F^-, in water also displays substantial variations. The source of most F^- is weathering of the mineral fluorapatite, $Ca_5(PO_4)_3F$. In many communities in which the F^- concentration in the drinking water source is low, a soluble fluoride salt is added artificially in order to bring the fluoride level up to about 1 ppm, that is, 5×10^{-5} M, which has been found optimum in strengthening children's teeth against decay while providing a margin of safety. If the fluoride level is substantially in excess of this value, as it is in some natural waters, deleterious effects on teeth such as mottling can occur. The addition of fluoride ion to public supplies of drinking water continues to be a controversial subject because at high concentrations fluoride is known to be poisonous and perhaps carcinogenic, and because some people feel that it is immoral to force everyone to drink water to which a substance has been added. However, for many people, the total amount of fluoride ion ingested from food exceeds that from water.

The maximum concentration of ions recommended for drinking water in the United States and in Canada are also listed in Table 8-1. The concentration of sodium ion, Na^+, in water is of interest since high consumption of it from water and salted food are believed to increase one's blood pressure, and the latter may lead to cardiovascular disease. Excessive sulfate, beyond 500 mg/L, may cause a laxative effect in some people. It is interesting to note that some varieties of bottled drinking water, which people presumably drink in preference to tap water due to

BOX **8-2** **SEA WATER**

The total concentration of ions in sea water is much higher than that in fresh water since it contains large quantities of dissolved salts. The predominant species in sea water are sodium and chloride ions, which occur at about one thousand times their average concentration in fresh water. Sea water also contains some Mg^{2+} and SO_4^{2-}, and lesser amounts of many other ions. If sea water is gradually evaporated, the first salt to precip- itate is $CaCO_3$ (present to the extent of 0.12 g/L), followed by $CaSO_4.H_2O$ (1.75 g/L), then NaCl (29.7 g/L), $MgSO_4$ (2.48 g/L), $MgCl_2$ (3.32 g/L), NaBr (0.55 g/L), and finally KCl (0.53 g/L). Due primarily to the operation of the CO_2/bicarbonate/car- bonate equilibrium system discussed previ- ously for fresh water, the average pH of sur- face ocean water is about 8.1 (see p. 338).

health concerns about the latter, exceed the recommended values for some ions. Several of the well-known bottled waters exceeded the drink- ing water standards for sulfate, fluoride, or chloride in a 1991 survey (see reference 1 of the Suggestions for Further Reading).

The nature of the ionic compounds that are present in sea water are discussed in Box 8-2.

THE pH OF RIVER AND LAKE WATER SATURATED WITH CO$_2$ AND CACO$_3$

According to our simplified analysis, calcareous water becomes neither acidic nor alkaline due to the dissolution of carbon dioxide and calcium carbonate. However, a more sophisticated determination of the pH of river and lake water saturated with carbon dioxide and calcium carbon- ate can be obtained by considering the acid-base properties of the bicar- bonate ion that is produced in the overall reaction (7) that describes the system. In particular, the simplification used previously—namely, that all the carbonate and carbon dioxide react to produce bicarbonate, leaving no excess hydrogen or hydroxide ions—can be corrected by considering that, once formed, the HCO_3^- can act as an acid or as a base. As an acid, its reaction is

$$HCO_3^- \rightleftharpoons H^+ + CO_3^{2-} \tag{8}$$

for which

$$K_a = 4.7 \times 10^{-11}$$

The K_6 for its reaction as a base,

$$HCO_3^- + H_2O \; \rightleftharpoons \; H_2CO_3 + OH^- \quad (9)$$

equals 2.2×10^{-8}, since the K_a value for the conjugate acid H_2CO_3 is 4.5×10^{-7}.

Since $K_b > K_a$, the predominant acid-base reaction of HCO_3^- is as a weak base; consequently, its action as an acid will be ignored in the present context. The equilibrium constant expression for the reaction of interest, (i.e., 9), is

$$K_b = [H_2CO_3]\,[OH^-]\,/\,[HCO_3^-]$$

From previous considerations in Chapter 3 (see solution to Problem 3-22) it is known that $[H_2CO_3]$ is determined by Henry's Law to be 1.2×10^{-5} M, so it follows from the reaction stoichiometry that

$$2.2 \times 10^{-8} = 1.2 \times 10^{-5}\,[OH^-]\,/\,(1.02 \times 10^{-3} - [OH^-])$$

The approximation $1.02 \times 10^{-3} >> [OH^-]$ is valid here, so the solution is readily obtained:

$$[OH^-] = 1.9 \times 10^{-6}\,M$$

This value gives pOH = 5.73, and so pH = 8.27. Thus river and lake water at 25°C whose pH is determined by saturation with CO_2 and $CaCO_3$ should be slightly alkaline, with a pH of about 8.3.

Typically, the pH of such calcareous waters lie in the range from 7 to 9, in reasonable agreement with our calculations. Due to the smaller amount of bicarbonate in noncalcareous waters, their pH values are usually close to 7. Of course if such waters are subject to acid rain, their pHs can become substantially lower since there is little HCO_3^- or CO_3^{2-} readily available with which to neutralize the acid. Lakes and rivers into which acid rain falls will have elevated levels of sulfate ion and perhaps of nitrate ion since the principal acids in the precipitation are H_2SO_4 and HNO_3 (see Chapter 3).

PROBLEM 8-6

Calculate the pH of water in equilibrium with air if it contains 1.0×10^{-2} M bicarbonate ion and no other significant acids or bases (other than H_2CO_3).

PROBLEM 8-7

In waters subject to acid rain, the pH is determined not by the CO_2/carbonate system but rather by the strong acid from the precipitation. Assuming that equilibrium with atmospheric carbon dioxide is in effect, calculate the concentration of HCO_3^- in natural waters with pH = 6, 5, and 4.

PROBLEM 8-8

Using algebraic expressions and numerical values for the K_a of both H_2CO_3 and HCO_3^-, calculate the pH values for which $[H_2CO_3] = [HCO_3^-]$ and for which $[HCO_3^-] = [CO_3^{2-}]$. From your answers, decide the pH domains in which the various carbon-containing species are dominant.

ALKALINITY INDICES FOR NATURAL WATERS

The actual concentrations of the cations and anions in a real water sample cannot simply be assumed to be the theoretical values calculated above for calcium, carbonate and bicarbonate for two reasons: first, the water may not be in equilibrium with either solid calcium carbonate or with atmospheric CO_2; and second, other acids or bases may also be present.

The index devised by analytical chemists to represent the *actual* concentration in water of the anions that are basic is provided by the **alkalinity** value for the sample. Alkalinity is a measure of the ability of a water sample to act as a base by reacting with protons; in practical use, the alkalinity of a body of water is a handy measure of the capacity of the water body to resist acidification when acid rain falls into it. From an operational viewpoint, alkalinity (more properly termed *total alkalinity*) is the number of moles of H^+ required to titrate one liter of a water sample to the equivalence point. For a solution containing carbonate and bicarbonate ions, as well as OH^- and H^+, by definition

$$\text{total alkalinity} = 2\,[CO_3{}^{2-}] + [HCO_3{}^-] + [OH^-] - [H^+]$$

The factor of two appears in front of carbonate ion concentration since in the presence of H^+ it is first converted to bicarbonate ion, and then the latter is converted to carbonic acid:

$$CO_3{}^{2-} + H^+ \rightleftharpoons HCO_3{}^- \tag{10}$$

$$HCO_3{}^- + H^+ \rightleftharpoons H_2CO_3 \tag{11}$$

Other, minor, contributors to the alkalinity of fresh water systems can include dissolved ammonia and the anions of phosphoric, boric, and silicic acids.

By convention in analytical chemistry, methyl orange is used as the indicator in titrations by which total alkalinity is determined. Methyl orange is chosen because it does not change color until the solution is slightly acidic ($pH = 4$); under such conditions, not only has all the carbonate ion in the sample been transformed to bicarbonate but virtually all the bicarbonate ion has been transformed to carbonic acid (see Problem 8-9).

Another index frequently encountered in the analysis of natural waters is the **phenolphthalein alkalinity,** which is a measure of the concentration of the carbonate ion and of other similarly basic anions. In order to titrate only $CO_3{}^{2-}$ and not $HCO_3{}^-$ as well, the indicator phenolphthalein or one with similar characteristics is used. Phenolphthalein changes color at a pH of about 10, that is, it provides a fairly alkaline endpoint. At this pH, only a negligible amount of the bicarbonate ion has been converted to carbonic acid, but the majority of $CO_3{}^{2-}$ has been converted to $HCO_3{}^-$. Thus,

$$\text{phenolphthalein alkalinity} = [CO_3{}^{2-}]$$

PROBLEM 8-9

Calculate the value of the ratios $[HCO_3{}^-]/[CO_3{}^{2-}]$ and $[H_2CO_3]/[HCO_3{}^-]$ at pH values of 4 and 10 to confirm the statements made above concerning the nature of the species present at the methyl orange and phenolphthalein endpoints of the titrations. (Hint: Use the equilibrium constant expressions and K values for reactions 2 and 4 or 8 and 9.)

PROBLEM 8-10

Calculate the value expected for the total alkalinity and for the phenolphthalein alkalinity of a 25°C saturated solution of calcium carbonate in water, and compare them to the values for a solution that also is in equilibrium with atmospheric carbon dioxide. Use the concentrations quoted in the text and in Box 8-1 for these two systems.

PROBLEM 8-11

Calculate the total alkalinity for a sample of river water whose phenolphthalein alkalinity is known to be 3.0×10^{-5} M, whose pH is 10.0, and whose bicarbonate ion concentration is 1.0×10^{-4} M.

PROBLEM 8-12

At first sight, one might think that pH would be a good measure of the ability of a lake to resist acidification, but in fact it is not as reliable a guide as is its total alkalinity. Deduce why this might be so.

The alkalinity value for a lake is sometimes used by biologists as a measure of its ability to support aquatic plant life, a high value indicating a high potential fertility. The reasons for such a state of affairs are often the following ones.

Normally, algae extract the carbon dioxide they need for photosynthesis from carbonic acid, H_2CO_3 (sometimes written as $CO_2 + H_2O$ by way of emphasizing the easy convertibility of the molecular acid species to its simpler components), in this fashion:

$$CO_2 + H_2O + \text{sunlight} \longrightarrow CH_2O \text{ polymer} + 0.5 \, O_2$$
carbonic acid   (as algae)

In the absence of sufficient carbonic acid in the lake, the bicarbonate ion in the water dissociates to form additional carbon dioxide:

$$HCO_3^- \;\rightleftharpoons\; CO_2 + OH^- \tag{12}$$

The algae readily exploit this CO_2 for their photosynthetic needs, at the cost of allowing a buildup of hydroxide ion to such an extent that the lake water becomes alkaline.

HARDNESS INDEX FOR NATURAL WATERS

As a measure of certain important cations that are present in samples of natural waters, analytical chemists often use the **hardness** index (see Box 7-2 for a discussion of hardness in water) since it measures the total concentration of the ions Ca^{2+} and Mg^{2+}, the two species that are principally responsible for hardness in water supplies. Chemically, the hardness index is defined in this way:

$$hardness = [Ca^{2+}] + [Mg^{2+}]$$

Experimentally, hardness can be determined by titrating a water sample with ethylenediaminetetraacetic acid (EDTA), a substance that forms very strong complexes with metal ions other than those of the alkali metals (see Chapter 9 for details). Traditionally, hardness is expressed not as a molar concentration of ions but as the mass in milligrams (per liter) of calcium carbonate that contains the same total number of dipositive (2+) ions. Thus for example, a water sample that contains a total of 0.0010 moles of $Ca^{2+} + Mg^{2+}$ per liter would possess a hardness value of 100 milligrams of $CaCO_3$, since the molar mass of $CaCO_3$ is 100 grams and thus 0.0010 moles of it weighs 0.1 g, or 100 milligrams.

Most calcium enters water from either $CaCO_3$ in the form of limestone or from mineral deposits of $CaSO_4$; the source of much of the magnesium is "dolmitic" limestone, $CaMg(CO_3)_2$. Hardness is an important characteristic of natural waters, since calcium and magnesium ions form insoluble salts with the anions in soaps, thereby forming a "scum" in washwater (see Box 7-2). Water is termed "hard" if it contains substantial concentrations of calcium and/or magnesium ions; thus calcareous water is "hard."

Many areas possess soils that contain little or no carbonate ion, and thus its dissolution and reaction with CO_2 to produce bicarbonate does not occur. Such "soft" water typically has a pH much closer to 7 than does hard water, since it contains few basic anions. However, there are lakes with little dissolved calcium or magnesium but relatively high concentrations of dissolved sodium carbonate, Na_2CO_3; such lakes have a very low degree of hardness but are high in alkalinity.

Interestingly, people who live in hard water areas are found to have a lower average death rate from heart disease than do people living in areas with very soft water. It is not clear whether the advantage of drinking hard water stems from its supply of magnesium ion to the body, or from the protection that hard water provides from the presence of other ions such as sodium and those of the heavy metals (see Chapter 9).

PROBLEM 8-13

Calculate the hardness, in milligrams of $CaCO_3$ per liter, of water that is in equilibrium at 25°C with carbon dioxide and calcium carbonate, using results from the calculations discussed in Box 8-1 on page 330. Is the calculated value greater or less than the median hardness value found for surface waters in the United States of 37 mg/L?

ALUMINUM IN NATURAL WATERS

The concentration of aluminum ions in natural waters normally is quite low, typically about 10^{-6} M. This low value is the consequence of the fact that in the pH range of 6 to 9 that is usual for natural waters, the solubility of the aluminum contained in rocks and soils to which the water is exposed is very small. The fact that aluminum is not very soluble in water is controlled by the insolubility of $Al(OH)_3$ (see Box 8-3). Given that the K_{sp} of the hydroxide is about 10^{-33} at usual water temperatures, then for the reaction

$$Al(OH)_3 \rightleftharpoons Al^{3+} + 3\,OH^-$$
$$\text{aluminum hydroxide} \qquad \text{aluminum ion}$$

it follows that
$$[Al^{3+}]\,[OH^-]^3 = 10^{-33}$$

Take, for instance, a sample of water whose pH is 6. Since the hydroxide concentration in such water is 10^{-8} M, it follows that

$$[Al^{3+}] = 10^{-33} / (10^{-8})^3 = 10^{-9}\ M$$

FURTHER ASPECTS OF THE CHEMISTRY OF ALUMINUM IN WATER

The aqueous chemistry of aluminum ion, Al^{3+}, like that of some other metal ions such as Pb^{2+} and Hg^{2+}, is complicated. The ions themselves do not exist in solution as uncomplexed species; in the absence of other ions or molecules, they are surrounded by a number of water molecules each of which is attracted by the metal's positive charge. The presence of the metal ion makes the water molecules more acidic than usual: in other words, the loss of H^+ occurs more readily. Aluminum ion, Al^{3+}, exists as $Al(H_2O)_6^{3+}$ only under highly acidic conditions. At pH values near 5, most such ions in solution lose H^+ from one of the water molecules (see Figure 8-2), yielding the ion $Al(H_2O)_5(OH^-)^{3+}$, which often is written for simplicity as $AlOH^{2+}$. Near a pH of 6, a second water molecule loses a proton, yielding $Al(H_2O)_4(OH^-)_2^{3+}$, or $Al(OH)_2^+$. The loss of a proton from a third water molecule yields the neutral species $Al(H_2O)_3(OH^-)_3^{3+}$, better known as the familiar $Al(OH)_3$, which occurs to a tiny extent as the dissolved material but in greater amounts as the gelatinous semi-solid. Under neutral and (especially) alkaline conditions, most dissolved aluminum exists as the anionic species with still another proton lost from a water molecule, namely $Al(H_2O)_2(OH_2)_4^{3+}$, usually written as $Al(OH)_4^-$. Indeed, the ready formation of this ion in alkaline conditions gives rise to a greater solubility for aluminum in alkaline waters than in those which are neutral. As a "rule of thumb," some of the aluminum precipitates from water if its pH lies in the range of 5 to 8; below 5, the solid dissolves to give soluble cations and above 8 it dissolves to yield the anion.

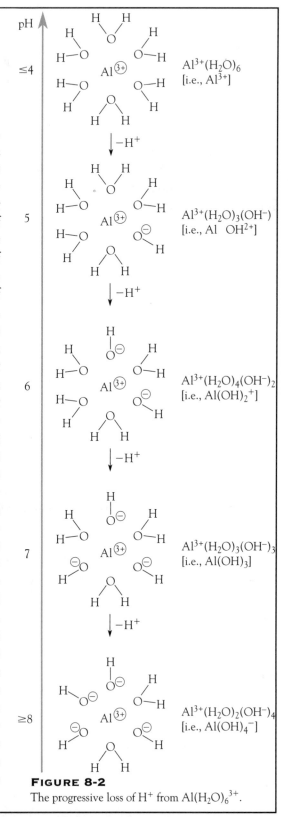

FIGURE 8-2

The progressive loss of H^+ from $Al(H_2O)_6^{3+}$.

Although this value is very tiny, for every decrease of the pH by one unit the concentration of aluminum ion increases by a factor of 10^3, and so it reaches 10^{-6} M at a pH of 5 and 10^{-3} M at a pH of 4. Thus aluminum is much more soluble in highly acidified rivers and lakes than in those whose pH values do not fall below 6 or 7. Indeed Al^{3+} is usually the principal cation in waters whose pH is less than 4.5, exceeding the concentrations of Ca^{2+} and Mg^{2+}, which are the dominant cations at pH values greater than 4.5.

It is thought that the principal deleterious effect of acid waters upon fish arises from the solubilization of aluminum from soil and its subsequent existence as a free ion in the acidic water. Unfortunately, the $Al(OH)_3$ then precipitates as a gel on contact with the less acidic gills of the fish, and the gel prevents the normal intake of oxygen from water, thus suffocating the fish.

It is also believed that aluminum mobilization in soils is one of the stresses that acid rain places on trees, and that results in the dieback of forests. Soils that contain limestone are usually considered to be "buffered" against much change in pH due to the ability of carbonate and bicarbonate ion to neutralize H^+, but over a period of decades surface soil may gradually lose its carbonate content due to a continual bombardment by acid rain. Thus soils receiving acid rain eventually become acidified. When the pH of the soil drops below about 4.2, aluminum leaching from soil and rocks becomes particularly appreciable. Such acidification has occurred in some regions in central Europe, including Poland, the former Czechoslovakia, and eastern Germany, and the resulting solubilization of aluminum may have contributed to the forest diebacks observed there in the 1980s.

PROBLEM 8-14

Calculate the pH value at which the aluminum ion concentration dissolved in water is 0.020 M, assuming that it is controlled by the equilibrium with solid aluminum hydroxide.

REVIEW QUESTIONS

1. What is the acid and what is the base that dominate the chemistry of most natural water systems, and whose interaction produces bicarbonate ion?

2. What is the source of most of the carbonate ion in natural waters? What name is given to waters that are exposed to this source?

3. Write the approximate net reaction between carbonate ion and water in a system which is *not* also exposed to atmospheric carbon dioxide. Is the resulting water acidic, alkaline, or neutral?

4. Write the approximate net reaction between carbonate ion and water in a system that *is* exposed to atmospheric carbon dioxide. Is the resulting water acidic, alkaline, or neutral?

5. If two equilibrium reactions are added together, what is the relationship between the equilibrium constants for the individual reactions and that for the overall reaction?

6. What is the natural source of fluoride ion in water? How and why is the fluoride level in drinking water artificially increased to about 1 ppm in many municipalities?

7. Define the *total alkalinity* index and the *phenolphthalein alkalinity* index for water.

8. Define the hardness index for water.

9. Which are the most abundant ions in fresh water?

10. Explain why aluminum ion concentrations in acidified waters are much greater than those in neutral water. How does the increased aluminum ion level affect fish and trees?

SUGGESTIONS FOR FURTHER READING

1. Allen, H. E., M. A. Henderson, and C. N. Haas, 1991. What's in the bottle of water? CHEMTECH (December): 738–742.

2. Stumm, W., and J. J. Morgan. 1970. *Aquatic Chemistry*. New York: Wiley-Interscience.

3. Carbonated Waters. 1992. *Consumer Reports*. September: 569–571.

HEAVY METALS AND THE CHEMISTRY OF SOILS

In chemistry, **heavy metal** refers not to a type of rock music but rather to a type of chemical element, many examples of which are poisonous to humans. The four discussed here—mercury (Hg), lead (Pb), cadmium (Cd), and arsenic (As)—are those that are of the greatest environmental hazard due to their extensive use, their toxicity, and their widespread distribution. None has yet pervaded the environment to such an extent as to constitute a widespread danger. However, each has been discovered to occur at toxic levels in certain locales in recent times. Metals differ from the toxic organic compounds we discussed in Chapter 6 in that they are totally nondegradable; since most elements cannot be transmuted except under truly extraordinary conditions, they are, practically speaking, indestructible, and so they accumulate in the environment.

The heavy metals occur near the bottom of the Periodic Table and their densities are high compared to those of other common materials. The densities of the four metals of interest here are collected in Table 9-1, as are contrasting values for water and two common "light" metals.

Although we think of heavy metals as water pollutants and as contaminants in our food, they are for the most part transported from place to place via the air, usually as species adsorbed on or absorbed in suspended

TABLE 9-1	DENSITIES OF SOME IMPORTANT HEAVY METALS AND OTHER SUBSTANCES

Element	Density (g/cm^3)
Hg	13.5
Pb	11.3
Cd	8.7
As	5.8
H$_2$O	1.0
Mg	1.7
Al	2.7

particulate matter. Thus for example, over half of the heavy metal input into the Great Lakes is due to deposition from air. Recent evidence from Sweden indicates that the deposition of lead into European lake sediments dates back to the time of the ancient Greeks, when silver was first mass-produced for use in coins. Apparently the rather substantial amount of lead contaminant in the crude silver escaped into the air during the refining of the metal.

TOXICITY OF THE HEAVY METALS

Although mercury vapor is highly toxic, the four heavy metals Hg, Pb, Cd, and As are not particularly toxic as the condensed free elements. However, all four are dangerous in the form of their cations and when bonded to short chains of carbon atoms. Biochemically, the mechanism of their toxic action arises from the strong affinity of the metal cations for sulfur. Thus "sulfhydryl" groups, —SH, which occur commonly in the enzymes that control the speed of critical metabolic reactions in the human body, readily attach themselves to ingested heavy metal cations or molecules that contain the metals. Because the resultant metal-sulfur bonding affects the entire enzyme, it cannot act normally and human health is adversely affected, sometimes fatally. The reaction of heavy metal cations M^{2+}, where M is Hg, Pb, or Cd, with the sulfhydryl units of enzymes R—S—H to produce stable systems such as R—S—M—S—R is analogous to their reaction with the simple inorganic chemical H$_2$S, with which they yield the insoluble solid MS.

$$M^{2+} + H—S—H \longrightarrow MS + 2 H^+$$
$$R—S—H + M^{2+} + H—S—R \longrightarrow R—S—M—S—R + 2 H^+$$

A common medicinal treatment for acute heavy metal poisoning is the administration of a compound that attracts the metal even more strongly than does the enzyme; subsequently the metal-compound combination is solubilized and excreted from the body. One compound used to treat mercury and lead poisoning is *British Anti-Lewisite* (BAL); its molecules contain two SH groups which together capture the metal.

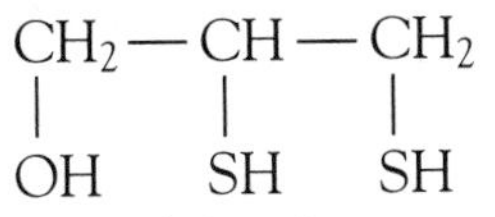

$$CH_2 - CH - CH_2$$
$$| \qquad | \qquad |$$
$$OH \quad SH \quad SH$$

British Anti-Lewisite

Also useful for this purpose is the calcium salt of ethylenediaminetetraacetic acid (EDTA), a well-known compound that extracts and solubilizes most metal ions. Its structure is illustrated below; the metal ions are complexed by the two nitrogens and the charged oxygens to form a chelate that is subsequently excreted from the body.

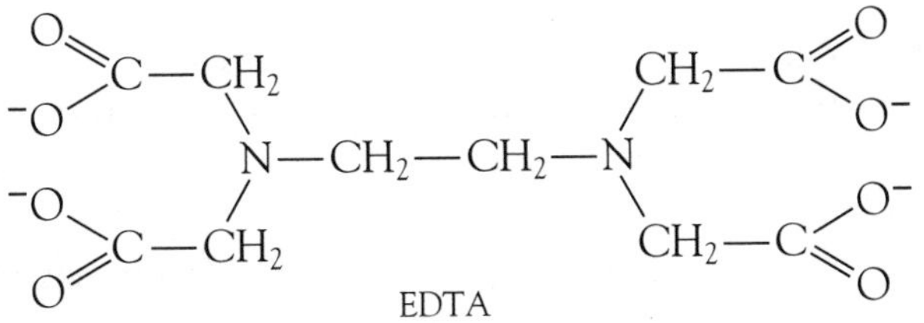

EDTA

Because EDTA is effective in chelating almost all +2 and +3 charged ions, it is usually administered as the calcium salt so that no net depletion of calcium in the body occurs during the therapy. Treatment of heavy metal poisoning by chelation therapy is best begun early, before neurological damage has occurred.

The toxicity for all four heavy metals depends very much on the chemical form of the element, that is upon its **speciation.** For example the toxicities of metallic lead, lead as the ion Pb^{2+}, and lead in the form of covalent molecules differ significantly. Forms which are almost totally insoluble pass through the human body without doing much harm. The most devastating forms of the metals are those that cause immediate sickness or death (as from a sufficiently large dose of arsenic oxide) so that therapy cannot exert its effects in time, and those which can pass through the membrane protecting the brain—the blood/brain barrier— or that protect the developing fetus. For some heavy metals such as mercury, the form that is the most toxic is that having alkyl groups attached to the metal, since many such compounds are soluble in animal tissue and can pass through biological membranes.

BIOACCUMULATION OF HEAVY METALS

Recall from Chapter 6 that some substances display the phenomenon of biomagnification: their concentrations increase progressively along an ecological food chain. The only one of the four heavy metals under consideration that is indisputedly capable of doing this is mercury. Many aquatic organisms do however bioconcentrate heavy metals. For example, oysters and mussels can contain levels of mercury and cadmium that are 100,000 times greater than those in the water in which they live.

The concentrations of most heavy metals that humans encounter in drinking water are usually small and cause no direct health problems; however, exceptions do occur and will be discussed later. As is the case with toxic organic chemicals, the amounts of metals that are ingested through our food supply are usually of much greater concern than is the intake attributable to drinking water. Paradoxically, the heavy metals in the fish that we ingest usually originate in fresh water.

The extent to which a substance accumulates in a human or in any other organism depends upon the rate R at which it is ingested from the source, for example the food supply, and the mechanism by which it is eliminated, that is, its sink. Commonly, the rate of elimination bears a simple relation to the organism's concentration C of the substance, namely, the rate is directly proportional to C. (This is known as a **first-order** relationship, since the power to which the independent variable is raised is unity.) If the rate constant for the process is defined as k, the rate of elimination is kC.

$$\text{Rate of intake} = R$$
$$\text{Rate of elimination} = kC$$

If none of the substance is initially present in an organism, that is, if $C = 0$, then initially the rate of elimination is zero and the concentration builds up solely due to its ingestion, as illustrated near the origin in Figure 9-1. However, as C rises, the rate of elimination also rises, and eventually meets the rate of intake if R is a constant. Once this equality is achieved, C does not vary thereafter; it is in a **steady state.** Since under these steady-state conditions

$$\text{Rate of elimination} = \text{Rate of intake}$$
$$kC = R$$

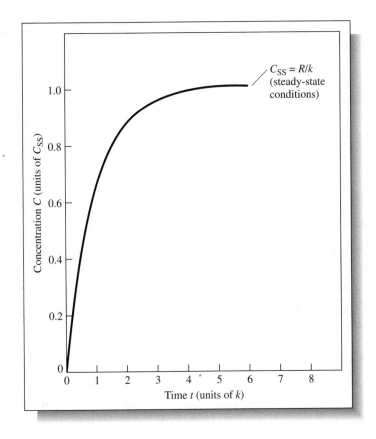

$$C_{SS} = R/k$$
(steady-state
conditions)

Concentration C (units of C_{SS})

Time t (units of k)

FIGURE 9-1
Increase in mercury concentrations with time to the steady-state level, C_{ss}.

It follows that the steady-state value for the concentration—C_{ss}—is

$$C_{ss} = R/k$$

Often the speed of elimination is discussed in terms of the half-life period, $t_{0.5}$, the length of time required for half of the reactant to react. From mathematical analysis of the kinetics of first-order reactions, it is known that $k = 0.69/t_{0.5}$. By substitution, it follows that

$$C_{ss} = R\, t_{0.5}/0.69$$
$$= 1.45\, R\, t_{0.5}$$

(This formula is similar to that developed on page 167 in Chapter 4 and which related C and R to T_{avg} the lifetime of a substance; clearly $T_{avg} = 1.45 t_{0.5}$ when formulas are compared.) Clearly, the longer the half-life of a substance, the higher its steady-state accumulation level. As an example of the above relationship, consider that the half-life in the human body of mercury in the form of Hg^{2+} is about 6 days; if one ingests one milligram of mercury per day, its steady-state accumulation would amount to 9 milligrams, since

$$C_{ss} = 1.45 \times 1 \ mg \ day^{-1} \times 6 \ days = 9 \ mg$$

The variations with time of concentrations and rates for systems of this type are illustrated in Figure 9-1 for the specific case where R, k, and hence C_{ss} is expressed in units of C_{ss}, and time t is expressed in units of k.

PROBLEM 9-1

If the half-life in the human body of the form of mercury called "methylmercury" is 70 days, what is its steady-state accumulation in a person who consumes daily one kilogram of fish containing 0.5 ppm methylmercury?

PROBLEM 9-2

Suppose a person consumes 100 μg a day of a substance whose steady-state level in the body is later established to be 10 milligrams. Once consumption stops, how long does it take for the level of this substance in the body to drop to 2.5 milligrams?

MERCURY

THE FREE ELEMENT

Because mercury is the only metal that is a liquid at room temperatures and because it expands uniformly with increases in temperature, it is widely used as the fluid in medical thermometers. Mercury's symbol, Hg, is derived from the Latin word *hydrargyrum*, which means *liquid silver,* and indeed the metal is sometimes called *quicksilver* (literally, *living silver*).

Elemental mercury is employed in hundreds of applications, many of which (e.g., electrical switches) take advantage of its unusual property of being a liquid that conducts electricity well. It is used in fluorescent light bulbs and in the mercury lamps employed for street lighting, since energized mercury atoms emit light in the visible wavelength region. In view of the contamination of the environment when mercury lamps are broken, there has recently been a shift toward the use of sodium vapor lamps, which present a lesser toxicity hazard and are more efficient light sources.

Mercury is the most volatile of all metals, and its vapor is highly toxic. Adequate ventilation is required whenever mercury is used in closed quarters, since the equilibrium vapor pressure of mercury is hundreds of times the maximum recommended exposure. Liquid mercury itself is not highly toxic, and most of that ingested is excreted. Nevertheless, children should not be allowed to play with droplets of the metal because of the danger from breathing the vapor. It diffuses from the lungs into the bloodstream, and then crosses the blood/brain barrier to enter the brain; the result is serious damage to the central nervous system, which is manifested by difficulties with coordination, eyesight, and tactile senses.

Large amounts of mercury vapor are released into the environment as a result of burning coal and fuel oil, both of which always contain trace amounts of the element (reaching several hundred ppm in some coals), and of incinerating solid waste that contain mercury in products such as batteries. This source of atmospheric mercury has increased substantially in the twentieth century, and now rivals the input from volcanoes, formerly the predominant source of airborne mercury. In air, the great majority of mercury is in the vapor (gaseous) state, with only a tiny fraction of it bound to airborne particles. Airborne gaseous mercury can travel long distances before being deposited on land or in waterways.

MERCURY AMALGAMS

Mercury readily forms **amalgams,** which are solutions or alloys with almost any other metal or combination of metals. For example, the solid "dental amalgam" used to fill cavities in teeth has a putty-like consistency initially and is prepared by combining approximately equal proportions of liquid mercury and a mixture that is mainly silver and tin. When first placed in a tooth, and whenever the filling is involved in the chewing of food, a tiny amount of the mercury is vaporized; some scientists believe that mercury exposure from this source causes long-term health problems in some individuals, although this matter has yet to be resolved one way or another. Mercury-free "amalgams" for use in dentistry are currently under development.

Until the nineteenth century, objects were plated ("gilded") with gold or silver by rubbing them with an amalgam of mercury and the precious metal, and then heating the object to boil off the mercury.

Similarly, mirrors were made by applying a tin-mercury amalgam to glass. In working some ore deposits, tiny amounts of elemental gold or silver are extracted from much larger amounts of dirt by adding elemental mercury to the mixture; this extracts the gold or silver by forming an amalgam, which is then heated to distill off the mercury. From 1570 until about 1900, this process was used to extract silver from ores in Central and South America; about one gram of mercury was lost to the environment for every gram of silver produced. Today, the extraction procedure is carried out on a large scale in Brazil to obtain gold, and results in substantial mercury pollution both in the air and, because of careless handling practices, in the Amazon river itself. The health hazards to workers using processes which involve the vaporization of mercury are substantial, since as mentioned above, the element is very toxic in its gaseous form. Currently there are some initiatives being undertaken by the European Commission to incorporate some additional but inexpensive technology into the process to prevent the massive release of mercury into the air and to the Amazon river during the extraction of gold.

Presumably on account of the toxicity of the vapor, the life expectancy of the slaves and convicts used by the Romans to mine mercury ore, HgS, ("cinnabar") was about six months. Mercury is obtained by roasting this ore in air, and condensing the mercury vapor released.

$$HgS(s) + O_2(g) \longrightarrow Hg(g) + SO_2(g)$$
$$\text{mercury (II) sulfide} \qquad\qquad \text{sulfur dioxide}$$

MERCURY AND THE CHLOR-ALKALI PROCESS

An amalgam of sodium and mercury is used in some industrial **chloralkali** plants in the process that converts aqueous sodium chloride into the commercial products chlorine and sodium hydroxide (and hydrogen) by electrolysis. In order to form a concentrated, pure solution of NaOH, flowing mercury is used as the negative electrode (cathode) of the electrochemical cell. The metallic sodium that is produced by reduction in the electrolysis combines with the mercury and is removed from the NaCl solution without having reacted in the aqueous medium.

$$Na^+ (aq) + e^- \xrightarrow{\ Hg\ } Na \text{ (in a Na-Hg amalgam)}$$

The need for this amalgamation process can be explained in the following way. When metals such as sodium are dissolved in amalgams, their reactivity is greatly lessened compared to that for the free state, so that the otherwise highly reactive elemental sodium in the Na-Hg amalgam does not react with the water in the original solution. Instead, the amalgam is removed, and later induced by the application of a small electrical current to react with water in a separate chamber, thereby producing sodium hydroxide that is free of salt. The mercury is then recovered and recycled back to the original cell.

The recycling of mercury is not complete, however, and some finds its way into the air and into the river from which the plant's cooling water is obtained and to which it is returned. Although liquid mercury is neither soluble in water nor in dilute acid, apparently it can be oxidized to soluble form by the intervention of bacteria that are present in natural waters. By this means the mercury becomes accessible to fish.

The mass of mercury lost to the environment from such chlor-alkali plants has decreased enormously since the problem was identified in the 1960s. Nevertheless, installations that use mercury electrodes are being phased out and replaced by those that use a membrane to separate the NaCl solution from the chloride-free solution at the negative electrode; the membrane is designed such that Na^+, but not anions, can pass through it. In both types of cells, the overall reaction is:

$$2\,NaCl(aq) + 2\,H_2O(\ell) \longrightarrow 2\,NaOH(aq) + Cl_2(g) + H_2(g)$$

The consumption of fish from Lake Saint Clair, a body of water that lies between Michigan and Ontario in the Great Lakes region and that received a heavy load of mercury from chlor-alkali plants in the 1960s, was restricted in the 1970s and 1980s in view of their contamination by mercury. Lately, the mercury levels in Great Lakes fish have fallen, since the mercury input into their waters has been sharply reduced. The reduction in the average mercury content of fish caught in Lake Saint Clair from concentrations exceeding 6 ppm in some fish in the early 1970s to less than 1 ppm today is illustrated in Figure 9-2.

IONIC MERCURY

Like its partners zinc and cadmium in the same subgroup of the Periodic Table, the common ion of mercury is the 2+ species, Hg^{2+}, the mercuric or mercury(II) ion. An example of a compound with mercuric ion is the ore HgS, $Hg^{2+}S^{2-}$. This salt is very insoluble in water; indeed, the wastewater at chlor-alkali plants is sometimes treated by adding a soluble

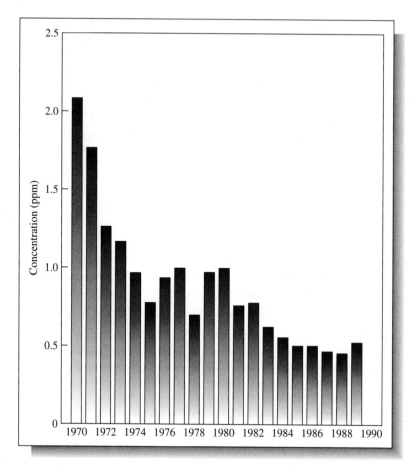

FIGURE 9-2

Annual variation of mercury concentrations in walleye fish from Lake Saint Clair.
(Source: Toxic chemicals in the Great Lakes and associated effects: Synopsis. Ministry
of Supply and Services Canada, 1991.)

salt such as Na_2S that contains the sulfide ion, since this action precipitates the mercury as HgS.

$$Hg^{2+} + S^{2-} \longrightarrow HgS(s)$$

mercuric   sulfide   mercuric
ion   ion   sulfide

PROBLEM 9-3

The solubility product, K_{sp}, for HgS is 3.0×10^{-53}. Calculate the
solubility of HgS in water in moles per liter, and transform your

answer into the number of monatomic mercury ions (Hg^{2+}) per liter. (As we shall see shortly, mercury readily forms diatomic ions.) According to this calculation, what volume of water in equilibrium with solid HgS contains a single Hg^{2+} ion?

The nitrate salt of Hg^{2+} is water soluble and was at one time used to treat the fur used for felt hats. Consequently, the workers in the felt trade often displayed nervous disorders: muscle tremors, depression, memory loss, paralysis, and insanity. Mercury vapor, and to a lesser extent mercury salts, attack the central nervous system, but the main target organs for Hg^{2+} are the kidney and the liver, where it can cause extensive damage.

Mercuric oxide is present in a paste in "mercury cell" batteries such as those used in hearing aids; when the battery operates and current flows, elemental mercury is produced:

$$Hg^{2+} + 2\,e^- \longrightarrow Hg(\ell)$$
$$\text{(as HgO)}$$

The practical advantage to using mercury is that no aqueous solutions are involved in the cell, and as a result its voltage remains constant until the reactants are consumed, at which point the voltage drops abruptly to zero. If the discarded spent batteries are subsequently incinerated as garbage, the volatile mercury can be released into the air. Indeed, the incineration of municipal garbage has become a major source of environmental mercury pollution. The mercury in ordinary flashlight batteries, present as a minor constituent in the zinc electrode to prevent its corrosion and thus to extend the shelf life of the product, has recently been drastically curtailed and may soon be completely eliminated. This action alone will eliminate half the mercury in garbage.

In natural waters, much of the Hg^{2+} is attached to suspended particulates, and so is eventually deposited in sediments. This topic is considered in further detail when soil and sediment chemistry is discussed (page 387).

METHYLMERCURY FORMATION

With anions that are more capable of forming covalent bonds than are nitrate, oxide, or sulfate ions, the mercuric ion Hg^{2+} forms covalent *molecules* rather than an ionic solid. For example, $HgCl_2$ is a volatile molecular compound. Just as Cl^- ions form a covalent compound with Hg^{2+}, so does the methyl anion CH_3^-, yielding the volatile molecular liquid dimethylmercury, $Hg(CH_3)_2$. The process of dimethylmercury formation occurs in the muddy sediments of rivers and lakes, especially under anaerobic conditions, when anaerobic bacteria and microorganisms convert Hg^{2+} into $Hg(CH_3)_2$. The active agent in the methylation process is a common constituent of microorganisms; it is a derivative of Vitamin B_{12} with a CH_3^- anion bound to cobalt and is called "methylcobalamin." Due to its volatility, dimethylmercury evaporates ("degasses") from water relatively quickly unless it is transformed by acidic conditions into the monomethyl form. The pathways for the production and fate of dimethylmercury and of other mercury species in a body of water are illustrated in Figure 9-3.

The less volatile "mixed" compounds CH_3HgCl and CH_3HgOH, often collectively written as CH_3HgX or somewhat misleadingly as CH_3Hg^+ (this molecular fragment may not exist as an ionic species),

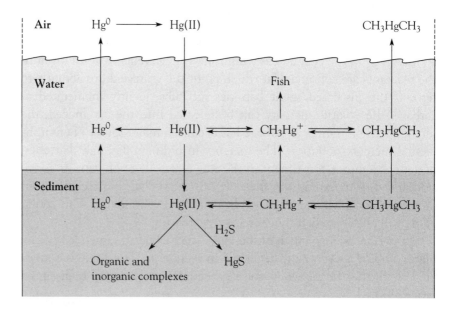

FIGURE 9-3

The cycling of mercury in freshwater lakes. (Source: Redrawn from M. R. Winfrey and J. W. M. Rudd, 1990. Environmental factors affecting the formation of methylmercury in low pH lakes. *Environmental Toxicology and Chemistry* 9:853–869. Copyright 1990. Reprinted with kind permission from Elsevier Science Ltd., The Boulevard, Langford Lane, Kidlington OX5 1AB, UK.)

"methylmercury" (or "monomethylmercury") are even more readily formed in the same way as dimethylmercury. Methylmercury production predominates over dimethylmercury formation in acidic or neutral aqueous solutions. Methylmercury is a more potent toxin than are salts of Hg^{2+} because it is soluble in fatty tissue in animals and bioaccumulates and biomagnifies there. Once ingested, the CH_3HgX compound is converted to compounds in which X is a sulfur-containing amino acid; in some of these forms it is soluble in biological tissue, and can cross both the blood-brain barrier and the human placental barrier, presenting a twofold hazard. Methylmercury is in fact the most hazardous form of mercury, followed by the vapor of the element. Mercuric ion, Hg^{2+}, itself is not readily transported across biological membranes. The other inorganic ion of mercury, Hg_2^{2+}, is not very toxic since it combines in the stomach with chloride ion to produce Hg_2Cl_2 (see page 362).

Most of the mercury present in humans is in the form of methylmercury, and almost all of it originates from the fish in our food supply; the mercury in fish is usually at least 80% in the form of methylmercury. Fish absorb methylmercury that is dissolved in water as it passes across their gills, and they also absorb it from their food supply. In contrast to organochlorines, which tend to predominate in the fatty portions of fish, methylmercury can bind to the sulfhydryl group in proteins and so is distributed throughout the fish; thus the mercury-containing part cannot be "cut away" before the fish is eaten. The ratio between methylmercury in fish muscle and that dissolved in the water in which the fish swims is often about one million to one, and can exceed ten million to one. The highest concentrations usually are found in large, long-lived predatory marine species such as shark and swordfish and in freshwater species such as bass and pike. Noncarnivorous species such as whitefish do not accumulate very much mercury since biomagnification by the food chain operates to a much lesser extent than in carnivorous fish. In lakes, the mercury content in fish is generally greater in acidic water, probably because the methylation of mercury is faster at lower pH. Thus, acidification of natural waters indirectly increases the exposure of fish-eaters to methylmercury.

METHYLMERCURY TOXICITY

The half-life of methylmercury compounds in humans, about 70 days, is much longer than that for Hg^{2+} salts. Consequently, methylmercury can accumulate in the body to a much higher steady-state concentration, even if on a daily basis a person consumes doses which individually would not be harmful. Most of the well-publicized environmental problems involving mercury have arisen in connection with the fact that in the methylated form it is a cumulative poison.

At the fishing village of Minamata, Japan, a chemical plant employing Hg^{2+} as a catalyst in a process that produced polyvinyl chloride discharged mercury-containing residues into Minamata Bay; the methylmercury that subsequently formed then bioaccumulated in the fish, which were the main component of the diet for many local residents. The fish contained up to 100 ppm mercury. By way of contrast, the current North American recommended limit for mercury in fish destined to be consumed by humans is 0.5 ppm. Thousands of people in Minamata were affected in the 1950s by mercury poisoning from this source; hundreds of them died from it. In fact, because the onset of symptoms in humans is delayed, the first signs of "Minamata disease" were observed in cats who ate discarded fish: they began jumping round and twitching, ran in circles, and finally threw themselves into the water and drowned. Symptoms in humans arise from dysfunctions of the central nervous system since the target organ for methylmercury is the brain; they include numbness in arms and legs, blurring and even loss of vision, loss of hearing and muscle coordination, and lethargy and irritability. Since methylmercury chloride can be passed to the fetus, children born to mothers poisoned even slightly by mercury showed severe brain damage, some to a fatal extent. The infants showed symptoms similar to those of cerebral palsy: mental retardation and motor disturbance, even paralysis. Just as in the case of high PCB levels in residents around Lake Michigan, the developing fetuses were much more affected by methylmercury than were the mothers themselves. The poisonings at Minamata must surely rank as one of the major environmental disasters of modern times.

In the town of Dryden in northern Ontario, Canada, the release of mercury from a chlor-alkali plant into the English-Wabigoon River in the 1960s and early 1970s led to the bioaccumulation of mercury in the river's fish to a level far beyond the legal limit of 0.5 ppm. As a result of bioaccumulation, the ratio of methylmercury in water, algae, and fish there is typically 1:3,000:20,000. Some members of the Indian band who regularly ate fish from these waters were found to have such elevated mercury levels that some clinical signs of mercury toxicity were observed. As a result of the mercury contamination, commercial fishing in the English-Wabigoon River was curtailed, as it also has been in some of the Great Lakes.

PROBLEM 9-4

What is the mass, in milligrams, of mercury in a 1.00 kilogram lake trout which just meets the Northern American standard of 0.50 ppm Hg?

PROBLEM 9-5

What mass of fish, each at the 0.50 ppm Hg level, would you have to eat in order to ingest a total of 100 milligrams of mercury?

OTHER SOURCES OF METHYLMERCURY

Organic compounds of mercury have been used as fungicides in agriculture and in industry, and enter the environment as a side effect of these applications. However, as a result of contact with soil, the compounds are broken down and the mercury becomes trapped as insoluble compounds by attachment to sulfur groups in clays and organic matter.

Hundreds of deaths in Iraq, and a few in the United States, have resulted in the past from the consumption of bread made from seed grain (intended for planting) that had been treated with methylmercury as a fungicide in order to reduce seedling losses from fungus attack. In Sweden and Canada, the use of methylmercury to treat seeds led to a significant reduction in the number of birds of prey which consumed the smaller birds and mammals that fed on the scattered seed. The use of methylmercury products in agriculture has now been curtailed in North America and western Europe.

Mercury is leached from rocks into water systems by natural processes, some of which are accelerated by human activities. In recent years, it has been established that flooding of vegetated areas can release mercury into water. For example, after the flooding of huge areas of northern Quebec and Manitoba to produce hydroelectric power dams, the newly submerged surface soils (and to a lesser extent the vegetation) released a considerable quantity of soluble methylmercury, formed from the "natural" mercury content of these media. The additional methylmercury resulted from contact of soil-bound Hg^{2+} with anaerobic bacteria produced by the decomposition of the immersed organic matter; in this way, previously insoluble inorganic mercury was converted to methylmercury, which readily dissolved in the water. The methylmercury subsequently entered the food chain through its absorption by fish, and native persons who ate fish from these flooded areas now have substantially elevated levels of mercury in their bodies. Indeed, the methylmercury concentration in fish, 5 ppm or more, from these areas approaches that previously associated only with regions of industrial mercury pollution.

With respect to methylmercury concentrations in human populations remote from regions affected by the sorts of mercury pollution discussed above, it is somewhat reassuring that both the direct effects of methylmercury on humans and the prenatal effects probably have thresholds below which no effects are observed. Currently the daily methylmercury intake of 99.9% of Americans lies below the World Health Organization's "safe limit"; however if prenatal health is the main consideration and if a safety factor of ten is applied, a substantial fraction of the population of the United States would exceed the limit.

OTHER FORMS OF MERCURY

Salts of the phenylmercury ion, $C_6H_5Hg^+$, have been used to preserve paint while in the can, and to prevent mildew after application of latex paint, particularly in humid areas. The phenylmercury salts are not as toxic to humans as are methylmercury compounds, since they break down into compounds of the less toxic Hg^{2+}. However, mercury compounds have been banned from indoor latex paints in the United States since 1990 because some ingestion of the element from this source is inevitable. Phenylmercury compounds were also formerly used as slimicides in the pulp and paper industry in order to prevent the growth of slime on wet pulp; because this practice has now been curtailed and because mercury-containing wastes are now usually treated, mercury releases from such sources have greatly decreased. Paper produced using mercury-containing slimicides had previously been banned in the United States for use in storing food. Because of their antiseptic and preservative qualities, however, some mercury compounds are still used in pharmaceuticals and cosmetics.

Mercury also forms a series of compounds containing the species Hg_2^{2+}. Because this ion forms the insoluble dichloride in chloride-rich environments such as the hydrochloric acid found in the human stomach, it is not a particularly hazardous form of the element. However, the use of Hg_2Cl_2 as a "teething powder" for children in England in the 1940s resulted in several deaths because, as we now realize, some humans are idiosyncratically sensitive to the compound.

PROBLEM 9-6

A quantity of a mercury-chlorine compound is included in a shipment of waste to a toxic waste disposal dump. Before it can be disposed of properly, the owners of the dump need to know whether the compound is $HgCl_2$, Hg_2Cl_2, or some other compound. They send a sample of it for analysis, and find that it contains 26.1% chlorine by mass. What is the formula of the compound?

LEAD

Lead (Pb), like mercury, cadmium and arsenic, is a heavy metal of environmental concern; it is the most abundant and the most widely used and widely dispersed of the four. Although its concentration is still increasing in some parts of the world, those uses that result in uncontrolled dispersion have been greatly reduced in the last two decades in many Western countries, and consequently its environmental concentration there has decreased.

THE FREE ELEMENT

Lead, like most metals, is soft when very pure. Its inherent luster is usually masked by a dull surface coating of its oxide. Lead's melting point of 327°C, rather low for a metal, allows it to be readily worked and shaped into pipes, etc. Indeed, lead has been known and used for millennia: some artifacts date from 3,800 B.C. Lead was used as a structural metal in ancient times (e.g., to brace masonry and statuary), and for weatherproofing buildings. The Romans used it in water ducts and for cooking vessels. Lead is still used in the building industry for roofing and flashing and for soundproofing. When combined with tin, it forms solder, the low-melting alloy used in electronics and in other applications (e.g., "tin cans") to make connections between solid metals. Various types of solder differ in their composition, from 50% to 95% lead.

Lead is also found in ammunition ("lead shot") used by hunters of water fowl. Many ducks and geese are injured or die from chronic lead poisoning after their ingestion of lead shot, which dissolves inside them. When birds that prey on some ducks and other waterfowl that have been shot by hunters but not harvested by them, these predators (such as bald eagles) became victims of lead poisoning. For these reasons, lead shot has been banned in the United States; the same problems continue, however, in Canada because lead shot is still available there. In both countries, many loons die because they swallow and are subsequently poisoned by lead sinkers used in sport fishing.

IONIC 2+ LEAD

Lead does not generally become an environmental problem until it dissolves to yield the ionic form. This behavior contrasts with that of mercury, for which the vapor is of environmental concern. In contrast, the boiling point of lead is 1740°C, compared to only 357°C for mercury, so the vapor pressure of lead is very much smaller than that of mercury. The stable ion of lead is the 2+ species. Consequently, lead forms the ionic sulfide PbS, $Pb^{2+}S^{2-}$; this compound is the metal-bearing component of the highly insoluble ore ("galena") from which almost all lead is extracted.

Lead does *not* react on its own with dilute acids. Indeed, elemental lead is stable as an electrode in the **lead storage battery,** even though it is in contact with fairly concentrated sulfuric acid. However, some lead in the solder that was used commonly in the past to seal "tin cans" will dissolve in the dilute acid of fruit juices and other acidic foods if air is present, that is, once the can has been opened, since lead is oxidized by oxygen in acidic environments:

$$2\,Pb + O_2 + 4\,H^+ \longrightarrow 2\,Pb^{2+} + 2\,H_2O$$

The Pb^{2+} produced by this reaction contaminates the contents of the can; for this reason such solder is not usually used any more for such purposes in North America.

Similarly, some lead used in the solder in the joints of domestic copper water pipes, and lead used in previous decades and centuries to construct the pipes themselves, can dissolve in drinking water, particularly if the water is quite acidic or if it is particularly "soft." Thus it is a good idea not to drink the "first draw" water that has been standing overnight in older drinking fountains or in the pipes of older dwellings; water in such plumbing systems should be allowed to run for 60 seconds before being used for human consumption. On the other hand, hard water contains carbonate ion, CO_3^{2-}, which together with oxygen forms an insoluble layer containing compounds such as $PbCO_3$ on the surface of the lead; this layer prevents the metal underneath from dissolving in the water that passes over it. In some regions of England, phosphates are added to drinking water in order to form a similar protective coating on the inside of lead pipes and so reduce the concentration of dissolved lead; about 4% of tap water samples in the United Kingdom exceed the old World Health Organization standards of 50 ppb (see Table 9-2). Drinking water in North America generally has very low levels of dissolved lead, and usually is a minor source of the element for humans compared to the amounts received from air and most importantly from food. It is true, however, that lead in water is more fully absorbed by the body than is lead in food. Now that many other sources of lead have been phased out, drinking water accounts for about one fifth of the collective lead intake of Americans. Some American cities, especially those in the northeast that have soft water and networks of old lead pipes, have started to add phosphates to their water supplies in order to reduce lead levels. Many domestic water treatment systems successfully remove the great majority of lead from drinking water.

There have also been incidents of serious lead poisoning among people who consume "moonshine", illegally produced whiskey that is often processed at high temperatures in apparatus containing lead solder. Another former source of Pb^{2+} to the environment was the use of lead arsenate, $Pb_3(AsO_4)_2$, as a pesticide.

TABLE 9-2	DRINKING WATER STANDARDS* FOR HEAVY METALS		
Metal	U.S. Environmental Protection Agency (EPA)	Canada	World Health Organization (WHO)
As	50 ppb (2 ppb)**	50 ppb (25 ppb)**	50 ppb (10 ppb)**
Cd	5 ppb	5 ppb	5 ppb
Pb	20 ppb	10 ppb	50 ppb (10 ppb)**
Hg	2 ppb	1 ppb	1 ppb

*Values in $\mu g/L$ are numerically identical to those listed for ppb.
**Revised standards reducing concentrations to these lower levels were under consideration during the writing of this book, and may now be in place.

PROBLEM 9-7

According to an informal 1992 survey, the drinking water in about one-third of the homes in Chicago had lead levels of about 10 ppb. Assuming that an adult drinks about 2 liters of water a day, calculate the total lead that residents of these Chicago homes obtain daily from their drinking water.

One form of the oxide PbO is a yellow solid that has been used at least as far back in history as Ancient Egypt to glaze pottery. In glazing, the material is fused as a thin film to the surface of the pottery in order to make it waterproof and to give it a brilliant high gloss. However, if applied incorrectly, some of the oxide will dissolve over a period of hours and days in acidic foods and acidic liquids stored in pottery containers, giving dissolved Pb^{2+}, sometimes in the range of hundreds or even thousands of parts per million:

$$PbO(s) + 2\,H^+ \longrightarrow Pb^{2+} + H_2O$$
lead (II) oxide

Indeed, lead-glazed dishware is still a major source of dietary lead, especially, but not exclusively, in developing countries.

Various salts of lead have been used as pigments for millenia, since they give stable, brilliant colors. Lead chromate, $PbCrO_4$, is the yellow pigment used in paints applied to school buses and for striping on roads. "Red lead," Pb_3O_4, is used in corrosion-resistant paints and has a bright red color. Lead pigments have been used to produce the colors used in glossy magazines and food wrappers. In past centuries, lead salts were used as coloring agents in various foods. "White lead," which is the compound $Pb_3(CO_3)_2(OH)_2$, was extensively used until the middle of the twentieth century as a major component of white indoor paint. However, when the paint peels off, small children may eat the paint flecks since Pb^{2+} has a sweet taste. Persons who renovate walls of old homes should ensure that dust from layers of old paint is properly contained. Children in inner-city slums, in which old coats of paint continue to peel off walls, are often found to have elevated blood levels of lead. In indoor paint, white lead has now been replaced by the pigment titanium dioxide, TiO_2. Though now banned from use in indoor paints, lead pigments continue to be used in exterior paints, with the result that soil around houses may eventually become contaminated. Some of this lead-contaminated soil may be ingested by small children because of its sweet taste.

THE SOLUBILIZATION OF "INSOLUBLE" LEAD SALTS

The presence of significant concentrations of lead in natural waters is unusual, given that both its sulfide, PbS, and its carbonate, $PbCO_3$, are highly insoluble in water.

$$PbS(s) \rightleftharpoons Pb^{2+} + S^{2-} \quad K_{sp} = 8.4 \times 10^{-28}$$

lead sulfide   sulfide ion

$$PbCO_3(s) \rightleftharpoons Pb^{2+} + CO_3^{2-} \quad K_{sp} = 1.5 \times 10^{-13}$$

lead carbonate   carbonate ion

In both salts, however, the anions behave as fairly strong bases. Thus, both of the above dissolution reactions are followed by the reaction of the anions with water:

$$S^{2-} + H_2O \rightleftharpoons HS^- + OH^-$$

sulfide   bisulfide
ion   ion

$$CO_3^{2-} + H_2O \rightleftharpoons HCO_3^- + OH^-$$

Thus, for reasons identical to those found for the dissolution of calcium carbonate in water (Chapter 8), the solubilities of PbS and $PbCO_3$ in water are increased by the reaction of the anion with water (see Problem 9-8).

PROBLEM 9-8

By adding the reaction involving PbS(s) to that for S^{2-}, determine the overall reaction when PbS dissolves and most of the resulting sulfide ion reacts with water. Calculate the solubility of PbS with and without the subsequent reaction of sulfide, given that for HS^-, $K_a = 1.3 \times 10^{-13}$.

If highly acidic water comes into contact with minerals such as PbS, the "insoluble" solid dissolves to a much greater extent than in neutral waters. This occurs because the sulfide ion initially produced is subsequently converted almost entirely to bisulfide ion, which in turn is converted to dissolved hydrogen sulfide gas since both S^{2-} and HS^- act as bases in the presence of acid:

$$S^{2-} + H^+ \rightleftharpoons HS^- \qquad K = 1/K_a(HS^-) = 7.7 \times 10^{12}$$

$$HS^- + H^+ \rightleftharpoons H_2S(aq) \qquad K' = 1/K_a(H_2S) = 1.0 \times 10^7$$

When these two reactions are added to that for the dissolution of PbS into Pb^{2+} and S^{2-}, the net reaction is

$$PbS(s) + 2 H^+ \rightleftharpoons Pb^{2+} + H_2S(aq)$$

for which $K_{overall} = K_{sp}KK' = 6.5 \times 10^{-8}$. From the Law of Mass Action,

$$K_{overall} = [Pb^{2+}][H_2S]/[H^+]^2$$

Under conditions in which no significant amount of the hydrogen sulfide gas is vaporized but which are sufficiently acidic that almost all the sulfur exists as H_2S rather than as S^{2-} or HS^-, then the stoichiometry of the above reaction allows us to write that $[Pb^{2+}] = [H_2S]$. By substitution of this relationship into the above equation, we obtain

$$[Pb^{2+}]^2 = 6.5 \times 10^{-8} [H^+]^2$$

or

$$[Pb^{2+}] = 2.5 \times 10^{-4} [H^+]$$

Thus the solubility of PbS increases linearly with the H^+ concentration in acidic water: at pH = 4 the solubility of PbS and the concentration of Pb^{2+} ion in water is 2.5×10^{-8} M, whereas at pH = 2, the solubility is 2.5×10^{-6} M. Thus we conclude that dangerous concentrations of lead ion can occur in highly acidic bodies of water that are in contact with "insoluble" lead minerals.

PROBLEM 9-9

By calculations similar to those for PbS above, deduce the relationship between the solubility of mercuric sulfide, HgS ($K_{sp} = 3.0 \times 10^{-53}$) and the hydrogen ion concentration in acidic water. Is the solubility of HgS increased by exposure to acid?

IONIC 4+ LEAD

In highly oxidizing environments, lead can form the 4+ ion. Thus the oxide PbO_2, written in ionic form as $Pb^{4+}(O^{2-})_2$, exists, as do the "mixed" oxides Pb_2O_3 and Pb_3O_4, both of which are combinations of PbO and PbO_2. The latter mixed oxide is known as "red lead" and has been widely used as a paint pigment, especially as a coating for iron since it forms a surface layer which prevents rusting.

The elemental lead and the lead oxide PbO_2 employed as the two electrodes in storage batteries in almost all vehicles constitute the major use of the element. When the battery functions to supply electricity, especially to start the motor, some of the lead electrode is converted to Pb^{2+} in the form of the insoluble sulfate, $PbSO_4$:

$$Pb(s) + SO_4^{2-} \longrightarrow PbSO_4(s) + 2\,e^-$$
$$\text{lead (II) sulfate}$$

A lead ion in the oxide electrode acquires the two electrons, converting it from Pb^{4+} to Pb^{2+}, also in the form of the sulfate:

$$PbO_2(s) + SO_4^{2-} + 4\,H^+ + 2\,e^- \longrightarrow PbSO_4(s) + 2\,H_2O(aq)$$

Overall the net reaction that produces power corresponds to

$$Pb(s) + PbO_2(s) + 2\,H_2SO_4(aq) \longrightarrow 2\,PbSO_4(s) + 2\,H_2O(aq)$$
$$\text{lead (IV) oxide} \qquad\qquad\qquad \text{lead (II) sulfate}$$

This reaction is reversed, and the reactants are thereby regenerated, during the recharging process which occurs after the engine starts.

The majority, though not all, of used lead storage batteries are recycled for their lead content; during this operation, lead can be expelled into the environment if careful controls are not maintained. Indeed, such recycling operations often constitute urban "hot spots" of lead emission into the surrounding communities. Those storage batteries that are not recycled constitute the main source of lead in municipal waste.

TETRAVALENT ORGANIC LEAD

Whereas the compounds of its 2+ ion are ionic, most of lead's tetravalent compounds are covalent molecules rather than ionic compounds of Pb^{4+}. In this respect, tetravalent lead is similar to the corresponding form of the other elements (C, Si, Ge, Sn) in its group of the Periodic

Table. Commercially and environmentally, the most important covalent compounds of lead are those formed with the methyl group, CH_3, and the ethyl group, CH_2CH_3, namely tetramethyllead, $Pb(CH_3)_4$, and tetraethyllead, $Pb(C_2H_5)_4$. In the past, both compounds found widespread use as additives to gasoline to give **leaded gasoline;** this practice now has been phased out in North America and in parts of Europe. The function in automobile engines of the lead additives was to prevent premature combustion, that is, before the arrival of the flame front, of the air-gasoline mixture in that part of the cylinder furthest from the spark plug. Such premature combustion is in effect a detonation that causes engine "knocking" by imparting a sharp mechanical shock to the engine, and can eventually cause serious damage. Engine knocking is prevalent when the gasoline is composed predominantly of straight-chain (i.e., unbranched) alkanes, the isomers that are present in greatest abundance in petroleum. The tetraalkyl lead compounds PbR_4 operate by decomposing at engine temperatures into atomic lead and alkyl free radicals. A few grams of the lead tetraalkyl compound per gallon of gas is sufficient to prevent engine knocking.

The atoms of lead liberated by the combustion of the tetraalkyl compounds must be removed before they can form metallic deposits and damage the vehicle's engine. In order to convert the combustion products into volatile forms which can leave the engine in the exhaust gases, small quantities of ethylene dibromide and ethylene dichloride are also added to the leaded gasoline. As a result, the lead is removed from the engine and enters the atmosphere from the tailpipe in gaseous form as the mixed dihalide $PbBrCl$ and as $PbBr_2$ and $PbCl_2$. Subsequently, under the influence of sunlight, these compounds form PbO. The transformations of tetraalkyl lead to inorganic lead are summarized conveniently as follows:

$$PbR_4 \longrightarrow [Pb] \longrightarrow PbBrCl \longrightarrow PbO$$

The lead oxide exists in particulate form as an aerosol in the atmosphere for hours or days, and consequently not all of it is deposited in the immediate surroundings of the roadway; it can thus enter the food chain at more distant sites if it is deposited upon vegetables or on fields used by grazing animals

PROBLEM 9-10

The object of this problem is to estimate the mass of lead that would have been deposited annually on each square meter of land near a typical busy six-lane freeway from the lead compounds emitted by cars using the roadway. Use reasonable estimates for the number of cars passing a point each day, and their average mileage per liter or gallon of gasoline. Assume that the gasoline contained about one gram of lead per gallon, or 0.2 g per liter, and make the approximation that half the lead was evenly deposited within 1000 meters on each side of the freeway.

Since tetraalkyl lead compounds PbR_4 are volatile, they evaporate to some extent from gasoline and enter the environment in gaseous form. They are not water-soluble, but they are readily absorbed through the skin. In the human liver, PbR_4 molecules are converted into the more toxic PbR_3^+ ions, which, like methylmercury ions, are neurotoxins because they can cross the blood-brain barrier. In substantial doses, these organic compounds of lead cause symptoms which mimic psychosis; it is not clear what the effects may be, if any, of chronic low-level exposure. At very high exposures, tetraalkyllead compounds are fatal, as was discovered many years ago upon the deaths of several employees of the companies that originally produced these compounds.

LEAD IN THE ENVIRONMENT

In contrast to mercury, little or no methylation of inorganic lead occurs in nature. Thus almost all the tetraalkylated lead in the environment probably originates from gasoline.

A high proportion of environmental lead in many parts of the world comes from that emitted from vehicles, and occurs mainly in inorganic form. The conversion to nonleaded gasoline in North America and Europe, the initial impetus for which was the interference of lead in exhaust gases with the proper functioning of catalytic converters, has had the welcome side effect of decreasing the average amount of lead ingested by urban inhabitants of these regions; as illustrated in Figure 9-4, the phase-out has led to a substantial decline in the lead levels in the blood of these people. Indeed, the noted environmentalist Barry Commoner has called the elimination of lead from gasoline "one of the (few) environmental success stories." The use of a manganese compound to replace lead additives in gasoline is explored in Box 9-1.

In many countries of the world the use of leaded gasoline continues unabated. In these areas, air is the major source of lead ingested by humans, as it was in the past in North America and Europe. Some of the

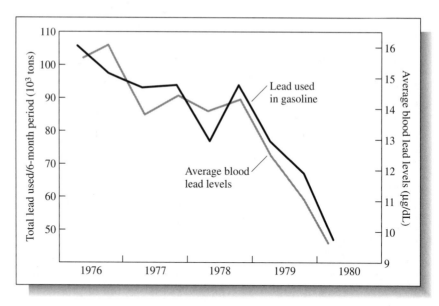

FIGURE 9-4
Annual variation in lead concentrations in human blood and lead usage in gasoline for selected U.S. cities. (Source: M. Lippmann. 1990. Lead and human health: Background and recent findings. *Environmental Research* 51:1–24.)

gasoline-based lead enters the body directly from inhaled air, and some enters indirectly from food into which lead has been incorporated; airborne lead oxide eventually settles on soil, water, fruits, or leafy vegetables, and can thereby enter the food chain since soluble lead is absorbed by plants. Microorganisms do bioconcentrate lead, but in contrast to mercury, lead does *not* undergo biomagnification in the food chain.

Most ingested lead in humans is initially present in the blood, but that amount eventually reaches a plateau and the excess next enters the soft tissues, including the organs, particularly the brain. Eventually the lead becomes deposited in bone, where it replaces calcium because Pb^{2+} and Ca^{2+} ions are similar in size. Indeed, lead absorption by the body increases in persons having a calcium deficiency and is much higher in children than in adults. The toxicity of lead is proportional to the amount present in the soft tissues, not to that in blood or bone. Lead remains in human bodies for several years, and thus it can accumulate in the body. The dissolving of bone, as occurs with old age or illness, results in the remobilization of bone-stored lead back into the bloodstream where it can produce toxic effects.

<table>
<tr><td>BOX
9-1</td><td># THE MANGANESE-BASED
REPLACEMENT FOR TETRAALKYL LEAD</td></tr>
</table>

The compound whose structure is shown below has the complicated name *methylcyclopentadienyl manganese tricarbonyl* and is usually called simply *MMT*. It has been adopted in some countries, including Canada but not the United States, as a replacement in gasoline for tetramethyllead and tetraethyllead to prevent engine knocking.

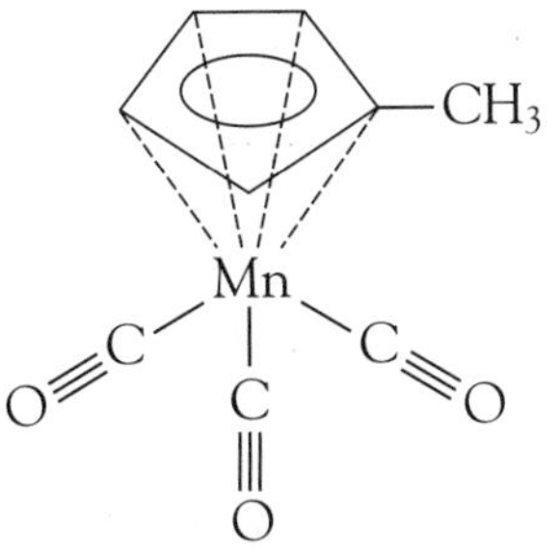

methylcyclopentadienyl manganese
tricarbonyl (MMT)

Like the lead compound, only a small amount of it (less than a tenth of a gram) needs to be added to a liter of gasoline to be effective. Upon combustion in gasoline engines, it is converted into oxides of manganese, principally Mn_3O_4, that become associated with suspended particulates in air.

The compound MMT itself is highly toxic, but the amount of additional manganese oxide its combustion produces does not significantly increase the total amount of manganese that we already take into our bodies from food, water, and air. Nevertheless, there has been opposition to the presence of this compound in gasoline since it does represent the use of a toxic material. Young children and iron-deficient adults are susceptible to increased ingestion of manganese. In the United States, the lead compounds in gasoline have not been replaced by another metal-organic compound. Instead, the proportion of aromatic hydrocarbons such as toluene, $C_6H_5CH_3$, and the xylenes, $C_6H_4(CH_3)_2$, in the gasoline mixture has been increased substantially. The presence of such benzene derivatives in appreciable amounts has the same advantageous mechanical effect as did the lead compounds, though at a higher cost of production. Recently, another application has been made to the Environmental Protection Agency to allow the use of MMT in the United States, since it has been used in Canada for 15 years without adverse health effects and without a measurable increase in manganese levels in urban soil.

At high levels, inorganic lead (Pb^{2+}) is a general metabolic poison. The effects of lead poisoning were known to the ancient Greeks, who realized that drinking acidic beverages from containers coated with lead-containing substances could result in illness. This information was not available to the Romans. Indeed, they sometimes deliberately adulterated overly acidic wine with sweet lead salts to improve the flavor. The concentration of lead in the bones of Romans is almost 100 times that found for modern North Americans. Some historians have hypothesized that chronic lead poisoning, from wine and other sources, of upper-class Romans contributed to the eventual downfall of the Roman Empire because of the metal's effects on the neurological and reproductive systems. The latter effects include dysfunctional sperm in males and an inability to bring the fetus to term in females. Due primarily to the contamination of beverages by lead from the distillation of alcohol in lead vessels, episodes of colic and gout due to lead poisoning were recorded through the Middle Ages and even until recent times.

Biochemically, lead interferes with the creation of hemoglobin by interfering with the enzymes involved in this process; at high levels, anemia can occur due to the lack of this oxygen-carrying component in blood. High lead levels also produce kidney dysfunction and cause permanent brain damage. Among adults, there is evidence that increased blood lead levels give rise to small increases in blood pressure.

The human groups most at risk from low levels of Pb^{2+} are fetuses and children under the age of about seven years; both are more sensitive to lead than are adults, a state of affairs due in part to the facts that they absorb a greater percentage of dietary lead and that their brains are growing rapidly. The metal readily crosses the placenta and thus is passed from mother to unborn child; because of the immaturity of the fetus's blood-brain barrier, there is little to prevent the entry of lead into its brain. Indeed, in the past women who worked in the lead industry suffered higher-than-average rates of miscarriages and stillbirths. In addition, lead is transferred postnatally from the mother in her breast milk, and/or from the tap water used to prepare formula for bottle-fed babies.

The principal risk to children from lead is interference with the normal development of their brains. A number of studies have found small but significant neuropsychological impairment in young children due to environmental lead absorbed either before or after birth. In particular, lead appears to have deleterious effects on children's behavior and attentiveness, and possibly also on their IQs. This is illustrated in Figure 9-5, where a mental development score is plotted as a function of age for groups of young children differentiated by the amount of lead in their umbilical cord at birth. A recent study of children in a lead-smelting community in Australia indicates that children with a blood lead level of 30 μg/100 g had an average IQ 4 to 5 points lower than those whose level was 10 μg/100 g. A survey in 1976–1980 of American children

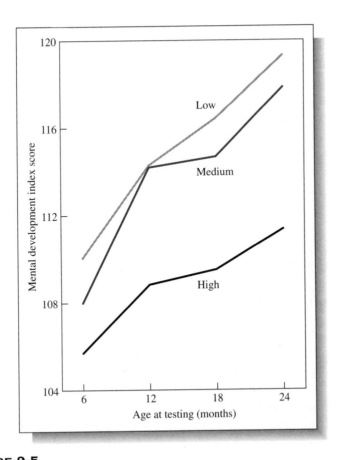

FIGURE 9-5

The effect of prenatal exposure to lead upon the mental development of infants. Lead exposure is measured by its concentration in the blood of the umbilical cord. "Low" corresponds to <3 μg/dL, "medium" to an average of 6.7 μg/dL, and "high" to > 10 μg/dL. (Source: Reprinted by permission of the *New England Journal of Medicine* 136 (1987): 1037–1043.)

aged from 6 months to 5 years found that about 4% of them had lead levels in excess of 30 μg per 100 g of blood, and that an additional 20% had levels over 20 μg/100 g; these concentrations represent two of the cutoffs that had been proposed in the past as "safe" levels. It now appears that there may be no level at which lead does not produce a deleterious effect in young and unborn children. In 1986, the Environmental Protection Agency concluded that "there is a national health problem associated with exposure to environmental lead for the general population and, in particular, pre-school children."

In summary, on an atom-for-atom basis, lead is not as dangerous as mercury. However, the general population is exposed to lead from a greater variety of sources and generally at higher levels than those associated with mercury. Overall, more people are adversely affected by lead, though on average to a lesser extent, than those fewer individuals exposed to mercury. Both metals are more toxic in the form of their organic compounds than as the simple inorganic cations. In terms of its environmental concentration, lead is much closer—within a factor of ten—to the level at which overt signs of poisoning become manifest than is any other substance, including mercury. Thus it is appropriate that society continues to take steps to further reduce human exposure to lead.

CADMIUM

THE FREE ELEMENT

Cadmium, Cd, lies in the same subgroup of the Periodic Table as zinc and mercury, but it is more similar to the former than the latter. Like zinc, the only common ion of cadmium is the 2+ species. In contrast to mercury, cadmium's compounds with simple anions such as chloride are ionic salts rather than molecules.

Most cadmium is produced as a byproduct of zinc smelting, since the two metals usually occur together. Some environmental contamination by cadmium often occurs in the areas surrounding zinc, lead, and copper smelters. As is the case for the other heavy metals, coal burning introduces cadmium into the environment. The disposal by incineration of waste materials which contain cadmium is also an important source of the metal in the environment.

In the past, the main use for elemental cadmium has been for the electroplating of other metals such as steel, since it is very resistant to corrosion and is especially useful when the plated object is to come into contact with sea water; this use for the metal is now in decline. Cadmium is also used as a stabilizer in PVC plastics (polyvinyl chloride; see page 198 of Chapter 5).

An increasingly important use of cadmium is as an electrode in rechargeable "nicad" (nickel-cadmium) batteries used in calculators, and similar devices. When current is drawn from the battery, the solid metal cadmium electrode partially dissolves to form insoluble cadmium hydroxide, $Cd(OH)_2$, by incorporating hydroxide ions from the medium into which it dips. When the battery is being recharged, the solid hydroxide, which was deposited on the metal electrode, is converted back to cadmium metal:

$$Cd(s) + 2\,OH^- \;\rightleftharpoons\; Cd(OH)_2(s) + 2\,e^-$$
$$\text{cadmium (II) hydroxide}$$

Each battery contains about 5 g of cadmium, much of which is volatilized and released into the environment if the spent batteries are incinerated as a component of garbage. The metallic cadmium tends to condense onto the smallest particles in the incinerator smokestream, which are precisely the ones that are difficult to capture by pollution control devices placed in the gas stack. In order not to release airborne cadmium into the environment upon combustion, some municipalities now require nicad batteries to be separated from other garbage. The recycling of metals from such batteries has also begun in some areas. However, some American states and European countries are taking steps to make the use of nicad batteries illegal due to the potential environmental contamination by cadmium; battery manufacturers hope to replace nicad batteries soon with those that do not contain cadmium.

ENVIRONMENTAL CADMIUM

In ionic form, the main use of cadmium is as a pigment. Because the color of cadmium sulfide depends very much on the size of the particles, cadmium pigments of many hues can be prepared. Both CdS and CdSe have been used extensively to color plastics. The latter salt is also used in photovoltaic devices (such as photoelectric cells) and in TV screens. Painters have used cadmium sulfide pigments in paints to produce brilliant yellow colors for 150 years and oppose any ban on them since at present there are no suitable replacements; van Gogh could not have painted his famous "Sunflowers" canvas without cadmium yellows, though it is speculated by some that cadmium poisoning may have contributed to the painter's anguished mental state.

Cadmium is released into the environment upon incineration of plastics and other materials which contain it as a pigment or as a stabilizer. Release to the atmosphere occurs as well when cadmium-plated steel is recycled, since the element is fairly volatile when heated. The analysis of low levels of cadmium and of other trace elements in environmental samples is described in Box 9-2.

BOX 9-2

THE ANALYSIS OF HEAVY METALS IN ENVIRONMENTAL SAMPLES

The concentration of metals, mercury for instance, in water, soil, and food samples, is usually so low that classical chemical analytical techniques that employ titration or precipitation are impractical for its determination. The most important analytical technique used to determine the total concentration of specific elements in such samples is **atomic absorption spectrometry.**

In atomic absorption spectrometry, samples to be analyzed are first vaporized by a flame operating at such high temperatures that all components of the sample dissociate to free, unbonded, electrically neutral atoms. A light beam is then shone through the vaporized sample, and the extent of its absorption by the gas is recorded. The wavelength of light in the beam is chosen to be precisely equal to one that is absorbed only by atoms of the element being determined; consequently, the fraction of light absorbed, rather than transmitted, by the sample is a direct measure of the concentration of that element. For example, since cadmium ions uniquely absorb light of wavelength 228.8 nm (among others)—no other gaseous element absorbs light having precisely this wavelength—the concentration of cadmium atoms in a mixture of many other substances can readily be determined by using this wavelength in the experiment.

Although Cd^{2+} is rather soluble in water unless sulfide ions also are present to precipitate the metal as CdS, humans usually receive only a small proportion of their cadmium directly from drinking water or from air. Exceptions arise for individuals who live near mines and smelters, particularly ones which process zinc. Indeed, there are isolated areas in which the cadmium concentration in drinking water is so high that the usual daily intake is doubled. Smokers are also exposed to cadmium that is absorbed from soil and irrigation water by tobacco leaves and released into the smokestream when a cigarette is burned; heavy smokers have approximately double the cadmium intake that is ingested from all other sources by nonsmokers.

For most of us, however, the greatest proportion of our exposure to cadmium comes from our food supply. Seafood and organ meats, particularly kidneys, have higher levels—100 μg or more per kilogram—than do most other foods. However, the majority of cadmium in the diet usually comes from potatoes, wheat, rice, and other grains, since most people consume so much more of them than seafood and kidneys. An exception is the Inuit people in Canada's Northwest Territories; a prize component of their diet is caribou kidneys, organs which are highly contaminated by cadmium that has reached the Arctic regions on the wind from industrial regions in Europe and North America.

Due to its similarity to zinc, plants absorb cadmium from irrigation water. The spreading on agricultural fields of phosphate fertilizers, which contain ionic cadmium as a natural contaminant, and of sewage sludge contaminated with cadmium from industrial releases, increases the cadmium level in soil and subsequently in plants grown in it. In the future, cadmium may be removed from phosphate fertilizers before it is sold to the consumer (see also page 387). Soil also receives cadmium from atmospheric deposition. Since cadmium uptake in plants increases with decreasing soil pH, one effect of acid rain is to increase cadmium levels in food.

Historically, all serious episodes of cadmium contamination have resulted from pollution from nonferrous mining and smelting. The most serious environmental problem involving cadmium occurred in the Jintsu River Valley region of Japan, where rice for local consumption was grown with the aid of irrigation water drawn from a river that was chronically contaminated with dissolved cadmium from a zinc mining and smelting operation upstream. Hundreds of people in this area, particularly older women who had borne many children and who had poor diets, contracted a degenerative bone disease called *itai-itai* or "ouch-ouch," so named because it causes severe pain in the joints. In this disease, some of the Ca^{2+} ions in the bones are apparently replaced by Cd^{2+} ions since they have the same charge and virtually the same size. The bones slowly become porous, and can subsequently fracture and collapse. The intake of cadmium by *itai-itai* sufferers was estimated at about 600 micrograms per day, which is about ten times the average ingestion of North Americans. Even rice grown in other areas of Japan is often contaminated with rather high cadmium levels. As a consequence, the average dietary intake of cadmium by residents of Japan is substantially greater than for peoples of other developed countries.

Cadmium is acutely toxic; the lethal dose is about one gram. Humans are protected against chronic exposure to low levels of cadmium by the sulfur-rich protein metallothionein, the usual function of which is the regulation of zinc metabolism. Because it has many sulfhydryl groups, metallothionein can complex almost all ingested Cd^{2+}, and the complex is subsequently eliminated in the urine. If the amount of cadmium absorbed by the body exceeds the capacity of metallothionein to complex it, the metal is stored in the liver and kidneys. Indeed, there is evidence that chronic exposure to cadmium eventually leads to an increased chance of acquiring kidney diseases. The average cadmium burden in humans is increasing, and in Japan the average daily amount of ingested cadmium is beginning to approach the maximum level recommended by health authorities, though this limit has a large built-in safety factor relative to levels at which health effects would occur. Although cadmium is not biomagnified, it is a cumulative poison since if not eliminated quickly (by metallothionein, as discussed above), its lifetime in the body is several decades. The areas of greatest risk are Japan and central Europe; in both regions the pollution of the soil by cadmium is particularly high due to contamination from industrial operations.

ARSENIC

Arsenic compounds such as the oxide As_2O_3 were the common poison used for murder and suicide in Roman times and through the Middle Ages: less than 0.1 gram of the element is required to kill a human being. Before modern times it was difficult to detect its presence in a dead body. Arsenic compounds also found widespread use as pesticides before the modern era of organic chemicals. Although its use in these applications has decreased, arsenic contamination remains an environmental problem in some areas of the world.

Arsenic is a semimetal, or metalloid: its properties lie between those of metals and those of nonmetals. It occurs in the same group of the Periodic Table as nitrogen and phosphorus, and so has an s^2p^3 electron configuration in its valence shell. Loss of all three p electrons gives the 3+ ion, whereas sharing of the three electrons gives trivalent arsenic; collectively these forms are designated As(III). Alternatively, loss of all five valence shell electrons gives the 5+ ion, and sharing them all gives pentavalent arsenic; collectively these forms are designated As(V). Overall, arsenic acts much as does phosphorus, though it has more of a tendency to form ionic rather than covalent bonds since it is more metal-like.

Environmental sources of arsenic stem from the continuing use of its compounds as pesticides, from its unintended release during the mining of gold and lead (in whose ores it commonly occurs), and from the combustion of coal, of which it is a contaminant. The leachate from abandoned gold mines of previous decades and centuries can still be a significant source of arsenic pollution in water systems. Due to the similarity in properties, arsenic compounds coexist with those of phosphorus in nature and so often contaminate phosphate deposits and commercial phosphates. The common arsenic-based pesticides are the insecticide lead arsenate, $Pb_3(AsO_4)_2$, and the herbicide calcium arsenate, both of which contain As(V); and the herbicide sodium arsenite, Na_3AsO_3, and Paris Green, $Cu_3(AsO_3)_2$, both of which contain As(III). Some of the methylated derivatives of arsenic acids (discussed below) are also used as herbicides. The environmental consequences of using another heavy metal, tin, in a pesticide are discussed in Box 9-3.

Drinking water, especially ground water, is a major source of arsenic for most people. There are small background levels of arsenic in many foods, and indeed a trace amount of this element apparently is essential to good human health. Tiny amounts of arsenic act as growth stimulants and are used to fatten pigs and poultry. Provided that use of the stimulant is stopped at least a few days before slaughter, arsenic levels in the resulting meat usually do not exceed the legal limit.

Arsenic is known to be carcinogenic in humans. Lung cancer results from the inhalation of arsenic and probably also from its ingestion. Skin and liver cancer, and perhaps cancers of the bladder and kidneys, arise from ingested arsenic. There is evidence that smoking and simultaneous exposure to environmental arsenic act *synergistically* in causing lung cancer, that is, their effect taken together is greater than the sum their individual effects would be if each acted independently. The human health risk, if any, from the background concentrations of arsenic encountered in the environment is not known. Its lethal effect when consumed in an acute dose is due to gastrointestinal damage, resulting in severe vomiting and diarrhea. Apparently As(III) is more toxic than is As(V), though the latter is reduced to the former in the human body. It is thought that the greater toxicity of As(III) is due to its ability to be retained in the body longer since it becomes bound to sulfhydryl groups.

The most common organic forms of arsenic are not simple methyl derivatives but are water-soluble acids that can be excreted, and thus are less toxic than some inorganic forms. In particular, in water, arsenic occurs most commonly as the As(V) acid H_3AsO_4 or one of its deprotonated forms. Like its phosphorus analog, the structure of the acid is $(OH)_3AsO$. Biological methylation in the environment by methylcobalamin initially involves the replacement of one or more OH groups by CH_3; monomethylation by the human liver and kidneys converts most but not all ingested inorganic arsenic to $(CH_3)(OH)_2AsO$, which is then readily excreted.

BOX 9-3   ORGANOTIN COMPOUNDS

Although inorganic compounds of tin (Sn) are relatively nontoxic, the bonding of one or more carbon chains to the metal results in substances that are toxic. Such organotin compounds have some common uses such as additives to stabilize PVC plastics and as fungicides to preserve wood, and therefore are of environmental concern.

Tin forms a series of compounds of general formula R_3SnX which are ionic substances $(R_3Sn^+)(X^-)$, where R is a hydrocarbon group and X is a monatomic anion; corresponding compounds such as $(R_3Sn)_2O$ occur when the anion is doubly charged. These compounds are toxic to mammals when R is a very short alkyl chain; maximum toxicity occurs when R is the ethyl group, C_2H_5, and decreases progressively with increasing chain length.

For fungi, the greatest toxic activity is attained when each hydrocarbon chain has four carbons in an unbranched chain, that is, when R is the n-butyl group, $-CH_2CH_2CH_2CH_3$, or simply $n-C_4H_9$. Tributyltin oxide, $(R_3Sn)_2O$ where R is $n-C_4H_9$, and the corresponding fluoride, have both been used as fungicides; commonly they are incorporated as "anti-fouling" agents in the paint applied to docks, to the hull of boats, to lobster pots, to fishing nets, etc. in order to prevent the accumulation of slimy marine organisms such as the larvae of barnacles. In recent years, tributyltin has been incorporated into polymeric coatings for boat hulls; a thin layer of the compound subsequently forms around the hull. The tin compounds replaced copper(I) oxide, Cu_2O, in such applications since their effectiveness lasts longer than a single season.

Unfortunately, some of the tributyltin compound leaches into the surface waters that are in contact with the coatings or paint, particularly in harbors where the boats are moored, and subsequently enters the food chain via the microorganisms that live near the surface. This can lead to sterility or death for fish and some types of oysters and clams that feed upon these microorganisms. Some countries have restricted the use of tributyltin compounds to large ships, but this still allows the pollutant to concentrate in marine coastal regions such as the North Sea. Scientists are worried that the presence of tributyltin compounds in these waters could affect fish reproduction. Higher organisms have the enzymatic ability to break down tributyltin fairly rapidly, and thus the compounds are not very toxic to humans.

PROBLEM 9-11

Although (for the sake of simplicity) many authors often state that the predominant form of arsenic in water is the arsenate ion, the fact is that since the AsO_4^{3-} ion is basic, the forms $HAsO_4^{2-}$,

$H_2AsO_4^-$, and H_3AsO_4 will all be present. Given that for H_3AsO_4, the successive acid dissociation constants are 6.3×10^{-3}, 1.3×10^{-7}, and 3.2×10^{-12}, deduce the predominant form of arsenic in waters of pH = 4, 6, 8 and 10.

Although most daily exposure of arsenic by North American adults is due to food intake, especially of meat and seafood, much of the arsenic present in these sources occurs in the organic form and is therefore readily excreted. In seafood, the common forms of arsenic are either the $(CH_3)_4As^+$ ion or that with one methyl group replaced by $-CH_2CH_2OH$ or $-CH_2COOH$; these forms are rather nontoxic to humans. In contrast, the neutral As(III) compounds such as AsH_3 and $As(CH_3)_3$ are the most toxic forms of arsenic. Curiously, the trimethyl compound is produced by the action under humid conditions of molds in wallpaper paste on arsenic-containing pigments in wallpaper. In the past a number of instances of human "death by wallpaper" occurred due to chronic exposure to the $As(CH_3)_3$ gas released into rooms by this mechanism.

The average inorganic arsenic content of drinking water is about 2.5 ppb; since the adult daily consumption of water is about 1.6 liters, the average intake of inorganic arsenic from drinking water is about 4 micrograms. In a study of residents of Taiwan who were exposed to high levels of the element in their well water, a relationship between arsenic exposure and skin cancer incidence has been established. Based upon linear extrapolations of cancer incidence from populations exposed to high levels of arsenic, and assuming that no threshold exists, some scientists have estimated that there is a 1-in-a-1,000 lifetime risk of dying from cancer induced by normal background levels of arsenic. This estimate places arsenic on an almost equal level with environmental tobacco smoke and radon exposure as an environmental carcinogen. Other scientists are not convinced that these estimates are realistic, since it is by no means certain that the extrapolation of the cancer incidence from high arsenic levels to the concentrations encountered in the environment is valid. In any case, the cancer risk would be most significant for the small proportion of the population who drink water having high arsenic levels. Most current governmental standards for arsenic in drinking water are 50 ppb, although there is currently consideration being given to reducing these values (see Table 9-2 for details). About 1% of Americans consume drinking water that has arsenic levels of 25 ppb or more; some water supplies in Utah and California have been found to contain up to 500 ppb.

HEAVY METALS IN SOILS, SEWAGE, AND SEDIMENTS

The ultimate sink for heavy metals, and for many toxic organic compounds as well, is deposition and burial in soils and sediments. Heavy metals often accumulate in the top layer of the soil, and are therefore accessible for uptake by the roots of crops. For these reasons it is important to know the nature of these systems and their mechanism of functioning.

SOIL CHEMISTRY

Most soils are composed mainly of small particles of weathered rock that are **silicate minerals.** These minerals consist of polymeric structures in which the fundamental unit is a silicon atom surrounded tetrahedrally by four oxygen atoms; the latter in turn are each bonded to another silicon, and so on, the resulting structure being an extended network. There are many variations on the silicate structural theme; some networks have exactly twice as many oxygens (formally O^{2-}) as silicons (formally Si^{4+}) and correspond to electrically neutral SiO_2 polymers. In others, the ratio of oxygen to silicon is greater than two to one, since the extra negative charge is neutralized by the presence of cations such as H^+, Na^+, K^+, Mg^{2+}, Ca^{2+}, Al^{3+}, Fe^{2+}, and others. Some common silicon-oxygen structural units are illustrated in Figure 9-6.

Over time, the weathering of the silicate minerals from rocks can involve chemical reactions such as the substitution by aluminum for some of the silicon atoms. Eventually, these reactions yield substances that are important examples of a class of soil materials known as **clay minerals.** A mineral having a particle size less than 2 μm is defined as a component of the clay fraction of soil. Small colloidal clay particles possess an outer layer of cations which are bound electrostatically to an electrically charged inner layer, as illustrated in Figure 9-7; depending upon the concentration of cations in the water surrounding the clay particle, the cations on the particle are capable of being exchanged for these others. For example, when exposed to water which is rich in potassium ions but poor in other ions, K^+ ions will displace those ions bound to the surface of the clay particle (see Figure 9-7). If instead the soil is acidic, the metal ions on the surface will be displaced by H^+ ions and the metal ions will then enter the aqueous phase.

In addition to minerals, the other important components of soil are organic matter, water, and air; the concentrations of each vary greatly

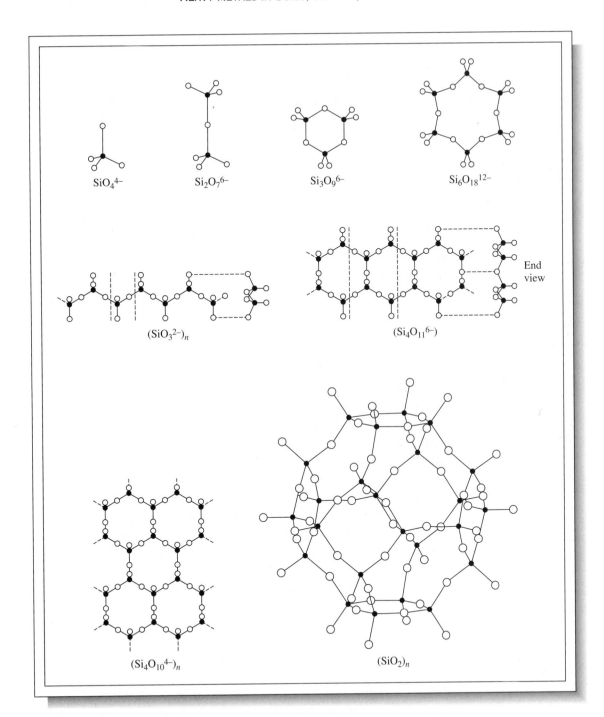

FIGURE 9-6

The common structural units in silicate minerals. (Source: Redrawn from R. W. Raiswell, P. Brimblecombe, D. L. Dent, and P. S. Liss. 1980. *Environmental Chemistry*. London: Edward Arnold Publishers.)

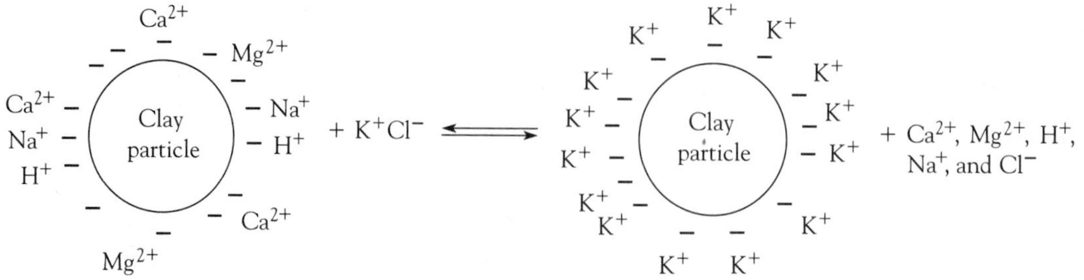

FIGURE 9-7

Ion-exchange equilibria on the surface of a clay particle. The addition of K^+ ions to the soil water displaces the exchange equilibrium to the right side, whereas removal of K^+ ions from solution displaces it to the left. (Source: Redrawn from R. W. Raiswell, P. Brimblecombe, D. L. Dent, and P. S. Liss. 1980. *Environmental Chemistry*. London: Edward Arnold Publishers.)

from one soil type to another. The organic matter, which gives soil its dark color, is primarily a material called **humus,** and is derived principally from photosynthetic plants, some components of which (such as cellulose and hemicellulose) have already been decomposed by organisms that live in the soil. The undecomposed plant material in humus is mainly protein and **lignin,** both of which are polymeric substances that are largely insoluble in water. A significant amount of the carbon in lignin exists in the form of six-membered aromatic benzene rings connected to each other by chains of carbon and oxygen atoms. As a result of the partial oxidation of some of the lignin, many of the resulting polymeric strands contain carboxylic acid groups, —COOH; this portion of humus is known as humic and fulvic acids, and is soluble in alkaline solutions due to the presence of the acid groups. Often these acid groups are complexed with clay minerals. Since some of the carbon in the original plant material is transformed to carbon dioxide and thus lost as a gas to the surroundings, humus is enriched in nitrogen relative to that in the original plants; its other main components are carbon, oxygen, and hydrogen.

BINDING OF HEAVY METALS TO SOILS AND SEDIMENTS

Humic materials have a great affinity for heavy metal cations, and extract them from the water that passes through them. The binding of metal cations occurs largely because of the formation of complexes with the metal ions by —COOH groups in the humic and fulvic acids. For example, in the case of fulvic acids the most important interactions probably involve a —COOH group and an —OH group on the adjacent carbon of a benzene ring in the polymeric structure

M = heavy metal

Humic acids normally yield water-insoluble complexes, whereas those of smaller fulvic acids are water-soluble.

Heavy metals are retained by soil in three ways: by adsorption onto the surfaces of mineral particles, by complexation by humic substances in organic particles, and by precipitation reactions. The precipitation processes for mercury ions and cadmium ions involve the formation of the insoluble sulfides HgS and CdS when the free ion in solution encounters sulfide ion, S^{2-}. Significant concentrations of aqueous sulfide ion occur near lake bottoms in summer months when the water is usually oxygen-depleted, as discussed in Chapter 7 (page 299). However, the concentration of mercury in the water of soils can exceed the limits set upon the Hg^{2+} content by the precipitating action of sulfide because some of the mercury will take the form of the moderately soluble molecular compound $Hg(OH)_2$.

In acidic soils, the concentration of Cd^{2+} can be substantial, since this ion adsorbs only weakly onto clays and other particulate materials. Above a pH of 7, however, Cd^{2+} precipitates as the sulfide, carbonate, or phosphate. Thus the liming of soil to increase its pH is an effective way of tying up cadmium ion and preventing its uptake by plants.

Like many other chemicals, heavy metals often are adsorbed onto the surfaces of particulates—especially organic ones—suspended in water rather than being simply dissolved in the water as free ions or as complexes with soluble biomolecules such as fulvic acids. For example, all the lead and two-thirds of the mercury, but only a tiny fraction of the cadmium, in Lake Huron is associated with the particulate phase. Of the Great Lakes, Erie and Ontario have the highest concentrations of heavy metals adsorbed onto suspended particulates. The particles eventually settle to the bottom of the lakes and become buried when other sediments accumulate on top of them. This "burial" represents an important sink for many water pollutants, and is a mechanism by which the water is cleansed. Before they are covered by later sediments, however, freshly deposited matter at the bottom of a body of water can recontaminate the water above it by desorption of the chemicals since adsorption and desorption establish an equilibrium. Furthermore, the adsorbed pollutants can enter the food web if the particles are consumed by bottom-growing organisms.

Although mercury in the form of Hg^{2+} is firmly bound to sediments and does not redissolve much into water, environmental problems have arisen in several bodies of water owing to the conversion of the metal into methylmercury and its subsequent release into the aquatic food chain. The overall cycling of mercury species between air, water, and sediments is illustrated in Figure 9-8.

As previously discussed, anaerobic bacteria methylate the mercuric ion to form $(CH_3)_2Hg$ and CH_3HgX, which then rapidly desorb from sediment particles and dissolve in water, and thereby enter the food web. Although the levels of methylmercury dissolved in water can be

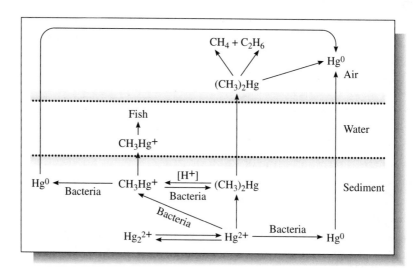

FIGURE 9-8

The biogeochemical cycle of bacterial methylation and demethylation of mercury in sediments. (Source: Redrawn with permission from AN ASSESSMENT OF MERCURY IN THE ENVIRONMENT. Copyright 1978 by the National Academy of Sciences. Courtesy of the National Academy Press, Washington, DC.)

extremely low, of the order of hundredths of parts per trillion, a biomagnification factor of 10^8 results in ppm range concentrations in the flesh of some fish. The devastating consequences of methylmercury poisoning have been already described (page 360).

Excavation and analysis of the sediments at the bottom of a body of water can yield a historical record of contamination by various substances. For example, the curves in Figure 9-9 show the levels of mercury and of lead in the sediments of the harbor in Halifax, Nova Scotia, as a function of depth and therefore also of year. For decades, raw sewage has been dumped into this harbor, and thus its sediments are a historical record of the levels of pollutants in sewage. Clearly, the metal pollution peaked about 1970 after having begun to increase dramatically about 1900. These trends are typical also of heavy metal levels in the Great

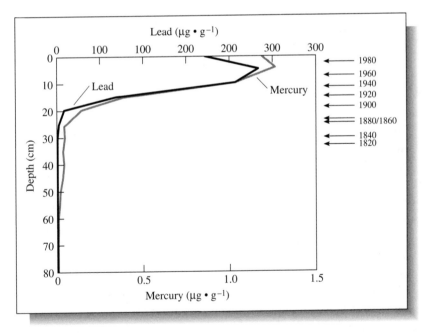

FIGURE 9-9

Lead and mercury concentrations in the sediments of Halifax Harbor versus depth (and therefore year). (Source: Redrawn from D. E. Buckley. 1992. Environmental geochemistry in Halifax Harbour. *WAT on Earth* Spring 1992:5–6.)

Lakes; the characteristic decreases of the past few decades are due to the imposition of pollution controls.

Heavy metal concentrations often are higher in sewage sludge than in soil: in built-up areas supplied with sewer systems, industrial wastes are sometimes released directly into sewage lines shared by households. For example, the lead level in municipal sludge can range from several hundred to several thousand parts per million, compared to an average of about 10 ppm in the Earth's crust. Some scientists have worried that food crops grown in soil fertilized by sewage sludge may incorporate some of the increased amounts of heavy metals. Control experiments indicate that vegetables vary greatly in the extent to which they will absorb increased amounts of the metals; for example the uptake of lead by lettuce is particularly large, but that by cucumbers is negligible. The concentration of arsenic in agricultural soils is greatly increased if arsenic pesticides are applied to them; crops planted on these soils subsequently absorb some of the adsorbed arsenic.

In summary, then, both soils and sediments act as vast sinks and reservoirs in the containment of heavy metals.

THE ANALYSIS AND REMEDIATION OF CONTAMINATED SEDIMENTS

We now realize that many river and lake sediments are highly contaminated by heavy metals and/or toxic organic compounds, and that such sediments act as sources to recontaminate the water that flows above them.

One way to determine the extent of contamination of a sediment is to analyze a sample of it for the total amounts of lead, mercury, and other heavy metals that are present. However, this technique fails to distinguish toxic materials that are either already dissolved in their actively toxic form or else potentially able to be resolubilized into the water, from those that are firmly bound to sediment particles and that could never be resolubilized. Thus a more meaningful test of the sediment involves extracting from a sample the substances that are soluble in water or in a weakly acidic solution, and analyzing the resulting liquid; thus, the permanently bound toxic agents can be left out of the account. Finally, the effect of sediments upon organisms that usually dwell in or on them can be determined by adding organisms to a sediment sample and seeing whether they survive and reproduce normally.

Several types of remediation have been used for sediments that were found to be highly contaminated. The simplest solution is often to simply cover the contaminated sediments with clean soil or sediment, thereby placing a barrier between the contaminants and the water system. In other instances, the contaminated sediments are dredged from the bottom of the water body to a depth below which the contaminant concentration is acceptable. If the sediments are high in organic content and inorganic nutrients, they are often used to enhance soil that is used for nonagricultural purposes. In some cases, the sediment can be used for cropland provided that its heavy metals and other contaminants will not enter the growing food. Cadmium is usually the heavy metal of greatest concern in such sediments; if the pH of the resulting soil is 6.5 or greater, most of the cadmium will not be soluble and so a higher total concentration often is tolerated.

Several chemical and biological methods of decontaminating sediments are in use. For example, treatment with calcium carbonate or lime increases the pH of the sediments and thereby immobilizes their heavy metals. In some situations, contaminated sediments are simply covered with chemically active solids, such as limestone (calcium carbonate), gypsum (calcium sulfate), iron (III) sulfate, or activated carbon that gradually detoxify the sediments. In other cases, the sediments are first dredged from the bottom of the water body and then treated. Heavy metals are often removed by acidifying the sediments or treating them with a chelating agent; in both cases the heavy metals become water-soluble and leach from the solid. For organic contaminants, extraction of toxic substances using solvents or destruction of them by either heat treatment of the solid or the introduction of microorganisms that consume them are the main options. The cleaned sediments can then be returned to the water body or spread on land. These techniques for removing metals and organics from sediments are also often useful on contaminated soils.

REVIEW QUESTIONS

1. What is a sulfhydryl group, and how does it become involved biochemically with heavy metals? How does the interaction affect processes in the body?

2. What principle underlies the usual cure for heavy metal poisoning?

3. Do heavy metals bioconcentrate? Do they biomagnify?

4. What equation relates the steady-state concentration of a heavy metal to its rate of ingestion and its half-life in an organism?

5. Is the liquid or the vapor of mercury more toxic? What are some important sources of airborne mercury?

6. What is an amalgam? Give two examples, and explain why they are used.

7. Explain how the chlor-alkali process leads to the release of mercury to the environment.

8. Name two uses for mercury in batteries.

9. Write the formulas for the methylmercury ion, for two of its common molecular forms, and for dimethylmercury. What is the principal source of exposure of humans to methylmercury?

10. Explain why mercury vapor and methylmercury compounds are much more toxic than other forms of the element.

11. What is meant by "Minamata disease"? Explain its symptoms and how it first arose.

12. List several uses for organic compounds of mercury. Which ones have been phased out?

13. What are the two common ionic forms of lead?

14. Explain how lead can dissolve—for example in canned fruit juice—even though it is insoluble in mineral acids.

15. Explain why lead contamination of drinking water by lead pipes is less common in hard water areas than in soft water areas.

16. Why are lead compounds used in paints? Why were mercury compounds used in paints?

17. Explain why heavy metal compounds such as PbS and $PbCO_3$ become much more soluble in acid water.

18. Write and explain the half-reactions by which a lead storage battery is charged and discharged.

19. What are the formulas and names of the two organic compounds of lead that were used as gasoline additives? What was their function? What was their fate upon combustion? In what form were they emitted from the tailpipe? Into what compound were they converted in air?

20. Discuss the toxicity of lead, especially with respect to its neurological effects. Which subgroups of the population are at particular risk from lead?

21. What are the main sources of cadmium in the environment?

22. Explain how nicad batteries operate. What other uses are made of cadmium?

23. What is the main source of cadmium to humans?

24. Describe what is meant by *itai-itai* disease, and relate where it arose and why.

25. What is metallothionein? What is its significance with respect to cadmium in the body?

26. What are some uses of arsenic that result in contamination of the environment?

27. What organic compounds of arsenic are of environmental significance? Why is arsenic in the form of $(CH_3)_4As^+$ not toxic to humans?

28. What are the main health concerns about arsenic in drinking water?

29. Describe the main inorganic constituents of soil.

30. What are the names and origins of the principal organic constituents of soil?

31. In what three ways are heavy metals retained by soil?

32. How can mercury stored in sediments be solubilized and enter the food chain?

33. Why is there concern about the use of sewage sludge as fertilizer?

34. How can sediments contaminated by heavy metals be remediated so they can be used or agricultural fields?

35. Describe two ways by which contaminated sediments can be treated without removing the sediments themselves.

SUGGESTIONS FOR

FURTHER READING

1. Kruus, P., M. Demmer, and K. McCaw, 1991. *Chemicals in the Environment*. Ch. VII. Morin Heights, Quebec: Polyscience Publications.

2. (a)Is there lead in your water? (b) Water-treatment devices. 1993. *Consumers Report* February: 73–78, and 79-82.

3. Ferguson, J.E., 1990. *The Heavy Elements: Environmental Impact and Health Effects*. Oxford: Pergamon Press.

4. Hutchinson, T. C., and K. M. Meema, eds. 1987. *Lead, Mercury, Cadmium and Arsenic in the Environment*. New York: John Wiley & Sons.

5. Spiro, T. G., and W. M. Stigliani. 1980. *Environmental Issues in Chemical Perspective*. Albany, NY: State University of New York Press.

6. Bunce, N. J. 1990. *Environmental Chemistry*. Winnipeg: Wuerz Publishing.

7. Harrison, R. M., ed. 1990. *Pollution: Causes, Effects and Control*. 2d ed. Cambridge: Royal Society of Chemistry.

8. Selinger, B. *Chemistry in the Marketplace*. 4th ed. Sydney, Australia: Harcourt, Brace, Jovanovich.

9. National Research Council of Canada Associate Committee on Scientific Criteria for Environmental Quality. 1979. Effects of mercury in the Canadian environment. NRCC.

10. Moore, J. W., and E. A. Moore. 1976. *Environmental Chemistry*. New York. Academic Press.

11. Kruss, P., and I. M. Valeriote, eds. 1979. *Controversial Chemicals: A Citizen's Guide*. Montreal: Multiscience Publications.

12. Krishnamurthy, S. 1992. Biomethylation and environmental transport of metals. *Journal of Chemical Education* 69(5):347–350.

13. Clarkson, T. W. 1992. Mercury: Major issues in environmental health. *Environmental Health Perspectives* 100: 31–38.

14. Tsuda, T. A., et al. 1992. Inorganic arsenic: A dangerous enigma for mankind. *Applied Organometallic Chemistry* 6:309–322.

15. Smith, A. H., et al. 1992. Cancer risks from arsenic in drinking water. *Environmental Health Perspectives* 97: 259–267.

16. Baghurst, P. A., et al. 1992. Environmental exposure to lead and children's intelligence at the age of seven years. *New England Journal of Medicine* 327(18): 1279–1284.

17. Goyer, R. A., 1993. Lead toxicity: Current concerns. *Environmental Health Perspectives* 100: 177–187.

18. Fitzgerald, W. F., and T. W. Clarkson. 1991. Mercury and monomethylmercury: Present and future concerns. *Environmental Health Perspectives* 96: 159–166.

19. Hoffmann, R. 1994. Winning gold. *American Scientist* 82: 15–17.

20. McBride, M. B. 1994. *Environmental Chemistry of Soils*. Oxford: Oxford University Press.

ENERGY PRODUCTION AND ITS ENVIRONMENTAL CONSEQUENCES

As discussed in preceding chapters, many environmental problems are the indirect result of producing and consuming energy, especially coal and gasoline. In the first part of this chapter, we explore the alternative fuels that are currently under development with respect to their production, use, and environmental consequences. During these discussions, it should be kept in mind that the world's reserves of coal are vastly greater than those of oil, natural gas and uranium combined; thus the use of coal to generate industrial energy will not only continue but will probably greatly increase, particularly in developing countries such as China and India that have vast supplies of it. Unfortunately, as we saw in Chapters 3 and 4, the burning of coal produces excessive amounts of sulfur dioxide and of carbon dioxide pollutants.

In the second part of this chapter, we explore nuclear energy, which is the production of useful energy, such as electricity, from the heat energy released when atomic nuclei themselves undergo change. Both current and future techniques for exploiting nuclear energy will be discussed.

LIQUID AND GASEOUS FUELS

We are familiar with the widespread use of gasoline (called "petrol" in many English-speaking countries outside North America) to power motor vehicles. We are also aware that supplies of the petroleum used to produce gasoline will eventually be depleted, and that most air pollution problems in cities stem from emissions from gasoline engines (Chapter 3). For these reasons, attention is turning to the development of alternative fuel sources that are cleaner and more abundant. In the material that follows, we discuss the nature and properties of gasoline and of the major contenders—chiefly alcohols, natural gas, and hydrogen—for "fuels of the future."

GASOLINE

Petroleum, or crude oil, is a complex mixture of thousands of compounds, most of which are hydrocarbons (see Chapter 5); the proportions of the various compounds vary from one oil field to another. The most abundant type of hydrocarbon is usually the alkane series which can be generically designated by the formula C_nH_{2n+2}. In petroleum the alkane molecules vary greatly, from the simple methane, CH_4 (i.e., $n = 1$), to molecules having almost one hundred carbons. Most of the alkane molecules found in crude oil are of two structural types: one type is simply a long, continuous chain of carbons; the other has one main chain and only short branches, for example, 3-methylhexane.

PROBLEM 10-1

Draw structural diagrams for n-hexane (C_6H_{14}) and for 3-methylhexane. Next, draw the structure of the isomer of the latter in which the longest continuous chain has four carbons.

Petroleum also contains cycloalkanes, mainly those with five or six carbons per ring, such as the C_6H_{12} systems methylcyclopentane and cyclohexane:

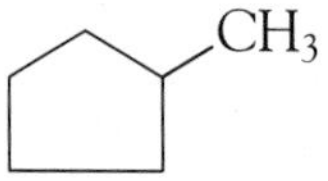

methylcyclopentane 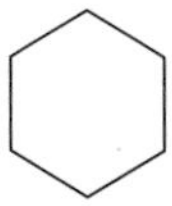   cyclohexane

Petroleum also contains some aromatic hydrocarbons such as benzene and its simple derivatives in which one or two hydrogen atoms have been substituted by methyl or ethyl groups. Recall from Chapter 3 that toluene is benzene with one hydrogen substituted by one methyl group, and the "xylenes" are the three isomers having two methyl groups. Collectively, the *benzene* + *toluene* + *xylene* component in gasoline is called BTX.

benzene   toluene   p-xylene

It is the BTX component of petroleum that is most toxic, chemically speaking, to shellfish and other fish when an oil spill occurs in an ocean, whether from an oil tanker or from an offshore oil well. The larger hydrocarbons form sticky, tarlike blobs that adhere to birds, sea mammals, and to rocks and other objects that the oil encounters.

In addition to hydrocarbons, petroleum also contains some sulfur compounds: hydrogen sulfide gas (H_2S), and organic sulfur compounds that are alcohol and ether analogs in which an S atom has replaced the oxygen. These substances are for the most part removed before the oil is sold for use. Small amounts of organic compounds containing oxygen or nitrogen also are present in crude oil.

Petroleum as it exits from the ground is useless for most applications. It must first undergo distillation, a process in which it is heated to about 400°C; at this temperature, its economically valuable constituents are vaporized. The components, or "fractions," of the vapor are then separated by condensing them from the hot vapor at different temperatures (see Figure 10-1). Some of the fractions that are recovered after distillation consist of molecules too large for any particular application; these are subjected to "cracking," a high-temperature process that reduces the size of the molecules by literally breaking them apart. Each fraction contains hydrocarbons with similar boiling points and therefore similar numbers of carbon atoms. The gases methane, ethane, and propane, which were dissolved in the oil, are collected and can be used as gaseous fuels or as "feedstock" for the petrochemical industry. The liquid

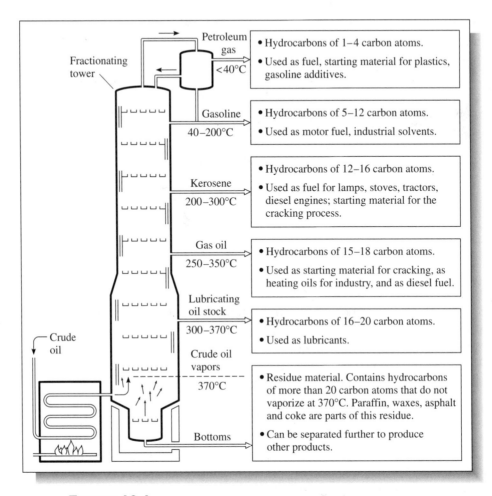

FIGURE 10-1

Petroleum distillation tower showing various fractions. (Source: Redrawn from *Chemistry in the Community* (*ChemCom*), 1st edition, copyright 1988 with permission of the American Chemical Society.)

components that have boiling points less than about 200°C are used as gasoline; this fraction consists mainly of hydrocarbons with 5 to 10 carbon atoms. Some of the higher-boiling components are altered chemically to reduce their molecular size before becoming components of gasoline. Regular gasoline contains predominantly C_7 and C_8 hydrocarbons. Common alkanes in gasoline are the colorless liquids *n*-pentane and 2-methylhexane, C_7H_{16}. Their structures and physical properties are shown below:

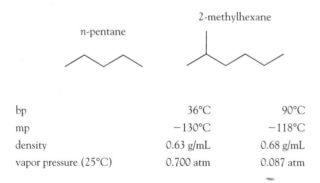

	n-pentane	2-methylhexane
bp	36°C	90°C
mp	−130°C	−118°C
density	0.63 g/mL	0.68 g/mL
vapor pressure (25°C)	0.700 atm	0.087 atm

Generally speaking, the more carbon atoms in the alkane, the higher its boiling point and the lower its vapor pressure—and hence the lesser its tendency to vaporize—at a given temperature. Thus, gasolines destined for warm summer conditions are formulated with smaller amounts of the smaller, more easily vaporized alkanes such as the butanes and pentanes, unlike those prepared for winters in cold climates. The presence of volatile hydrocarbons in gasoline is necessary in cold climates so that automobile engines can be started.

The next two higher-boiling fractions to be obtained from crude oil (Figure 10-1) are kerosene, used chiefly as aviation fuel, and gas oil; these fractions, spanning hydrocarbon molecules with about 12 to 18 carbon atoms, are also used as diesel fuel or as starting materials for the cracking process. The "residue" from crude oil, that is, that component which boils above 350°C, is used as lubricating fluid or is cracked so that it may be used as gasoline. This conversion is accomplished by subjecting the residue to chemical reactions that shorten the carbon chains, which originally contain 20 or more carbons. In general, the sulfur content (not including H_2S) of the fractions increases with boiling point, so that the diesel fuel fraction contains a higher percentage of sulfur than does gasoline, and the residue contains the highest concentration of sulfur of all. The residue also contains the highest levels of the metals vanadium and nickel, usually at levels of several parts per million. The sulfur that is present in fuels is usually converted during the fuel's combustion into sulfur dioxide, which is a serious pollutant if released into the air (Chapter 3).

Gasoline consisting primarily of straight-chain alkanes has poor combustion characteristics when burned in internal combustion engines. A mixture of air and vaporized gasoline of this type tends to ignite spontaneously in the engine's cylinder before it is completely compressed and ignited by the spark plug, so that the engine "knocks," with a resulting loss of power. Consequently, as noted in Chapter 9, all gasoline is formulated to contain substances which will prevent knocking.

In contrast to unbranched alkanes, highly branched ones such as the octane isomer 2,2,4-trimethylpentane, "isooctane" (illustrated below), have excellent burning characteristics. Unfortunately, highly branched alkanes do not occur naturally in significant amounts in crude oil. The ability of a gasoline to generate power without engine knocking is measured by its **octane number.** To define the scale, isooctane is given the octane number of 100, and *n*-heptane, which causes extremely severe knocking, is arbitrarily assigned a value of zero.

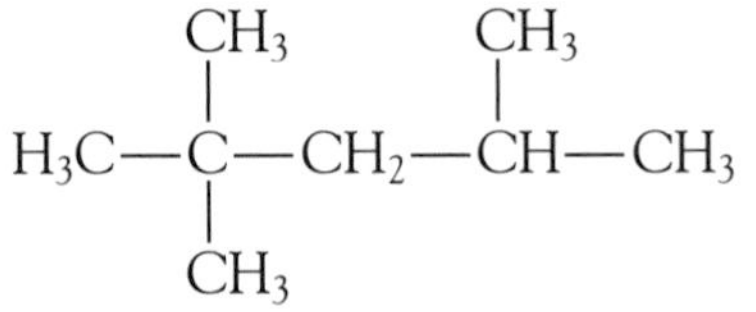

$$\underset{\text{2,2,4-trimethylpentane ("isooctane")}}{H_3C - \overset{\displaystyle CH_3}{\underset{\displaystyle CH_3}{\overset{|}{\underset{|}{C}}}} - CH_2 - \overset{\displaystyle CH_3}{\overset{|}{CH}} - CH_3}$$

When present in small amounts, the compounds tetramethyllead, $Pb(CH_3)_4$, and its ethyl equivalent prevent engine knocking and hence greatly boost the octane number of gasoline. For decades, they were added to all gasolines that consisted predominantly of unbranched alkanes. However, as mentioned in Chapter 9, these additives have by now been largely phased out in most developed countries in view of the pollution hazards they present.

In Canada and some European countries, lead compounds have been replaced by an organic compound of manganese (see Chapter 9). The alternative to using lead or manganese additives to boost octane ratings is to blend into gasoline significant quantities of organic substances which have high octane numbers, such as highly branched alkanes or MTBE (discussed later). A list of the common additives is shown in Table 10-1. Currently, most unleaded gasoline sold in the United States contains significant quantities of BTX or of ethanol to boost the octane number. Unfortunately, with respect to causing photochemical air pollution, the BTX hydrocarbons are more reactive than the alkanes that they replace, so that these additives have the unwelcome property of reducing lead pollution at the price of producing more smog. In addition, the use of BTX in unleaded gasoline in countries such as Great Britain, in which few cars are equipped with catalytic converters, has resulted in an increase in BTX concentrations in outdoor air. Benzene in particular is a worrisome air pollutant, since at higher levels it has been linked to increases in the incidence of leukemia.

The use of alcohols and compounds derived from them as gasoline additives or as "oxygenated" motor fuels in their own right is discussed in a later section.

NATURAL GAS AND OTHER ALKANES

In the developed world, the fuel called "natural gas" is used extensively. It consists principally of methane, but may contain small amounts of ethane and propane. Normally, the gas is transported by pipelines from its source to domestic consumers, who use it for cooking and heating, and to some utilities that burn it as an alternative to coal or oil in power plants to produce electricity:

$$CH_4\ (g) + 2\ O_2(g) \longrightarrow CO_2(g) + 2H_2O(l)$$

$$\Delta H = -890\ \text{kJ mol}^{-1}$$

$$= -55.6\ \text{kJ per gram of fuel}$$

Unfortunately, where gas pipelines do not exist, the natural gas that is produced as a byproduct of petroleum production in the field is simply wasted by venting it directly to the atmosphere or by "flaring" it off (i.e., allowing it to burn in the open air), with the result that the atmospheric burden of greenhouse gases is aggravated.

TABLE **10-1**	OCTANE NUMBERS OF COMMON GASOLINE ADDITIVES	
Compound	Octane number
Benzene	106
Toluene	118
p-Xylene	116
Methanol	116
Ethanol	112
MTBE	116

Highly **compressed natural gas** (CNG) is used to power some vehicles, especially in Canada, Italy, and New Zealand. Due to the cost of converting a gasoline engine to accept natural gas as a fuel, the current use of CNG in vehicles is largely restricted to taxis, commercial trucks, and other vehicles that are in constant service, since the additional capital cost of converting the fuel-system components to accept natural gas rather than gasoline is much less in the long run than the savings from the lower cost of the fuel. Because the compressed gas must be maintained at very high pressures (several thousand pounds per square inch) in order to allow it to be stored in reasonably small volumes, heavy fuel tanks with thick walls are required. Similar considerations apply to propane, C_3H_8 (also known as **liquefied petroleum gas,** or LPG), in its use as a gasoline replacement in vehicles. Currently in North America there are more propane-powered vehicles than those which use CNG. The heat energy produced per gram of propane burnt, 50.3 kilojoules, is not quite as high as that of 55.6 kJ for methane. The heat released by the burning of gasoline depends on the composition of the particular blend under consideration, but is generally slightly less than that for propane.

Compressed natural gas has both environmental advantages and disadvantages as a vehicular fuel when compared to gasoline. Since methane's molecules contain no carbon chains, neither organic particulates nor reactive hydrocarbons are formed and emitted into air as a result of its combustion. Thus, regional air quality is improved by the use of natural gas rather than gasoline or diesel oil. However, the release of methane gas from pipelines during its transport or from tailpipes of vehicles due to its incomplete combustion leads to increased global warming, since methane's atmospheric lifetime exceeds a decade and it is a potent greenhouse gas (as discussed in Chapter 4).

Some interesting proposals have been made recently to improve the performance of natural gas as a vehicular fuel. More efficient burning of the methane results if a small amount—about 15% by volume—of hydrogen gas is added to it. Alternatively, a smaller volume for storage of methane results if it is liquefied rather than simply compressed; however, additional energy is expended in the liquefaction process.

Although methane is useful as a fuel in its own right, more attention has been paid to its conversion to fuels which are liquids at atmospheric temperatures and pressures. Liquids have a greater "energy density" than gaseous fuels, in the sense of containing more combustion energy per unit of volume.

ALCOHOL FUELS: METHANOL AND ETHANOL

Methanol (CH_3OH) and ethanol (C_2H_5OH) are both colorless liquids which burn easily in air to produce heat. Many energy experts believe that alcohols will become increasingly important as vehicular fuels in the near future. Indeed, these alcohols have found niche domestic uses as fuels for many decades, and have recently found increasing acceptance as gasoline additives or gasoline replacements for vehicles. Alcohols have an advantage over hydrogen and natural gas since, as liquids at normal pressures and temperatures, they are "energy-dense" fuels. The physical properties of methanol and ethanol are compared in Table 10-2.

PROBLEM 10-2

Using the data in Table 10-2, calculate the heat released by methanol and by ethanol per milliliter. From your results, comment on the superiority of one alcohol or the other with respect to energy content on a weight and on a volume basis. Are these alcohols superior or inferior to methane as fuels in terms of energy content per gram (see text for additional data)?

Although small amounts of both alcohols can be produced by a variety of different chemical or biochemical reactions, for reasons of economics, the large-scale production of methanol for fuel purposes would have to start with a fossil fuel, either coal or natural gas. In contrast, ethanol for fuel is produced from plant material—in Brazil from sugarcane and in the United States and Canada from starchy grains such as corn.

TABLE 10-2   **SOME PROPERTIES OF METHANOL AND ETHANOL**

	Methanol	Ethanol
Melting point (°C)	−94	−117
Boiling point (°C)	65	78
Density (g/mL)	0.79	0.79
$\Delta H_{combustion}$ (kJ mol^{-1})	−726	−1367
$\Delta H_{combustion}$ (kJ g^{-1})	−22.7	−29.7

The conventional conversion of either coal (mostly carbon, C) or natural gas (mostly methane, CH_4) into methanol begins with the reaction of the fossil fuel with steam:

$$C(s) + H_2O(g) \longrightarrow CO(g) + H_2(g)$$

$$CH_4(g) + H_2O(g) \longrightarrow CO(g) + 3 H_2(g)$$

In the first process, steam is blown over white-hot coal, whereas in the second, methane gas is combined with steam that has been heated to about 1,000°C. In principle at least, an analogous mixture of hydrogen and carbon monoxide could also be obtained from heating renewable sources of biomass such as wood or garbage.

PROBLEM 10-3

Deduce the fraction of the CO (or H_2) produced by the reaction with steam of (a) coal, and (b) methane that must subsequently be converted to H_2 (or CO) by the "shift" reaction $CO + H_2O \longrightarrow CO_2 + H_2$ in order to obtain the 2:1 ratio of hydrogen to carbon monoxide required to produce methanol. Deduce also the net reactions for the conversion of coal and methane to methanol in both cases.

Methanol is synthesized from a 2:1 molar ratio of H_2 to CO by the following reaction, which occurs in the presence of a catalyst:

$$2 H_2 + CO \xrightarrow{\text{catalyst}} CH_3OH$$

Unfortunately, existing catalysts allow only a partial conversion of the gases into methanol, and the processes are energy-intensive and require high temperatures. Substantial research is currently underway to find a way to directly convert methane into methanol in a much more efficient manner. Most of the difficulty stems from the fact that methane is a very unreactive substance: its C—H bond dissociation energy is the highest of all the alkanes. Once one C—H bond is broken, however, the molecule becomes highly reactive and in the presence of oxygen tends to oxidize completely to carbon dioxide, rather than partially oxidize to give a useful intermediate product such as methanol. Some recently proposed schemes involve the catalyzed oxidation of CH_4 to CH_3^+. This process requires an oxidizing agent whose reduced form can be readily recycled back to its oxidized form using atmospheric oxygen. An example is sulfuric acid, whose reduced form is SO_2; the mechanism for the overall process is shown below:

$$CH_4 \longrightarrow CH_3{}^+ + H^+ + 2\,e^-$$

$$H_2SO_4 + 2\,H^+ + 2\,e^- \longrightarrow SO_2 + 2\,H_2O$$

followed by

$$SO_2 + 1/2\,O_2 + H_2O \longrightarrow H_2SO_4$$

The methyl cation, $CH_3{}^+$, bound in a compound such as methyl bisulfate, CH_3OSO_3H, is then reacted with water to produce methanol:

$$CH_3{}^+ + H_2O \longrightarrow CH_3OH + H^+$$

It is not yet clear whether this process will be cost-effective when carried out on a massive scale.

PROBLEM 10-4

By summing the above four reactions, deduce the overall reaction in the production of methanol from methane.

Methanol can also be produced by combining carbon dioxide and hydrogen gases in the presence of a suitable catalyst:

$$CO_2(g) + 3\,H_2(g) \longrightarrow CH_3OH(l) + H_2O(l)$$

Since this reaction is only slightly exothermic, most of the fuel energy originally associated with the hydrogen is still contained in the methanol product. Some of the massive quantities of CO_2 that are released annually into the atmosphere could be used as reactants in this process. Indeed, methanol produced in this manner could be considered a renewable fuel provided that the hydrogen is produced without consuming fuel (see below).

Ethanol, also called ethyl alcohol or grain alcohol, can be obtained on a massive scale by the enzyme-catalyzed fermentation of plant crops that consist predominantly of carbohydrates, whether starch, sugars, or cellulose. In the processing, starch or cellulose is first converted to sugars. Even waste paper has been proposed as a feedstock for the process.

As in the processes that produce beer and wine, the ethanol that results from this fermentation is a dilute solution whose major component is water. To convert this solution of ethanol into the almost-pure liquid that can be used as a fuel, the alcohol must be distilled off. If the heat for distillation is supplied from a fossil fuel, the cost of the energy required to produce the pure ethanol may be comparable to that recovered when the alcohol is sold as a fuel. This problem is circumvented, however, in some operations in which the heat needed to distill the ethanol is provided by waste biomass.

Currently, massive amounts of ethanol—over 10 billion liters annually—for vehicular fuel are produced from sugar cane in Brazil; unfortunately, massive water pollution results from disposing of the byproducts of this operation. Smaller quantities of ethanol are obtained from cane in Zimbabwe and from corn and grain in some midwestern American states and recently also in Canada. As of the mid-1990s, more than 3.5 billion liters of ethanol for use as fuel are being produced annually from the starch of surplus corn and grain in the Midwestern United States, and that amount is expected to increase substantially over the next few years. Many farmers in the United States and Canada have provided major political support for the production and use of ethanol in gasoline, and in particular for the government subsidies required to make it competitive with petroleum. A substantial increase in ethanol production in temperate climates could result if efficient methods were developed to convert cellulose, for example from hardwood trees, into sugar and then into ethanol in a fermentation process; research currently is under way to develop such a process.

As fuels for vehicles, alcohols can be used "neat", that is, in pure form, or as a component in a solution which includes gasoline. Often these fuels are referred to by either M (for methanol) or E (for ethanol) combined with a subscript which indicates the percentage of alcohol in the liquid; thus the pure alcohols are denoted M_{100} and E_{100}, respectively. Currently E_{100} is used only in Brazil, where about one-eighth of the cars have been designed with engines that permit pure ethanol to be used as fuel. About one-quarter of Brazil's vehicle fuel is ethanol. In North America, the "gasohol" currently sold is about 10% ethanol and 90% gasoline, that is, E_{10}.

One disadvantage to methanol blends is that the alcohol is only soluble to the extent of about 15% in gasoline, corresponding to M_{15};

greater amounts of methanol form a second layer rather than dissolve. The presence of water as a contaminant causes this unacceptable phase separation to occur with even smaller percentages of methanol. Additives such as *tert*-butyl alcohol (i.e., 2-methyl-2-propanol) that are soluble in both methanol and gasoline prevent such separations from occurring. Looking at things from the other direction, gasoline is moderately soluble in methanol, and so fuel blends such as M_{85} have been tested and are now on sale in limited quantities. "Cold starting" of engines in vehicles, especially in winter, using blends such as M_{85} and E_{85} is much less of a problem than with the pure alcohols, which have much lower vapor pressures than does gasoline.

One of the attractive features of "oxygenated" transportation fuels such as alcohols is that they result in lower emissions of many pollutants—specifically carbon monoxide, alkenes, aromatics, and particulates—compared to those generated by the use of pure gasoline or diesel fuel. To a large extent, however, the reduction in urban ozone that would result from the lowered emissions of CO and reactive hydrocarbons would be countered by increases due to the higher amounts of aldehydes and vaporized alcohols that would be emitted into the air. This is particularly true for urban areas in which ozone formation is NO_x-limited rather than hydrocarbon-limited (see p. 85 of Chapter 3). As a motor fuel, however, methanol gives rise to lower levels of NO_x emission than does ethanol, which in turn is less noxious in this respect than gasoline, presumably because the flame temperature—which governs the amount of nitric oxide originally formed—increases in the same order. Studies in Rio de Janeiro indicate that the concentration in air of the important pollutant PAN (see Chapter 3) has actually increased due to the use of ethanol fuel, but since Brazilian cars are not equipped with catalytic converters, this finding is not directly relevant to the North American situation. However there is recent evidence from measurements in Albuquerque, New Mexico that the use of ethanol as a gasoline additive has increased the concentrations of pollutants such as PAN in the air of that city.

PROBLEM 10-5

> Based on the principles of air chemistry developed in Chapter 3, and given that the O—H bond in methanol is stronger than the C—H bonds, predict the fate of CH_3OH molecules in an urban atmosphere.

PROBLEM 10-6

Neither methanol nor ethanol produce many hydrocarbon-based particulates during their combustion. Given their chemical structures, explain why this should be so, and therefore why the use of alcohol fuels would lead to a reduction in particulate pollution emissions from vehicles.

Some concern has been expressed about the safety of methanol for use as a vehicular fuel given the toxicity of the compound. Methanol-water solutions have been widely used as windshield-washer liquids in northern climates for many years without much environmental impact. However, the use of methanol as a fuel may be more dangerous, as it could involve a very high concentration of the alcohol. Ethanol, on the other hand, is much less toxic than either methanol or gasoline.

Another disadvantage of alcohols as fuels, and of methanol in particular, is that, liter for liter, they produce less energy than gasoline; to travel the same distance, fuel tanks for alcohols must be larger. To travel a given distance requires 25% more ethanol (on a volume basis) than gasoline. In addition, methanol cannot be used in conventional automobile engines because it reacts with and corrodes some engine and fuel tank components. However, alcohols also possess some advantages: they are inherently high-octane fuels, and indeed methanol is used to power all the cars at the Indy 500 races. Methanol has the added advantage that it does not produce a fireball when a tank-rupturing crash of racing cars occurs.

Methanol is also used to produce the oxygenated gasoline additive "MTBE", which stands for methyl *tert*-butyl ether, the structure of which is illustrated below:

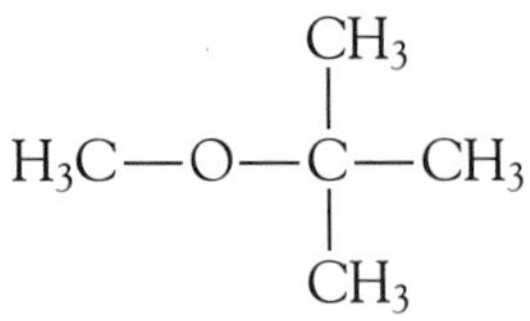

$$
\begin{array}{c}
CH_3 \\
| \\
H_3C-O-C-CH_3 \\
| \\
CH_3
\end{array}
$$

methyl *tert*-butyl ether ("MTBE")

Ethanol can be used to produce the corresponding ethyl *tert*-butyl ether, which also can be blended with gasoline. Both ethers have octane numbers (116 and 118, respectively) well in excess of 100. Currently MTBE is used in American and European unleaded gasoline blends not only to increase their overall octane number but also to reduce air pollution from carbon monoxide: like the alcohols, MTBE is an "oxygenated" fuel and generates less CO during its combustion than do hydrocarbons. Some gasolines for sale in American cities in the winter contain up to 15% MTBE, since the Clean Air Act of 1992 requires gasoline sold in winter months in 39 designated areas to contain at least 2.7% oxygen. The usage of MBTE is not without controversy, however. It has an objectionable odor, and residents of some cities like Fairbanks, Alaska, where it has been used in oxygenated fuels, complain that it actually makes them sick. In addition, like the alcohols, its combustion can also produce more aldehydes and other oxygen-containing air pollutants than result from hydrocarbon combustion. As a consequence of the use of MBTE in gasoline, it is now one of the top 25 industrial chemicals produced in North America. The advantages to using MBTE rather than ethanol to achieve the required oxygen level in gasoline are that it has a higher octane number and that it does not evaporate as readily.

PROBLEM 10-7

Calculate the mass of MTBE that is required in each kilogram of gasoline whose oxygen content is to be 2.7% by mass.

HYDROGEN—FUEL OF THE FUTURE?

Hydrogen gas can be used as a fuel in the same way as carbon-containing compounds, and some futurists believe that the world will eventually have a hydrogen-based economy. Hydrogen gas combines with oxygen gas to produce water, and in the process it releases a substantial quantity of energy in the form of heat:

$$H_2(g) + 1/2\ O_2(g) \xrightarrow{\text{spark}} H_2O(g) \quad \Delta H = -242\ \text{kJ mol}^{-1}$$

The idea that hydrogen would be the ultimate fuel of the future goes back at least as far as 1874, when it was mentioned by a character in the novel *Mysterious Island* by Jules Verne. Indeed, hydrogen has already found use as fuel in applications for which lightness is an important factor, namely in powering the Saturn rockets to the moon and the main engines of the U.S. Space Shuttles. Hydrogen gas can also be combined with oxygen to produce electricity in fuel cells (widely used in space vehicles) and to produce heat via low-temperature combustion in catalytic heaters. The main advantages of using hydrogen, as opposed to other fuels, are its low mass per unit of energy produced, and the smaller quantity of polluting gases produced by its combustion.

THE STORAGE OF HYDROGEN

In rocketry applications, hydrogen is stored as a liquid, as is oxygen. Since hydrogen's boiling point of only 20 K, or $-253°C$ at 1 atmosphere pressure, is so low, much energy must be expended in keeping it very cold even after energy has been employed to liquefy it. This drawback effectively limits the applications of liquid hydrogen to a few specialized situations in which its lightness is the most important factor.

Conceivably, hydrogen could be stored as a compressed gas, in much the same way as is currently done for methane in the form of natural gas. However, compared to CH_4, hydrogen has its drawbacks: a much greater amount of H_2 gas needs to be stored in order to release the same amount of energy. Compared with methane, the combustion of one mole of hydrogen consumes only one-quarter the oxygen, and consequently generates about one-quarter the energy (see Chapter 4), even though both occupy equal volumes under the same pressure (Ideal Gas Law). Thus the "bulky" nature of hydrogen gas limits its applications (see Problem 10-9). But hydrogen is superior to electricity in some ways, since its transmission over long distances does not consume large amounts of energy, and since no batteries are required to store the energy.

A more practical and safer way to store hydrogen for use in vehicles is in the form of a metal hydride. Many metals, including alloys, absorb hydrogen gas reversibly—as a sponge absorbs water—forming metal hydrides by incorporating the small atoms of hydrogen into "holes" in the crystalline structure. For example, titanium metal absorbs hydrogen to form the hydride TiH_2, a compound in which the density of hydrogen is four times that of liquid hydrogen! Heating the solid gradually releases the hydrogen as the molecular gas H_2 which then can be burned in air or oxygen to power the vehicle. Both Mercedes-Benz and Mazda have produced prototype cars which use hydrogen as fuel. In order to improve acceleration, which is sluggish on account of the low rate of hydrogen's release from the vehicle's magnesium hydride fuel source, the Mazda vehicle uses a battery for supplemental power. Research continues in the effort to find a light metal alloy which can efficiently store hydrogen without making the vehicle excessively heavy. Even existing metal-hydride systems, however, are lighter than are the pressurized tanks needed to store liquid hydrogen.

Calculate the mass of titanium metal required to absorb each kilogram of hydrogen and form TiH_2 in a "tankful" of hydrogen. Repeat the calculation for magnesium given that the hydride has the formula MgH_2. Which metal is superior for the storage of hydrogen from a weight standpoint?

Assuming that the energy released by the combustion of H_2 is proportional to the amount of oxygen it consumes in the same manner as discussed previously (Chapter 4) for carbon-based fuels, estimate the ratio of heat released by one mole of methane compared to one mole of hydrogen gas.

Using the thermochemical information in the equation above (page 409), calculate the enthalpy (heat) of combustion of hydrogen per gram, and by comparing it to that of methane (see page 401) decide which fuel is superior on a weight basis. By comparing the energy released by combustion per mole of gas—and hence per molar volume—decide which fuel is superior on a volume basis.

Although it is sometimes stated that hydrogen combustion produces only water vapor and no pollutants, this in fact is not true. Since combustion involves a flame, some of the nitrogen from the air that is used as the source of oxygen reacts to form nitrogen oxides, NO_x, although the lower flame temperature for the combustion of H_2, compared to that for fossil fuels, inherently produces less NO_x. Some hydrogen peroxide, H_2O_2, is released as well. Thus, hydrogen-burning vehicles are not really zero-emission systems. The nitrogen oxide release can be eliminated only by using pure oxygen rather than air to burn the hydrogen; alternatively, it can be minimized by passing the emission gases over a catalytic converter or by lowering the flame temperature as much as possible.

Even electric vehicles are not really pollution-free if a fossil fuel is burned to generate the electricity to charge the battery, since the fossil fuel's combustion in any power plant yields NO_x that is released into the atmosphere (see Chapter 3).

One of the practical difficulties in using hydrogen as a fuel is its tendency to react over time with the metal in the pipelines or storage containers in which it is present. This reaction embrittles the metal and it deteriorates, eventually forming a powder. Recent progress has been made in overcoming this difficulty by using composite materials rather than simple metals in the fabrication of storage and transport facilities for hydrogen.

THE PRODUCTION OF HYDROGEN

It is important to realize that hydrogen is not an energy source, since it does not occur as the free element on the Earth's crust. Hydrogen gas is an energy "vector" or "carrier" only; it must be produced, usually from water and/or methane as starting materials, by the expenditure of large amounts of energy derived from other fuels. The most expensive way to produce hydrogen is by electrolysis of water, using electricity generated by some energy source:

$$2 H_2O(l) \xrightarrow{\text{electricity}} 2 H_2(g) + O_2(g)$$

Unfortunately, about half the energy associated with the electricity is unavoidably converted to waste heat. An alternative is to use excess hydroelectric or nuclear power—such power sources typically generate more power than can be immediately used, and this excess cannot readily be stored as such for future use—to produce hydrogen.

The hope for the future is that either wind power or solar energy collected via photovolatic collectors (i.e., "solar cells" that convert light directly into electricity) will be an economically viable means of providing electricity to generate hydrogen. Even better would be the direct decomposition of water into hydrogen and oxygen by sunlight, though no practical, efficient method has yet been devised to effect this transformation. One of the difficulties in using sunlight to decompose water is that H_2O does not absorb light in the visible or UV-A regions; thus some substance must be found that can absorb sunlight, transfer the energy to the decomposition process, and finally be regenerated. The substances devised to date for this purpose are very inefficient in converting sunlight, and since they are not 100% recoverable in the cycle, they must be continuously resupplied.

One of the chief merits of solar power, whether it is employed to produce hydrogen or to produce electricity directly, is that it has a low environmental impact. It can also be part of a decentralized power system, so that the need for extensive transmission lines can be eliminated or minimized. In fact, prototype plants are already in operation in Germany and Saudi Arabia which use electricity from solar collectors to produce hydrogen; the energy stored in this form is later released by reacting the hydrogen with oxygen.

The disadvantages of solar power include its intermittent nature, its low density, and its high initial capital costs. However, the manufacturing costs of photovoltaic cells has fallen substantially in recent years. This is due to the fact that amorphous silicon rather than the costly crystalline form of the element is now being used as the cell material. Indeed, it now is cheaper to install solar cells to generate electricity than to extend power lines to some remote sites in North America.

PROBLEM 10-11

Determine the longest wavelength of light that has photons capable of decomposing liquid H_2O into H_2 and O_2 gases, given that ΔH for this process equals $+285.8$ kJ per mole of water. In which region of the spectrum does this wavelength lie?

Hydrogen gas can be produced by reacting a fossil fuel such as coal or natural gas with water so as to form hydrogen and carbon dioxide; the energy value of the fuel is transferred from carbon to the hydrogen atoms in water. The net reactions (assuming the coal in the first reaction to be graphite), are as follows:

$$C + 2\,H_2O \longrightarrow 2\,H_2 + CO_2$$

$$CH_4 + 2\,H_2O \longrightarrow 4\,H_2 + CO_2$$

Notice that as much carbon dioxide is produced in this way as would be obtained by combustion of the fossil fuels in oxygen. The actual conversions occur in two steps: first the fossil fuel reacts with steam to yield carbon monoxide and some hydrogen, as discussed on page 404. Then the

CO/H_2 mixture and additional steam are passed over a suitable catalyst to obtain additional hydrogen and complete the oxidation of the carbon by the "water gas shift reaction":

$$CO + H_2O \xrightarrow{\text{catalyst}} CO_2 + H_2$$

Finally, it should be mentioned that hydrogen is considered to be a dangerous fuel on account of its high flammability and explosiveness; furthermore, it ignites more easily than do most conventional fuels. On the positive side, however, spills of liquid hydrogen rapidly evaporate and the vapor rises high into the air.

CONCLUSIONS CONCERNING LIQUID AND GASEOUS FUELS

In summary, technology is currently being developed to produce and use hydrogen, natural gas, and alcohols as new vehicle fuels. Each of these fuels possesses some advantages and some disadvantages compared to gasoline; these are conveniently summarized for several fuels in Table 10-3. At present the main disadvantage to using these alternative fuels is their cost of production; hydrogen is particularly expensive.

NUCLEAR ENERGY

While most of the energy used by humans originates as the heat generated by the combustion of carbon-containing fuels, heat in commercial quantities can also be produced indirectly when certain processes involving atomic nuclei occur; this energy source is called **nuclear energy.** There are two processes by which energy is obtained from atomic nuclei: fission and fusion. In *fission*, the collision of a certain type of heavy nuclei (all of which have many neutrons and protons) with a neutron results in the splitting of the nucleus into two similarly sized fragments. Since together the fragments are more stable energetically than was the original heavy nucleus, energy is released by the process. The combination of two very light nuclei to form one combined nucleus is called *fusion* and also results in the release of huge amounts of energy, again since the combined nucleus is more stable than the lighter ones. Since nuclear forces are much greater than chemical bond forces, the energy released in nuclear reactions is immense compared to those obtained in combustion reactions.

TABLE 10-3	ALTERNATIVE FUELS' ENVIRONMENTAL, ECONOMIC, AND PERFORMANCE CHARACTERISTICS COMPARED WITH THOSE OF GASOLINE MADE FROM CRUDE OIL				
	Methanol	Ethanol	Compressed natural gas	Propane (LPa)	Electricity
Feedstock size/diversity	++	−	++	−	++
Environmental impacts	++	++	++	++	++
Vehicle cost	0	0	−	−	− −
Vehicle utility (range, luggage space)	0	0	− −	0	− −
Vehicle performance	0/+	0/+	− −	−	− −
Current fuel operating cost (low demand)	−	− −	0	0	0/+
Future fuel operating cost (high demand)	++	−	+	0	0/+
Refueling convenience (time, complexity)	0	0	− −	− −	− −

Codes: ++   much better than gasoline
    +   somewhat better than gasoline
    0   similar to gasoline
    −   somewhat worse than gasoline
    − −   much worse than gasoline

Reproduced from C. L. Gray, Jr. and J. A. Alson. 1989. The Case for Methanol. *Scientific American.* (November) 1989: 108–114.

FISSION REACTORS

The most economically useful example of fission is the sort that is induced by the collision of a $^{235}_{92}U$ nucleus with a neutron. The products of the decomposition of the unstable combination of these two particles are typically a nucleus of barium, $^{142}_{56}Ba$, one of krypton, $^{91}_{36}Kr$, and three neutrons:

$$^{1}_{0}n + ^{235}_{92}U \longrightarrow ^{142}_{56}Ba + ^{91}_{36}Kr + 3\,^{1}_{0}n$$

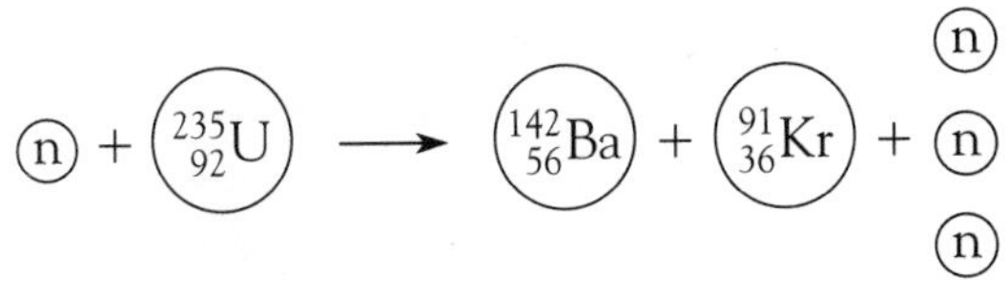

$$\boxed{n} + \left(^{235}_{92}U\right) \longrightarrow \left(^{142}_{56}Ba\right) + \left(^{91}_{36}Kr\right) + \boxed{n}\;\boxed{n}\;\boxed{n}$$

Not all the uranium nuclei that absorb a neutron form exactly the same products, but the process always produces two nuclei of about these sizes together with several neutrons. The two new nuclei are very fast-moving, as are the neutrons; it is the thermal, or heat, energy due to this excess kinetic energy that is used to produce electrical power. Indeed, the generation of electricity by nuclear energy and by the burning of fossil fuels both involve using the energy source to produce steam, which is used to turn large turbines that produce the electricity.

An average of about three neutrons are produced per ^{235}U nucleus that reacts; one of these neutrons can be used to produce the fission of another ^{235}U nucleus, and so on, yielding a **chain reaction.** In atomic bombs, the extra neutrons are used to induce a very rapid fission of all the uranium in a small volume, and so energy is released explosively, rather than gradually, as in a nuclear power reactor.

The only naturally occurring uranium isotope that can undergo fission is ^{235}U, which constitutes only 0.7% of the native element. The remaining uranium is ^{238}U (99.3% natural abundance). A neutron produced by the fission of a ^{235}U nucleus can subsequently be absorbed upon collision by a ^{238}U nucleus. The resulting ^{239}U nucleus is radioactive and emits a beta (β) particle (i.e., a fast-moving electron, as discussed in Chapter 3 and summarized in Table 10-4), as also does the heavy product (^{239}Np) of this process; consequently a nucleus of plutonium, ^{239}Pu, is produced:

$$\ce{^{1}_{0}n} + \ce{^{238}_{92}U} \longrightarrow \ce{^{239}_{92}U} \xrightarrow{-\beta} \ce{^{239}_{93}Np} \xrightarrow{-\beta} \ce{^{239}_{94}Pu}$$

$$\text{n} + \left(\ce{^{238}_{92}U}\right) \longrightarrow \left(\ce{^{239}_{92}U}\right) \longrightarrow \beta + \left(\ce{^{239}_{93}Np}\right) \longrightarrow \beta + \left(\ce{^{239}_{94}Pu}\right)$$

Thus, as a result of the operation of nuclear power reactors, ^{239}Pu is produced as a byproduct. The uranium in reactors is contained in a series of enclosed bars called fuel rods. When the uranium fuel in a rod is "spent," that is, when its ^{235}U content becomes too low for it to continue to be useful as a fuel, it is removed from the reactor. The plutonium-239 that has formed in the fuel rods is fissionable and can be removed and used as fuel for other reactors or to make atomic bombs; this "reprocessing" of fuel from commercial reactors is done in France but not in the United States or Canada. Some reactors called "breeders" are designed to maximize the production of plutonium and actually produce more fissionable material than they consume! Breeder reactors are in operation in France, Great Britain, and Russia.

TABLE 10-4	SUMMARY OF SMALL PARTICLES PRODUCED RADIOACTIVELY		
Particle symbol and name	Chemical symbol	Comment	Effect on nucleus by particle emission
α (alpha)	^4_2He	nucleus of a helium atom	atomic no. reduced by 2
β (beta)	$^0_{-1}\text{e}$	fast-moving electron	atomic no. increased by 1
γ (gamma)	none	high-energy photon	none

Unfortunately, both the pair of nuclei into which the ^{235}U nucleus fissions and the ^{239}Pu byproduct are highly radioactive substances; thus the spent fuel is much more radioactive than was the original uranium. Most of the common fission products of uranium emit both β and γ rays (see Table 10-4), as in this example:

$$^{142}_{56}\text{Ba} \longrightarrow \beta + ^{142}_{57}\text{La}$$

After the barium decays (half-life of 11 minutes), the product, a lanthanum isotope, subsequently also decays by β emission. Although many of the fission products decay rapidly by β emission, others have half-lives of many years. After ten years, most of the radioactivity in spent fuel rods (often stored under water in what look like swimming pools) is due to strontium-90, $^{90}_{38}\text{Sr}$ (half-life of 28 years), and to cesium-137, $^{137}_{55}\text{Cs}$ (half-life of 30 years). The dispersal of radioactive strontium and cesium into the environment would constitute a serious environmental problem, since they would be readily incorporated into the body. This is because strontium and cesium readily replace chemically similar elements which are normally integral parts of the bodies of animals including humans. For this reason, the radioactive waste from the spent fuel rods of nuclear power plants must be carefully monitored and will eventually have to be deposited in a secure environment from which it cannot escape. Scientists have as yet not developed any clearly foolproof methods for the long-term storage of such waste.

The plutonium-239 isotope that is produced during uranium fission is an α particle emitter (see Table 10-4) and has a long half-life: 24,000 years. After 1,000 years, the main sources of radioactivity from fuel rods will be plutonium and other such very heavy elements, since the medium-sized nuclei produced in fission, having much shorter half-lives than 1,000 years, will have decayed to a great extent by this time. Thus the long-term radioactivity of the spent fuel rods can be greatly reduced by chemically removing the very heavy elements from it, and using the plutonium as discussed above. Such reprocessing has the disadvantage that it converts the solid fuel rods into a radioactive aqueous solution from which the water must be extracted before disposal. Contamination may pose further difficulties: unfortunately, the safety record in handling radioactive materials at many reprocessing facilities has not been good, and some have been shut down since the environment around them has been heavily contaminated by radioactive isotopes (see Box 10-1 for more detail).

BOX 10-1 RADIOACTIVE CONTAMINATION BY PLUTONIUM PRODUCTION

Plutonium has been deliberately produced for more than fifty years in order to provide the readily fissionable material needed for nuclear weapons. In the United States, most plutonium production and processing was carried out at Hanford, Washington. Given the huge quantities of radioactive waste that were produced, stored, and disposed of at this facility, the surrounding environment is now so heavily polluted that it has been called the "dirtiest place on Earth." About 190,000 m^3 of highly radioactive solid waste and 760 million liters of moderately radioactive liquid waste and toxic chemicals were deposited in the ground at this site. As much as a metric ton of plutonium may be contained within the masses of solid wastes buried there. The cleanup of the wastes will cost between 50 and 200 billion dollars, and will not be completed earlier than the year 2020. One proposed—though expensive—technique for immobilizing the radioactive wastes involves passing a strong electrical current through the contaminated soil for a period of days; the electricity would fuse the soil and sand into a glassy rock from which the contaminants could not escape.

Plutonium was used as the explosive material in some atomic bombs themselves and as a "trigger" for hydrogen bombs, forcing the reactants together and thereby initiating the thermonuclear explosion. Consequently, about 100 metric tons of plutonium must be removed from nuclear weapons as tens of thousands of them are dismantled by the United States and Russia over the next few decades. Aside from questions of health and safety, a major problem associated with handling this material involves the enforcement of security measures intended to pre-

PROBLEM 10-12

Given that the half-life of ^{239}Pu is 24,000 years, how many years will it take for the level of radioactivity from plutonium in a sample of it to decrease to 1/128 (i.e., about 1%) of its original value?

Although nuclear energy has been used to generate electricity now for many decades, there is still no consensus among scientists and policy-makers concerning the best procedure for the long-term storage of the radioactive wastes from the plants. Proposals have included their burial in stable, dry geological formations such as salt domes once the radioactive substances have been encapsulated and immobilized in a glass or ceramic form, that is, in a "synthetic rock."

vent its slipping into the hands of terrorists and governments who want to fashion their own bombs. One disposal option is to use the plutonium, mixed with uranium, as fuel in power plants. Another expedient is to mix it with more highly radioactive wastes and seal the mixture as glass rods that would be stored underground. Alternatively, the plutonium could be buried in deep holes in the ground, though there must be the assurance that the plutonium would never leach into ground water.

The disposal of "weapons-grade" uranium is somewhat easier than of plutonium. Uranium that is highly enriched in ^{235}U can be mixed with an excess of ^{238}U, making it no more accessible or valuable than uranium ore, since it would require a major effort to separate the heavy isotopes from each other. In contrast, there is no naturally occurring plutonium with which the fissionable ^{239}Pu can be diluted. The separation of plutonium from other elements, even highly radioactive ones, can be accomplished chemically by exploitation of the differing solubilities of the elements, and so is much less energy-intensive than is isotope separation. Only four or five kilograms of weapons-grade plutonium (93% ^{239}Pu) is required to produce an atomic bomb; about seven kilograms of reactor-grade plutonium, which after reprocessing still contains more contaminant isotopes and elements than the weapons grade, is needed for a bomb. Given that the world's current stockpile of plutonium exceeds 1000 metric tons, a huge number of atomic bombs could be fashioned by terrorists or rogue governments if even a tiny fraction of this material fell into their hands.

During the mining of uranium, contamination of the environment by radioactive substances commonly occurs. Since naturally occurring uranium slowly decays into other substances that also are radioactive (recall the sequence involving radon in Chapter 3), uranium ore contains a variety of radioactive elements. Consequently the large volume of waste material that remains after the uranium is chemically extracted from the ore is itself radioactive. Once the radioactive waste is released from the original rock ore, which essentially immobilized it, it occurs as a liquid and as a powder, both called "tailings." The liquid tailings are normally held in special ponds until the solids separate, but pollution of the local ground water can occur if these ponds leak or overflow. In addition, the solid tailings, exposed to the weather, are partially dissolved by rainfall and can then contaminate local water supplies. The use of the solid tailings as landfill on which buildings are constructed can also lead to problems, since the radon produced by the radioactive decay of the radium in the tailings is quite mobile (see Chapter 3). Radon is a particular hazard for uranium ore miners, since the radioactive gas is always present in the ore and is released into the air of the mine. Indeed, the incidence of lung cancer among such miners was particularly high until mine ventilation was improved to permit more frequent changes in air and thus a more efficient clearing out of accumulated radon gas.

In most but not all nuclear power reactors (the Canadian CANDU system being the main exception), the uranium fuel must be "enriched" in the fissionable ^{235}U isotope; that is, the abundance of this isotope must be increased to 3.0%, compared to 0.7% in the naturally occurring element. The extent of enrichment required for use of uranium in bombs is much greater still; uranium sufficiently enriched for this purpose is called "weapons-grade" material and is 90% or more ^{235}U. Enrichment is a very expensive, energy-intensive process since it requires physical rather than chemical means of separation, given the fact that all isotopes of a given element behave identically chemically.

FUSION REACTORS

The optimum energetic stability per nuclear particle (protons and neutrons) occurs for nuclei of intermediate size, such as iron. It is for that reason that the fission of a heavy nucleus into two fragments of intermediate size releases energy. Similarly, the fusion of two very light nuclei to produce a single one also releases substantial quantities of energy.

Indeed, fusion reactions are the sources of the energy in stars, including our own Sun, and in "hydrogen bombs."

Unfortunately, fusion reactions all have huge activation energies; thus it is difficult to initiate and sustain a controlled fusion reaction such that it provides more energy than it consumes. The fusion reactions that have the greatest potential as producers of useful commercial energy involve the nuclei of the heavier isotopes of hydrogen, namely deuterium, ^{2}H, and tritium, ^{3}H. A sample reaction follows; note that two different sets of products can be produced.

$$\,_1^2H + \,_1^2H \longrightarrow \,_2^3He + \,_0^1n \quad \text{or} \quad \,_1^3H + \,_1^1H$$

In other terms,

$$D + D \longrightarrow He + n \quad \text{or} \quad T + H$$

deuterium + deuterium $\longrightarrow$ helium + a neutron, or tritium and regular hydrogen

There is an abundant supply of deuterium available, since it is a nonradioactive, naturally occurring isotope (constituting 0.015% of hydrogen) and thus is a natural component of all water. A somewhat lower activation energy is required for the reaction of deuterium with tritium:

$$\,_1^2H + \,_1^3H \longrightarrow \,_2^4He + \,_0^1n$$

However, because tritium is a radioactive element (a β emitter) with a short half-life (12 days), it is not a significant component of naturally occurring hydrogen and would have to be synthesized by the fission of the relatively scarce element lithium.

The environmental consequences of generating electrical power from fusion reactors should be less serious than those associated with fission systems. The only radioactive waste produced in quantity would be tritium. Although the β particle that it emits is not sufficiently energetic to penetrate the outer layer of human skin, tritium is dangerous since biological systems incorporate it as readily as they do normal hydrogen (^{1}H or ^{2}H) as a result of inhalation, absorption through the skin, or ingestion of water or food. Currently, tritium in drinking water (some of which results from artificial sources) constitutes the source of about 3% of our exposure to radioactivity.

REVIEW QUESTIONS

1. What are the three main classes of hydrocarbons present in crude oil? Draw the structural diagram for a representative member of each class.

2. What is meant by a "fraction" of petroleum? Name two of the fractions, and indicate the range of carbon atoms found in the molecules of each fraction.

3. What types of sulfur compounds are present in petroleum? How are the nongaseous sulfur compounds apportioned between fractions?

4. What is meant by engine "knocking"?

5. How is the octane number rating scale for fuels defined? Name a few fuels that have octane numbers greater than 100.

6. List several ways in which the octane number of fuels are increased by the addition of other compounds to straight-chain alkane mixtures.

7. What is the main component of "natural gas"? Write out the balanced chemical equation illustrating its combustion.

8. What are the advantages and disadvantages to fueling vehicles with CNG?

9. Describe the methods by which methanol can be produced in volume for use as a fuel. What does the notation M_{85} mean?

10. Describe the method used in producing ethanol in volume for use as a fuel. What is the very energy-intensive step involved? Describe the composition of E_{10} fuel.

11. What are the advantages and disadvantages of using alcohol fuels in regard to air pollution?

12. Describe the three ways in which hydrogen can be stored in vehicles for use as a fuel, and discuss briefly the disadvantages of each method.

13. Does the burning of hydrogen really produce no pollutants? Under what conditions do no pollutants form?

14. What is the difference between an energy source and an energy carrier (vector)? Into which category does H_2 fall?

15. Describe the production of hydrogen gas by electrolysis. Can solar energy be used for this purpose? Why isn't water decomposed directly by absorption of sunlight?

16. Define fission, and write the reaction in which a ^{235}U nucleus is fissioned into typical products.

17. Write the nuclear reaction that produces plutonium from ^{238}U in a fission reactor.

18. What is a "breeder" reactor? Why is it useful to breed fissionable fuel? What is meant by *reprocessing*?

19. Why are the spent fuel rods from fission reactors more radioactive than the initial fuel? How long does their radioactivity last?

20. Describe why the mining of uranium ore often pollutes the local environment.

21. Define fusion, and give two examples of fusion processes (i.e., reactions) that may be used in power reactors of the future.

22. Describe the nature of any radioactive byproducts produced by the operation of fusion reactors.

SUGGESTIONS FOR

FURTHER
READING

1. Emsley, J. 1994. Energy and Fuels. *New Scientist,* Jan. 15, 1994 (suppl.): 1–4.

2. Lynd, L. R., et al. 1991. Fuel Ethanol from Cellulosic Biomass. *Science* 251:1318–1323.

3. Gray, C. L., Jr. and J. A. Alson. 1989. The Case for Methanol. *Scientific American.* November, 1989: 108–114.

4. Miller, G. T., Jr. 1990. *Living in the Environment,* 6th ed. Belmont, CA: Wadsworth Publishing Company.

5. Kruus, P., M. Demmer, and K. McCaw. 1991. *Chemicals in the Environment.* Morin Heights, Quebec: Polyscience Publications. The authors elaborate upon nuclear energy and uranium mining.

INTERCHAPTER:
CLEAN-UP OF VALDEZ OIL SPILL

AN INTERVIEW WITH GREGORY DOUGLAS

Gregory Douglas received his Ph.D. from the University of Rhode Island Graduate School of Oceanography in 1986. In 1989, as an employee of the Batelle Corporation, he was called to work on the clean-up of the Exxon Valdez oil spill in Prince William Sound the day after the spill occurred. He is now the Unit Manager for Environmental Monitoring and Analysis for Arthur D. Little, where he is an expert in hydrocarbon fingerprinting.

How does PAH analysis allow one to determine the origin of oil samples found in the marine environment of Prince William Sound?

Oil associated with the sediment particles can be detected in the deep subtidal sediments of Prince William Sound. Hydrocarbon fingerprint analysis of polycyclic aromatic hydrocarbons (PAH) is one way to evaluate the chemical signature of this oil. The relative concentrations of some PAH compounds are source-specific and can be used to separate different oil signatures. For example, C3-phenanthrenes (C3-P) and C3-dibenzothiophenes (C3-D) are degraded at similar rates in the environment. Therefore, as the oil is degraded by evaporation, solubilization, and biodegradation, the characteristic source ratio is maintained. For example, the C3-D/C3-P source ratio for regional crude oils and the deep subtidal sediments in Prince William Sound is 0.16 ± 0.07 and the C3-D/C3-P source ratio for Exxon Valdez crude is 1.19 ± 0.08.* This is strong forensic chemical evidence that there has not been any measurable contamination of the deep subtidal sediments with Exxon Valdez crude oil. In addition to the PAH data, a second class of diagnostic hydrocarbons called biomarkers (for example, triterpenes) was measured in the sediment and used to confirm the source of the hydrocarbons.

What did you discover to be the sources of hydrocarbons in the Prince William Sound/Gulf of Alaska area?

Based on the hydrocarbon fingerprinting results, we discovered that, in addition to the Exxon Valdez oil, there were many natural and anthropogenic sources of hydrocarbons in the Prince William Sound subtidal sediments.

*Page, D. S., P. D. Boehm, G. S. Douglas, and A. E. Bence. 1993. Identification of hydrocarbon sources in the benthic sediments of PWS and the Gulf of Alaska following the Exxon Valdez oil spill. In: *Third Symposium on Environmental Toxicology and Risk Assessment: Aquatic, plant and terrestrial.* ASTM STP (D) 1. American Society for Testing and Materials.

These included diesel fuel, diesel soot, creosote, combustion related PAHs from forest and camp fires (for example, benzo(a)pyrene), naturally produced PAHs (for example, perylene), and oil seeps.

Were most of the oil samples on the sea floor from the Exxon Valdez spill? If not, what was the origin of the samples and where did the Exxon oil go?

The chemical fingerprinting of evidence combined with the Landsat MSS image from the Gulf of Alaska shoreline showing the extensive particle transport into Prince William Sound, and the general ocean circulation pattern in the Gulf of Alaska provide compelling evidence that suspended sediments from seep areas in the Gulf of Alaska are the vehicle for transport and sedimentation of seep hydrocarbons in Prince William Sound.

Oil present on high-energy shorelines (exposed beaches) is generally removed during winter storm events and dispersed at sea. Some Exxon Valdez tanker oil in low energy environments (enclosed bays and estuaries, embayments) will persist for longer periods until it is dispersed, buried, or biodegraded. Most of the sea floor in Prince William Sound has no detectable oil from the Exxon Valdez oil spill.

In what other ways can we apply hydrocarbon fingerprint analysis to the study of environmental problems?

These chemical fingerprinting techniques are being used to provide data to hydrogeologists and environmental chemists to evaluate the source, transport, and fate of spilled petroleum products in soil and groundwater systems to allocate responsibility in multiple spill situations.

These included diesel fuel, diesel soot, creosote, combustion related PAHs from forest and camp fires (for example, benzo(a)pyrene), naturally produced PAHs (for example, perylene), and oil seeps.

Were most of the oil samples on the sea floor from the Exxon Valdez spill? If not, what was the origin of the samples and where did the Exxon oil go?

The chemical fingerprinting of evidence combined with the Landsat MSS image from the Gulf of Alaska shoreline showing the extensive particle transport into Prince William Sound, and the general ocean circulation pattern in the Gulf of Alaska provide compelling evidence that suspended sediments from seep areas in the Gulf of Alaska are the vehicle for transport and sedimentation of seep hydrocarbons in Prince William Sound.

Oil present on high-energy shorelines (exposed beaches) is generally removed during winter storm events and dispersed at sea. Some Exxon Valdez tanker oil in low energy environments (enclosed bays and estuaries, embayments) will persist for longer periods until it is dispersed, buried, or biodegraded. Most of the sea floor in Prince William Sound has no detectable oil from the Exxon Valdez oil spill.

In what other ways can we apply hydrocarbon fingerprint analysis to the study of environmental problems?

These chemical fingerprinting techniques are being used to provide data to hydrogeologists and environmental chemists to evaluate the source, transport, and fate of spilled petroleum products in soil and groundwater systems to allocate responsibility in multiple spill situations.

WORLD SCIENTISTS' WARNING TO HUMANITY*

INTRODUCTION

Human beings and the natural world are on a collision course. Human activities inflict harsh and often irreversible damage on the environment and on critical resources. If not checked, many of our current practices put at serious risk the future that we wish for human society and the plant and animal kingdoms, and may so alter the living world that it will be unable to sustain life in the manner that we know. Fundamental changes are urgent if we are to avoid the collision our present course will bring about.

THE ENVIRONMENT

The environment is suffering critical stress:

THE ATMOSPHERE

Stratospheric ozone depletion threatens us with enhanced ultraviolet radiation at the earth's surface, which can be damaging or lethal to many

---

* Sponsored by the Union of Concerned Scientists, 26 Church Street, Cambridge, MA 02238. Reprinted by permission.

life forms. Air pollution near ground level, and acid precipitation, are already causing widespread injury to humans, forests, and crops.

WATER RESOURCES

Heedless exploitation of depletable ground water supplies endangers food production and other essential human systems. Heavy demands on the world's surface waters have resulted in serious shortages in some 80 countries, containing 40% of the world's population. Pollution of rivers, lakes, and ground water further limits the supply.

OCEANS

Destructive pressure on the oceans is severe, particularly in the coastal regions which produce most of the world's food fish. The total marine catch is now at or above the estimated maximum sustainable yield. Some fisheries have already shown signs of collapse. Rivers carrying heavy burdens of eroded soil into the seas also carry industrial, municipal, agricultural, and livestock waste—some of it toxic.

SOIL

Loss of soil productivity, which is causing extensive land abandonment, is a widespread by-product of current practices in agriculture and animal husbandry. Since 1945, 11% of the earth's vegetated surface has been degraded—an area larger than India and China combined—and per capita food production in many parts of the world is decreasing.

FORESTS

Tropical rain forests, as well as tropical and temperate dry forests, are being destroyed rapidly. At present rates, some critical forest types will be gone in a few years, and most of the tropical rain forest will be gone before the end of the next century. With them will go large numbers of plant and animal species.

LIVING SPECIES

The irreversible loss of species, which by 2100 may reach one-third of all species now living, is especially serious. We are losing the potential they hold for providing medicinal and other benefits, and the contribution that genetic diversity of life forms gives to the robustness of the world's biological systems and to the astonishing beauty of the earth itself.

Much of this damage is irreversible on a scale of centuries, or permanent. Other processes appear to pose additional threats. Increasing levels of gases in the atmosphere from human activities, including carbon dioxide released from fossil fuel burning and from deforestation, may alter climate on a global scale. Predictions of global warming are still uncertain—with projected effects ranging from tolerable to very severe—but the potential risks are very great.

Our massive tampering with the world's interdependent web of life—coupled with the environmental damage inflicted by deforestation, species loss, and climate change—could trigger widespread adverse effects, including unpredictable collapses of critical biological systems whose interactions and dynamics we only imperfectly understand.

Uncertainty over the extent of these effects cannot excuse complacency or delay in facing the threats.

POPULATION

The earth is finite. Its ability to absorb wastes and destructive effluent is finite. Its ability to provide food and energy is finite. Its ability to provide for growing numbers of people is finite. And we are fast approaching many of the earth's limits. Current economic practices which damage the environment, in both developed and underdeveloped nations, cannot be continued without the risk that vital global systems will be damaged beyond repair.

Pressures resulting from unrestrained population growth put demands on the natural world that can overwhelm any efforts to achieve a sustainable future. If we are to halt the destruction of our environment, we must accept limits to that growth. A World Bank estimate indicates that world population will not stabilize at less than 12.4 billion, while the United Nations concludes that the eventual total could reach 14 billion, a near tripling of today's 5.4 billion. But, even at this moment, one person in five lives in absolute poverty without enough to eat, and one in ten suffers serious malnutrition.

No more than one or a few decades remain before the chance to avert the threats we now confront will be lost and the prospects for humanity immeasurably diminished.

WARNING

We the undersigned, senior members of the world's scientific community, hereby warn all humanity of what lies ahead. A great change in our stewardship of the earth and the life on it is required, if vast human misery is to be avoided and our global home on this planet is not to be irretrievably mutilated.

WHAT WE MUST DO

Five inextricably linked areas must be addressed simultaneously:

1. *We must bring environmentally damaging activities under control to restore and protect the integrity of the earth's systems we depend on.* We must, for example, move away from fossil fuels to more benign, inexhaustible energy sources to cut greenhouse gas emissions and the pollution of our air and water. Priority must be given to the development of energy sources matched to Third World needs—small-scale and relatively easy to implement.

   We must halt deforestation, injury to and loss of agricultural land, and the loss of terrestrial and marine plant and animal species.

2. *We must manage resources crucial to human welfare more effectively.* We must give high priority to efficient use of energy, water, and other materials, including expansion of conservation and recycling.

3. *We must stabilize population. This will be possible only if all nations recognize that it requires improved social and economic conditions, and the adoption of effective, voluntary family planning.*

4. *We must reduce and eventually eliminate poverty.*

5. *We must ensure sexual equality, and guarantee women control over their own reproductive decisions.*

The developed nations are the largest polluters in the world today. They must greatly reduce their overconsumption, if we are to reduce pressures on resources and the global environment. The developed nations have the obligation to provide aid and support to developing nations, because only the developed nations have the financial resources and the technical skills for these tasks.

Acting on this recognition is not altruism, but enlightened self-interest: whether industrialized or not, we all have but one lifeboat. No nation can escape from injury when global biological systems are damaged. No nation can escape from conflicts over increasingly scarce resources. In addition, environmental and economic instabilities will cause mass migrations with incalculable consequences for developed and undeveloped nations alike.

Developing nations must realize that environmental damage is one of the gravest threats they face, and that attempts to blunt it will be overwhelmed if their populations go unchecked. The greatest peril is to become trapped in spirals of environmental decline, poverty, and unrest, leading to social, economic, and environmental collapse.

Success in this global endeavor will require a great reduction in violence and war. Resources now devoted to the preparation and conduct of war—amounting to over \$1 trillion annually—will be badly needed in the new tasks and should be diverted to the new challenges.

A new ethic is required—a new attitude towards discharging our responsibility for caring for ourselves and for the earth. We must recognize the earth's limited capacity to provide for us. We must recognize its fragility. We must no longer allow it to be ravaged. This ethic must motivate a great movement, convincing reluctant leaders and reluctant governments and reluctant peoples themselves to effect the needed changes.

The scientists issuing this warning hope that our message will reach and affect people everywhere. We need the help of many.

We require the help of the world community of scientists—natural, social, economic, political;

We require the help of the world's business and industrial leaders;

We require the help of the world's religious leaders; and

We require the help of the world's peoples.

We call on all to join us in this task.

A P P E N D I X

II

STEADY-STATE ANALYSIS FOR CONCENTRATIONS

The discussion in Chapter 3 of the chemical processes found to occur in clean air indicated that there are many reactions that proceed simultaneously in such a mixture, especially in the daytime. Several possible pathways exist for the reaction of many of the substances present in such a system. One might imagine that such great complexity would rule out any simple mathematical analysis of reaction rates, and indeed fast computers and sophisticated programs are required to analyze such systems exactly. However, some of the main features characterizing stable air masses can be deduced by applying the steady-state concept introduced in Chapter 2. In this section, the principles of this approach are developed by application to the kinetics of some important reactions in clean air.

As a simple illustration of the approach, consider first an atmosphere into which methane is released—due, say, to biological decay processes—at a constant rate R. The atmospheric concentration of methane will not rise without limit since it is destroyed by its reaction with the hydroxyl radical:

$$OH^\bullet + CH_4 \longrightarrow CH_3^\bullet + H_2O$$

433

According to the Collision Theory of Reaction Rates, it follows that the rate of this process is proportional to the CH_4 concentration, and so the destruction rate will increase as the CH_4 level rises:

$$\text{Rate of } CH_4 \text{ destruction reaction} = k\,[OH^\bullet][CH_4]$$

Here k is the rate constant for this particular reaction step. Eventually the rate of destruction increases to such an extent (since the CH_4 concentration is increasing) that it becomes equal to R; once this point is reached, this equality is maintained thereafter and the methane concentration remains constant—that is, the level of methane has achieved a steady state, since the rates of its production and destruction are equal. The value of $[CH_4]$ under steady-state conditions can be deduced by equating the expressions for the rates of the production and destruction processes:

$$\text{Rate of destruction of } CH_4 = \text{Rate of production of } CH_4$$

$$k[OH^\bullet][CH_4] = R$$

Solving for $[CH_4]$, we obtain

$$[CH_4]_{SS} = R/k\,[OH^\bullet]$$

In other words, the concentration of methane rises to $R/k[OH^\bullet]$, the value of which we can determine from numerical values for R, k, and $[OH^\bullet]$. Clearly, doubling the rate R of release of CH_4 into the atmosphere will lead to a doubling of its steady-state concentration; this conclusion is, however, based on the assumption that changes in the methane level would not significantly deplete the supply of hydroxyl radical, an assumption which is not completely valid.

Before proceeding to a more complex example, we should emphasize that a species is in a steady state whenever its *total* rate of destruction—by the sum of the rates of all reactions which consume it—is equal to the *total* rate of its creation from all the reactions which produce it. Thus the steady-state analysis is mathematically more complicated the more reactions for a given chemical that occur simultaneously.

An example of a chemical species that is produced and consumed by several different reactions is the hydroxyl radical, $OH^\bullet$. Recall that it is created by the photochemical decomposition of ozone into O_2 and atomic oxygen; some of the oxygen atoms react subsequently with H_2O

molecules to produce 2 OH$^\bullet$. Let us define as R_p the rate at which hydroxyl radicals are generated from ozone in a sunlight-irradiated troposphere:

$$\text{Rate of production of OH}^\bullet \text{ from ozone} = R_p$$

The most important *destruction* reaction for OH$^\bullet$ is its reaction with carbon monoxide:

$$OH^\bullet + CO \longrightarrow HCO_2^\bullet \xrightarrow{O_2} HOO^\bullet + CO_2 \qquad (1)$$

According to the Collision Theory, the rate of reaction 1 is $k_1[OH^\bullet][CO]$. However we know that OH$^\bullet$ is efficiently regenerated from HOO$^\bullet$ by reaction of the latter with nitric oxide in air:

$$HOO^\bullet + NO^\bullet \longrightarrow OH^\bullet + NO_2^\bullet \qquad (2)$$

The rate of this reaction is $k_2[HOO^\bullet][NO^\bullet]$. Thus we can write

$$\text{Total rate of OH}^\bullet \text{ production} = R_p + k_2[HOO^\bullet][NO^\bullet]$$

whereas

$$\text{Total rate of OH}^\bullet \text{ destruction} = k_1[OH^\bullet][CO]$$

Since at the steady state for OH$^\bullet$ the total rates of production and consumption are equal, it follows that

$$k_1[OH^\bullet][CO] = R_p + k_2[HOO^\bullet][NO^\bullet] \qquad (I)$$

Solving for [OH$^\bullet$], we obtain

$$[OH^\bullet] = \frac{R_p + k_2[HOO^\bullet][NO^\bullet]}{k_1[CO]}$$

As it stands, this equation is of little practical use for estimating the steady-state value of [OH$^\bullet$] since it requires knowledge of the value for [HOO$^\bullet$], another reactive species. This mathematical bottleneck can be eliminated by realizing that HOO$^\bullet$ will also be in a steady state. It is produced by reaction 1, and is consumed not only by reaction 2 but also by the bimolecular collision of 2 HOO$^\bullet$ radicals, to yield hydrogen peroxide:

$$2\,HOO^\bullet \longrightarrow H_2O_2 + O_2 \qquad (3)$$

The rate of consumption of HOO$^\bullet$ radicals by reaction 3 is $2\,k_3[HOO^\bullet]^2$. Thus we can write that

Rate of HOO$^\bullet$ production $= k_1[OH^\bullet][CO]$
Rate of HOO$^\bullet$ destruction $= k_2[HOO^\bullet][NO^\bullet] + 2\,k_3[HOO^\bullet]^2$

Thus at the steady state,

$$k_1[OH^\bullet][CO] = k_2[HOO^\bullet][NO^\bullet] + 2\,k_3[HOO^\bullet]^2$$

Inspection of equations I and II reveals that they are two simultaneous equations in two unknowns, [OH$^\bullet$] and [HOO$^\bullet$], assuming that the concentrations of CO and NO$^\bullet$ are known through measurement. The pair of equations can be solved most readily by recognizing that their left-hand sides are identical, so their right-hand sides can be equated to obtain an equation in which [OH$^\bullet$] has been eliminated:

$$R_p + k_2\,[HOO^\bullet][NO^\bullet] = k_2\,[HOO^\bullet][NO^\bullet] + 2\,k_3[HOO^\bullet]^2$$

After striking out the term common to both sides, we obtain

$$R_p = 2\,k_3\,[HOO^\bullet]^2$$

from which a simple expression for [HOO$^\bullet$] is obtained:

$$[HOO^\bullet] = (R_p/2\,k_3)^{1/2}$$

Substituting this expression for [HOO$^\bullet$] into equation I, we obtain an expression in which [OH$^\bullet$] is the sole unknown:

$$k_1[OH^\bullet][CO] = R_p + k_2[NO^\bullet](R_p/2\,k_3)^{1/2}$$

By rearrangement, we can solve at last for [OH$^\bullet$]:

$$[OH^\bullet] = \frac{R_p + k_2[NO^\bullet](R_p/2\,k_3)^{1/2}}{k_1\,[CO]}$$

Thus, the level of $OH^\bullet$ in air increases as the $NO^\bullet$ concentration increases (since an increase in $[NO^\bullet]$ speeds up its reformation from $HOO^\bullet$) and is inversely proportional to the CO concentration in air (since increasing [CO] increases the rate of reaction of CO with $OH^\bullet$).

Another important reaction in which $OH^\bullet$ is consumed is the reaction with CH_4 shown in Problem A-1 below. To include this reaction in our analysis would complicate the algebra so much that a simple analytical solution could not be derived, and a computer simulation would be needed to explore the system. An algebraic steady-state analysis is useful and illuminating only when there are but a few dominant creation and destruction processes for a given radical.

PROBLEM A-1

As discussed previously, the methyl radical in air is produced by the reaction of methane with hydroxyl radical:

$$CH_4 + OH^\bullet \xrightarrow{k_1} CH_3^\bullet + H_2O$$

and is subsequently destroyed by its reaction with oxygen in this way:

$$CH_3^\bullet + O_2 \xrightarrow{k_2} CH_3OO^\bullet$$

Develop an expression for the steady-state concentration of $CH_3^\bullet$ in terms of k_1, k_2, $[OH^\bullet]$, $[O_2]$, and $[CH_4]$. Is $[CH_3^\bullet]_{SS}$ directly proportional to $[CH_4]$?

PROBLEM A-2

Consider an atmosphere in which the only elements present are hydrogen and oxygen. Ozone molecules undergo photochemical decomposition to yield O_2 and oxygen atoms

$$O_3 \longrightarrow O_2 + O^*$$

Some of the O* atoms react with water molecules to form hydroxyl radicals:

$$O^* + H_2O \longrightarrow 2\,OH^{\bullet}$$

Under these circumstances, the dominant process in which the $OH^{\bullet}$ radicals take part is to react with each other, forming hydrogen peroxide:

$$2\,OH^{\bullet} \longrightarrow H_2O_2$$

Perform steady-state analyses for [O*] and [$OH^{\bullet}$] to deduce expressions for them as a function of rate constants and the concentrations of O_3, H_2O, O_2, and H_2O_2.

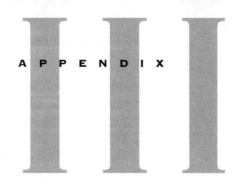
REVIEW OF ACID-BASE EQUILIBRIUM PRINCIPLES

An *acid* is a substance that increases the concentration of hydrogen ions, H^+, in water. A few acids, such as hydrochloric acid, HCl, are called *strong* acids since they ionize completely in water:

$$HCl(aq) \longrightarrow H^+ + Cl^-$$

Most acids, typically denoted in general as HA, however, are *weak* since they release only a limited portion of their hydrogen as H^+ in water. The reason for this limitation is that the ions H^+ and A^- exist in equilibrium with the dissolved but unionized form HA:

$$HA(aq) \rightleftharpoons H^+ + A^-$$

According to the Law of Mass Action, the ratio of the arithmetic product of the concentrations of the chemical products to that of the reactants is a constant K that is characteristic of the reaction equilibrium at a given temperature; this ratio for a weak acid is called its *acid dissociation (or ionization) constant* and is symbolized K_a:

$$K_a = [H^+][A^-]/[HA]$$

439

The concentrations [⋯] in these expressions must be equilibrium concentrations or their ratio would not be equal to the equilibrium constant. However, when only a small fraction of the acid exists in ionized form, the numerical value for the equilibrium concentration [HA] is very similar to the initial, pre-ionization value.

A *base* B is a substance that increases the concentration of hydroxide ions, OH^-, in aqueous solution. For weak bases, this usually occurs by the reaction of B with a water molecule:

$$B + H_2O \rightleftharpoons BH^+ + OH^-$$

The *base dissociation (ionization) constant* K_b is defined as the ratio

$$K_b = [BH^+][OH^-]/[B]$$

By convention, the concentration of water, $[H_2O]$, does not appear in such equilibrium expressions since its value is close to that for pure water and can be incorporated into K_b.

The anion, A^-, of a weak acid itself acts as a weak base; it can acquire a hydrogen ion from a water molecule to form the acid HA and release a hydroxide ion:

$$A^- + H_2O \rightleftharpoons HA + OH^-$$

HA and A^- are called a *conjugate acid-base pair*; similarly BH^+ and B are a conjugate pair since BH^+ can act as an acid, producing hydrogen ions in water:

$$BH^+ \rightleftharpoons B + H^+$$

In the chemistry of natural waters, we encounter H_2CO_3 and HCO_3^- as one acid-base conjugate pair, and HCO_3^- and CO_3^{2-} as another such pair.

For any acid-base conjugate pair, the product K_aK_b of their ionization constants is equal to K_w, the equilibrium constant for the *self-ionization reaction* of water:

$$H_2O \rightleftharpoons H^+ + OH^-$$

for which

$$K_w = [H^+][OH^-]$$

At 25°C, the numerical value for K_w is 1.0×10^{-14}. Thus, for example,

$$K_b \text{ (for } A^-) = K_w/K_a \text{ (for HA)}$$
$$= 10^{-14}/K_a$$

The concentration of hydrogen ion in aqueous solutions is often reported in terms of the *p*H of the solution, where pH equals the negative of the base 10 logarithm of $[H^+]$:

$$pH = -\log_{10} [H^+]$$

Similarly we define

$$pOH = -\log_{10} [OH^-]$$

Given that $[H^+][OH^-] = 10^{-14}$, it follows that for aqueous solutions that the pH and pOH values are interrelated simply as

$$pH + pOH = 14$$

Although the discussion above implies that a given substance is either an acid or a base (and not both), in fact some substances behave in either fashion depending upon their environment. Substances that can act either as acids or as bases are said to be **amphoteric.** The most important amphoteric species in environmental chemistry is the bicarbonate ion, HCO_3^-:

Action as an acid: $\qquad\qquad HCO_3^- \rightleftharpoons H^+ + CO_3^{2-}$

Action as a base: $\qquad\qquad HCO_3^- + H_2O \rightleftharpoons OH^- + H_2CO_3$

PERIODIC TABLE

1	2											3	4	5	6	7	8
1 **H** 1.008																	2 **He** 4.003
3 **Li** 6.941	4 **Be** 9.012											5 **B** 10.81	6 **C** 12.01	7 **N** 14.01	8 **O** 16.00	9 **F** 19.00	10 **Ne** 20.18
11 **Na** 22.99	12 **Mg** 24.31											13 **Al** 26.98	14 **Si** 28.09	15 **P** 30.97	16 **S** 32.06	17 **Cl** 35.45	18 **Ar** 39.95
19 **K** 39.10	20 **Ca** 40.08	21 **Sc** 44.96	22 **Ti** 47.90	23 **V** 50.94	24 **Cr** 52.00	25 **Mn** 54.94	26 **Fe** 55.85	27 **Co** 58.93	28 **Ni** 58.70	29 **Cu** 63.55	30 **Zn** 65.38	31 **Ga** 69.72	32 **Ge** 72.59	33 **As** 74.92	34 **Se** 78.96	35 **Br** 79.90	36 **Kr** 83.80
37 **Rb** 85.47	38 **Sr** 87.62	39 **Y** 88.91	40 **Zr** 91.22	41 **Nb** 92.91	42 **Mo** 95.94	43 **Tc** (98)	44 **Ru** 101.1	45 **Rh** 102.9	46 **Pd** 106.4	47 **Ag** 107.9	48 **Cd** 112.4	49 **In** 114.8	50 **Sn** 118.7	51 **Sb** 121.8	52 **Te** 127.6	53 **I** 126.9	54 **Xe** 131.3
55 **Cs** 132.9	56 **Ba** 137.3	71 **Lu** 175.0	72 **Hf** 178.5	73 **Ta** 180.9	74 **W** 183.9	75 **Re** 186.2	76 **Os** 190.2	77 **Ir** 192.2	78 **Pt** 195.1	79 **Au** 197.0	80 **Hg** 200.6	81 **Ti** 204.4	82 **Pb** 207.2	83 **Bi** 209.0	84 **Po** (209)	85 **At** (210)	86 **Rn** (222)
87 **Fr** (223)	88 **Ra** (226.0)	103 **Lr** (260)	104 **Unq**	105 **Unp**	106 **Unh**	107 **Uns**	108	109 **Une**									

Lanthanide series

57 **La** 138.9	58 **Ce** 140.1	59 **Pr** 140.9	60 **Nd** 144.2	61 **Pm** (145)	62 **Sm** 150.4	63 **Eu** 152.0	64 **Gd** 157.3	65 **Tb** 158.9	66 **Dy** 162.5	67 **Ho** 164.9	68 **Er** 167.3	69 **Tm** 168.9	70 **Yb** 173.0

Actinide series

89 **Ac** (227)	90 **Th** 232.0	91 **Pa** (231)	92 **U** 238.0	93 **Np** (244)	94 **Pu** (242)	95 **Am** (243)	96 **Cm** (247)	97 **Bk** (247)	98 **Cf** (251)	99 **Es** (252)	100 **Fm** (257)	101 **Md** (258)	102 **No** (259)

ANSWERS TO SELECTED PROBLEMS

CHAPTER 2

2-1 a. 199 kJ mol^{-1}

b. 60 kJ mol^{-1}

2-2 1140 nm; IR region

2-3 390.7 nm; 127.5 nm

2-4 406 nm

2-5 Location of the unpaired electron:

a. O   b. C   c. C

d. terminal O   e. O   f. terminal O

g. O   h. C

2-7 No

2-8 $O^* + H_2O \longrightarrow 2\,OH^\bullet$

2-9 Yes

2-10

$$O_3 + Cl^\bullet \longrightarrow ClO^\bullet + O_2$$
$$O_3 + Br^\bullet \longrightarrow BrO^\bullet + O_2$$
$$\underline{ClO^\bullet + BrO^\bullet \longrightarrow Cl^\bullet + Br^\bullet + O_2}$$
$$\text{Overall}\quad 2\,O_3 \longrightarrow 2\,O_3$$

2-11 The dichloroperoxide reaction would become relatively more important.

2-13 a. CF_2Cl_2

b. $C_2F_3Cl_3$

c. $C_2HF_3Cl_2$

d. $C_2H_2F_4$

2-14 a. 140

b. 10

c. 141

2-15 Reaction is highly endothermic and therefore too slow to be significant.

2-16 a.

$$F^\bullet + O_3 \longrightarrow FO^\bullet + O_2$$
$$\underline{FO^\bullet + O \longrightarrow F^\bullet + O_2}$$
$$O_3 + O \longrightarrow 2\,O_2$$

b. The reaction $OH^\bullet + HF \longrightarrow F^\bullet + H_2O$ is highly endothermic so negligibly slow; thus fluorine will never be reactivated from HF.

2-17 Reactions to abstract H will be faster for propane and butane than for methane.

2-18 Presumably they have efficient tropospheric sinks and short atmospheric lifetimes.

2-19 b, c, and d

2-20 a. O_3, $ClO^\bullet$, $BrO^\bullet$, $HOO^\bullet$

b. O_3, $ClO^\bullet$, $BrO^\bullet$, $HOO^\bullet$, and $NO_2^\bullet$

c. all of them

d. $ClOOCl$, $(NO_2)_2$, and perhaps $BrOOBr$

e. two $HOO^\bullet$ and two O_3

2-21 a. $BrO^\bullet + O \longrightarrow Br^\bullet + O_2$

b. $BrO^\bullet + ClO^\bullet$
$$\longrightarrow Br^\bullet + Cl^\bullet + O_2$$

c. $2\,BrO^\bullet \longrightarrow BrOOBr$

d. $BrO^\bullet + UV \longrightarrow Br^\bullet + O$

2-22 a. $ClO^\bullet + NO_2^\bullet \longrightarrow ClONO_2$

b. $2\,ClO^\bullet \longrightarrow ClOOCl$

c. reaction of UV with $ClO^\bullet$ or with $ClOOCl$, or reaction of $ClO^\bullet$ with O or $NO^\bullet$.

CHAPTER 3

3-1   a. 0.032 ppm
       b. 7.9×10^{11} molecules cm^{-3}
       c. 1.3×10^{-9} M

3-2   24 ppm; 1.0×10^{-6} M

3-3   a. 9.1×10^{11} molecules cm^{-3}
       b. $74.2 \ \mu g/m^3$

3-4   99.2%; 92.3%; yes

3-5   One million to one

3-6   3:1 ratio; area bigger with small particles

3-7   16%

3-8 1.4×10^{-17} M; 0.35 ppt

3-10   Very slow

3-11   Quite endothermic

3-12   Overall
$$H_2 + O_2 + NO^\bullet \longrightarrow H_2O + NO_2^\bullet$$

3-13   Overall $H_2S + 4 O_2 + 4 NO^\bullet$
$$\longrightarrow H_2SO_4 + 4 NO_2^\bullet$$

3-15   Overall
$$CO + O_2 + NO^\bullet \longrightarrow CO_2 + NO_2^\bullet$$

3-16   Overall
$$H_2CO + 2 NO^\bullet + 2 O_2 + sunlight$$
$$\longrightarrow CO_2 + H_2O + 2 NO_2^\bullet$$

3-18   Overall
$$CH_3(H)CO + 7 O_2 + 7 NO^\bullet$$
$$+ sunlight \longrightarrow 2 CO_2$$
$$+ 7 NO_2^\bullet + 4 OH^\bullet$$

3-19   Same as for 3-16

3-21 $NO^\bullet + O_3 \longrightarrow NO_2^\bullet + O_2$, when $[HOO^\bullet]$ is low

3-22   pH=5.63 at 25°C, and 5.61 at 15°C

3-23   3.88

3-24   0.59 ppm

3-26   +1 power for H_2SO_3, -1 power for $SO_3{}^{2-}$

3-27   a. $^{218}_{84}Po$
       b. $^{214}_{84}Po$
       c. α
       d. $^{238}_{92}U$

CHAPTER 4

4-1 H_2, Cl_2 cannot. Vibrations must be outside thermal range.

4-2   0.44 metric ton

4-3   a. oil; $CH_3OH + 1.5 O_2$
$$\longrightarrow CO_2 + 2 H_2O$$
       b. $3 C + 4 H_2O$
$$\longrightarrow CO_2 + 2 CH_3OH$$
       c. no, amount of CO_2 is same in both cases.

4-4 10^8 kilograms

4-5 2.5×10^{15} grams per year

4-6   1.4%

4-7   480 Tg per year

4-8   8%

4-9   129 grams

4-10   6.0 Tg

4-11   No; yes; no

4-13   b > c > a; most nonlinear at 7 and 16 μm

4-14   Linear for 0.001; square root for 0.9

CHAPTER 5

5-1 a.

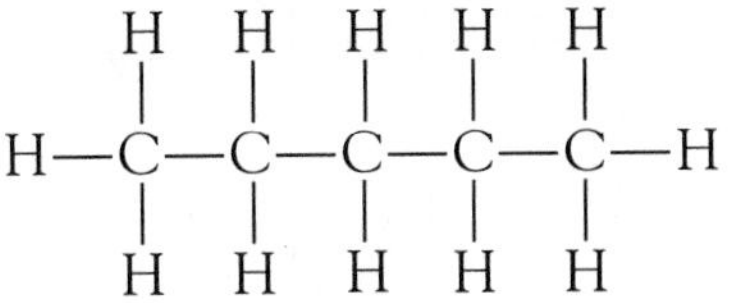

$CH_3CH_2CH_2CH_2CH_3$

b.

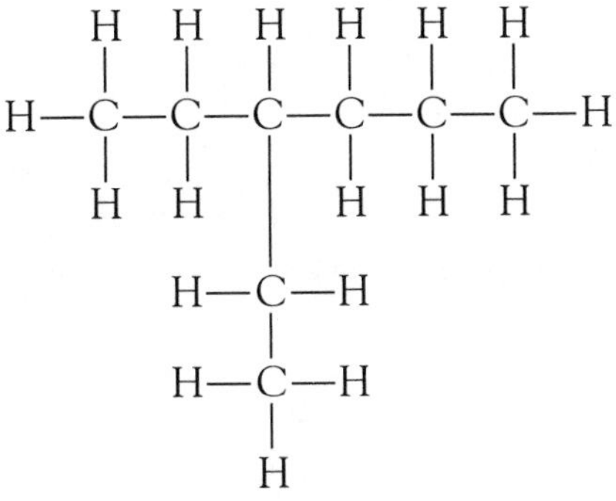

$(CH_3CH_2)_2CHCH_2CH_3$

c.

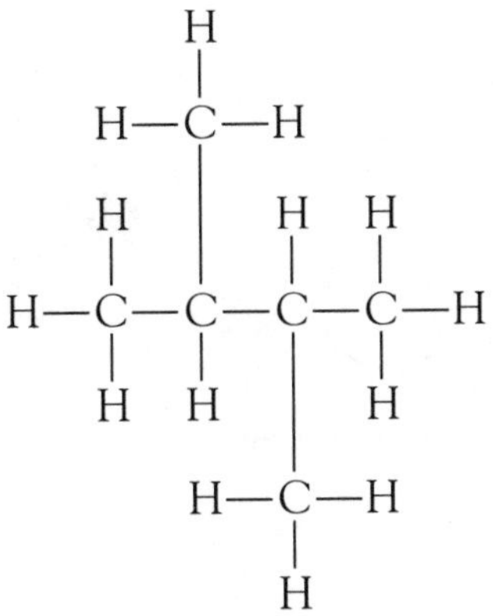

$(CH_3)_2CHCH(CH_3)_2$

5-2 a.

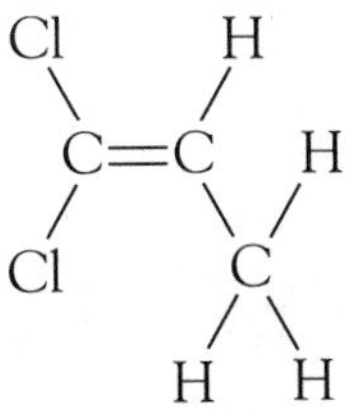

5-2 b.

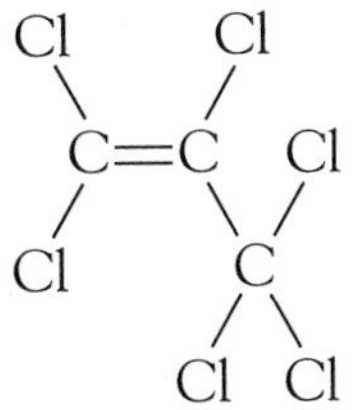

c.

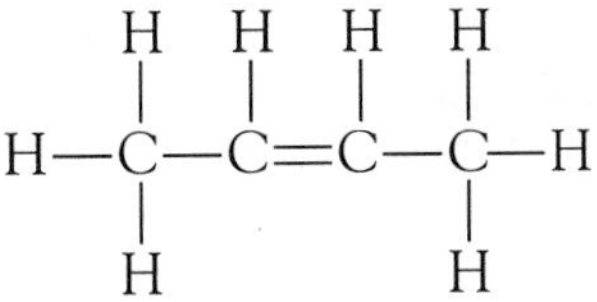

5-3 a. 1,1,2,2-tetrachloroethane

b. 1-butene

c. 1,4-butadiene

5-4 a.

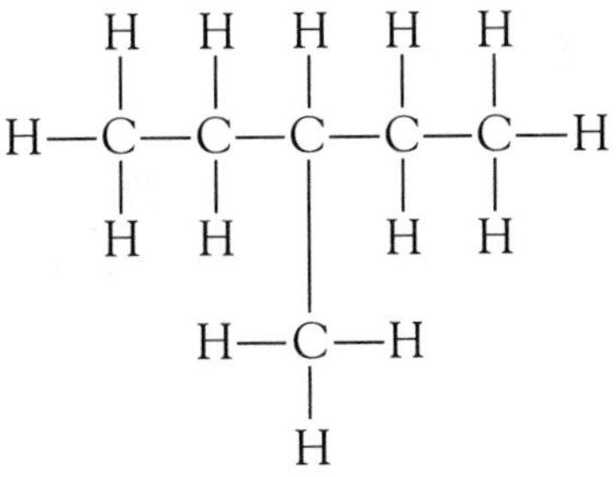

b.

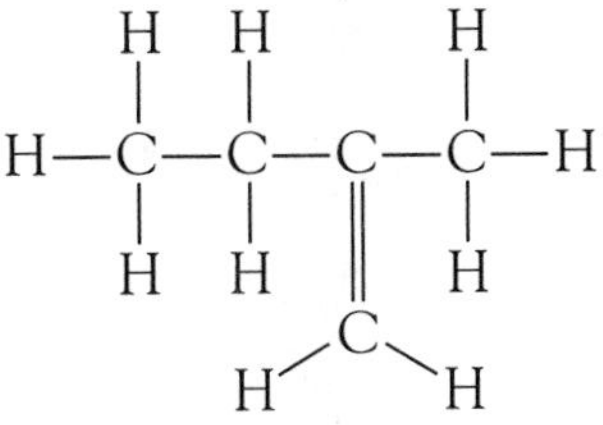

c.

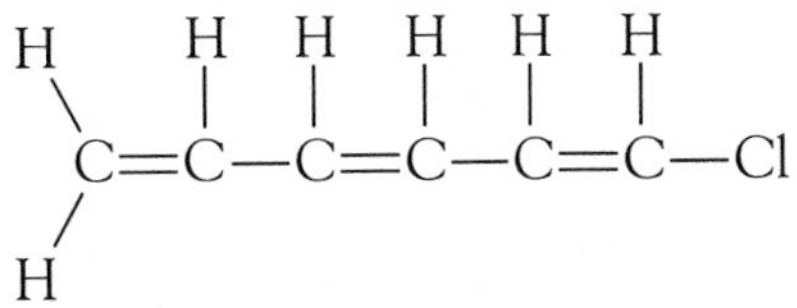

5-5   a.

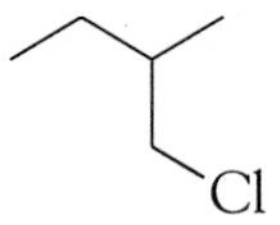

b.

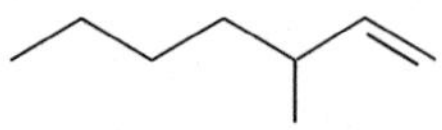

c.

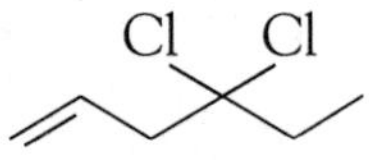

5-6   a.

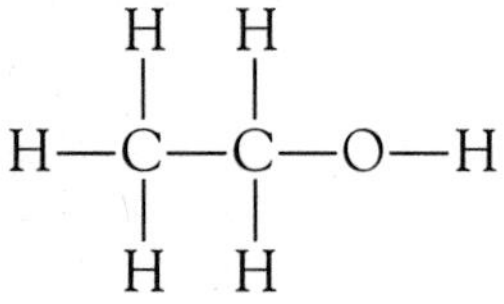

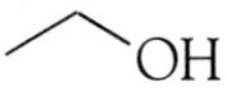

b.

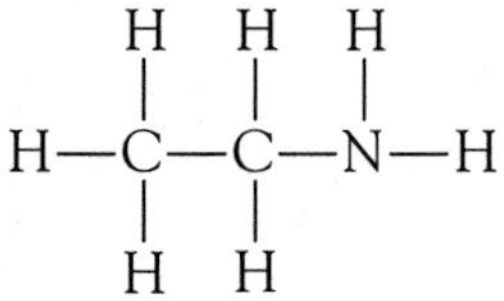

c.

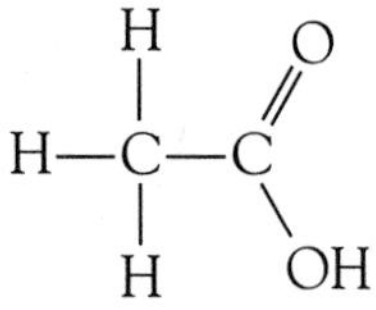

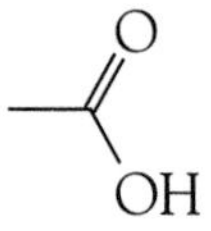

5-7   a.

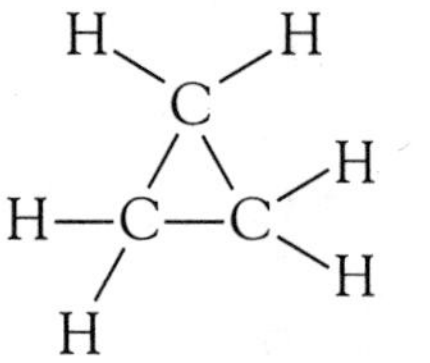

b.

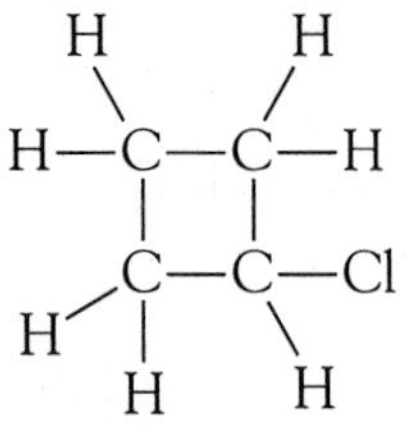

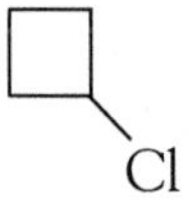

5-7   c.

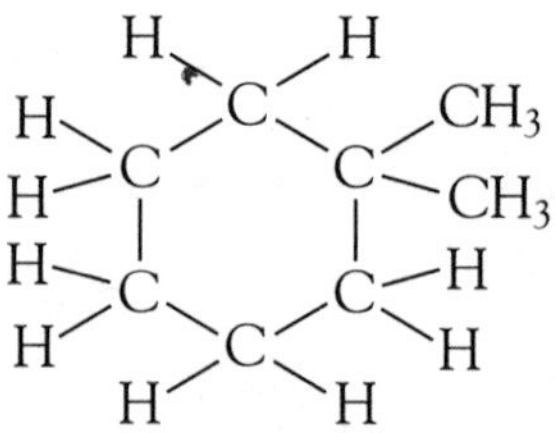

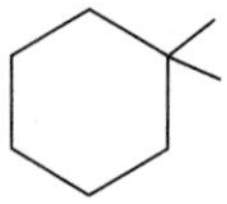

5-8

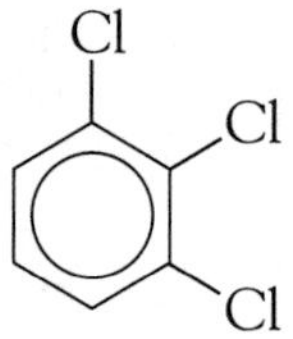

1,2,3-trichlorobenzene

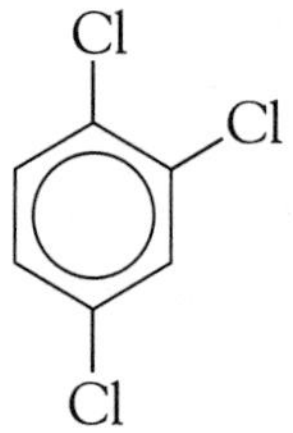

1,2,4-trichlorobenzene

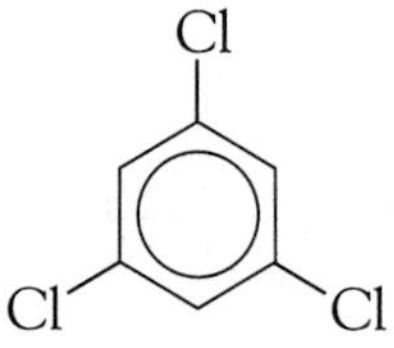

1,3,5-trichlorobenzene

CHAPTER 6

6-1   a. 4.0×10^{-5} ppm; 0.04 ppb

     b.   3.0 μg/L

6-2   2.0 ppm

6-4   a. yes   b. no

6-5   No, the same. Yes. The unique ones are 1,2; 1,3; 1,4; 1,6; 1,7; 1,8; 1,9; 2,3; 2,7; 2,8

6-6 a. 2,3,5,6- and 2,3,4,5-tetrachloro-phenols

   b. Two 2,3,5,6-tetrachlorophenols give the 1,2,4,6,7,9-dioxin

   Two 2,3,4,5-tetrachlorophenols give the 1,2,3,6,7,8-dioxin

   Two 2,3,4,6-tetrachlorophenols give the 1,2,3,6,8,9-dioxin and the 1,2,4,6,7,9- and the 1,2,3,6,7,8-dioxins

   c. i. One 2,3,4,5- and one 2,3,4,6-tetrachlorophenols

   ii. One 2,3,5,6- and one 2,3,4,6-tetrachlorophenols

   iii. One 2,3,4,6- and either one 2,3,4,5- or one 2,3,5,6-tetra-chlorophenol

6-7 a. 2,7-dichlorodibenzo-*p*-dioxin

   b. Octachloro, and 1,2,3,4,6,7,9- and 1,2,3,4,6,7,8-heptachlorodibenzo-*p*-dioxins, and 1,2,3,6,8,9- and 1,2,3,6,7,8- and 1,2,4,6,7,9-hexa-chlorodibenzo-*p*-dioxins

6-8 6.26

6-9 2,3; 2,4; 2,5; 2,6; 3,4; 3,5; 2,2'; 2,3'; 2,4'; 2,5'; 2,6'; 3,3'; 3,4'; 3,5'; 4,4'. With rotation, 2,2' and 2,6' intercon-vert, as do 2,3' and 2,5', as well as 3,3' and 3,5'.

6-10 From that at left, 1-chloro- and 1,9-, 1,4-, and 1,6-dichloro-dibenzofurans. From that at right, 1- and 4-chlorodi-benzofuran and dibenzofuran itself.

6-11 No additional ones from the structure on the left. 1,9-dichlorodibenzofuran from that on the right.

6-12 3-chloro; 2,3- and 2,3'- and 2,2'-dichloro; 2,3,2'-trichlorobiphenyl

6-13 1,2; 1,3; 1,4; 2,3; 2,4; 3,4; 1,6; 1,7; 1,8; 1,9; 2,6; 2,7; 2,8; 3,6; 3,7; 4,6 dichloro-dibenzofurans

6-14 1,2,3,7,8 > 2,3,7 > 1,2,3

6-15 28.6 pg

6-16 a. $C_{12}H_6Cl_4 + 12.5\ O_2 \longrightarrow$

   $12\ CO_2 + H_2O + 4\ HCl$

   b. $C_{12}H_6Cl_4 + 23\ H_2 \longrightarrow$

   $12\ CH_4 + 4\ HCl$

6-20 No for naphthalene and anthracene; yes for phenanthrene and benzo[ghi]perylene

CHAPTER 7

7-3 $NH_3 + 2\ O_2 + OH^- \longrightarrow$

   $NO_3^- + 2\ H_2O$

7-4 16 mg/L

7-5 $SO_4^{2-} + 10\ H^+ + 8\ e^- \longrightarrow$

   $H_2S + 4\ H_2O$

   $SO_4^{2-} + 2\ H^+ + 2\ CH_2O$

   $\longrightarrow\ H_2S + 2\ CO_2 + 2\ H_2O$

7-6 7.9×10^{-3} M and 6.3×10^{-20} M; 8.32 and 2.51

7-7 $Ca(HCO_3)_2 + Ca(OH)_2$

   $\longrightarrow\ 2\ CaCO_3 + 2\ H_2O$; 1:1 ratio

7-8 0.79, 0.54, 0.27, 0.10

7-9 Ammonium at the bottom, nitrate at the top

7-10 $NH_4^+ + 3\ H_2O \longrightarrow$

   $NO_3^- + 10\ H^+ + 8\ e^-$

   $NO_2^- + 8\ H^+ + 6\ e^- \longrightarrow$

   $N_2 + 4H_2O$

CHAPTER 8

8-1 8.3×10^{-5} M; increases.

8-2 3.2×10^{-6} M; 3.9×10^{-7} M; yes

8-3 5.5×10^{-3} M; 0.001%; no

8-4 7.9×10^{-4} M

8-5 1.9×10^{-4} M; 1.9×10^{-4} M; 7.6×10^{-4} M

8-6 9.26

8-7 5.5×10^{-6} M; 5.5×10^{-7} M; 5.5×10^{-8} M

8-8 6.3, 10.3. H_2CO_3 is dominant for pH < 6.3; HCO_3^- is dominant between pH 6.3 and 10.3; CO_3^{2-} is dominant for pH > 10.3.

8-9 2.1×10^6 and 220 at pH = 4; 2.1 and 2.2×10^{-4} at pH = 10

8-10 2.4×10^{-4} M, 3.5×10^{-5} M for saturated $CaCO_3$. 1.04×10^{-3} M, 9.6×10^{-6} M if in equilibrium with CO_2.

8-11 2.6×10^{-4} M

8-13 51 mg $CaCO_3$ per liter; slightly greater

8-14 3.6

CHAPTER 9

9-1 0.051 g

9-2 138 days

9-3 5.5×10^{-27} M; 0.0033 ions/L; 300 L

9-4 0.5 mg

9-5 2.0×10^5 g

9-6 $HgCl_2$

9-7 2×10^{-5} g

9-8 $PbS(s) + H_2O \rightleftharpoons Pb^{2+} + HS^- + OH^-$

9-9 $[Hg^{2+}] = 4.8 \times 10^{-17} [H^+]$; yes

9-11 $H_2AsO_4^-$ at pH = 4 and 6; $HAsO_4^{2-}$ at pH = 8 and 10

CHAPTER 10

10-2 17.9 kJ/mL; 23.5 kJ/mL; ethanol is superior to methanol on both counts, but they are inferior to methane on energy per mass basis

10-3 a. one-third; $3 C + 4 H_2O \longrightarrow 2 CH_3OH + CO_2$

   b. one-ninth; $3 CH_4 + CO_2 + 2 H_2O \longrightarrow 4 CH_3OH$

10-4 $CH_4 + 0.5 O_2 \longrightarrow CH_3OH$

10-5 Converted to formaldehyde, which then oxidizes to CO_2

10-8 23.7 kg; 12.0 kg; magnesium

10-9 4:1

10-10 120 kJ g^{-1} for H_2 vs. 55.6 for methane, so H_2 superior on weight basis. On molar and volume basis, CH_4 is superior (890 vs 242) to H_2

10-11 419 nm; visible light

10-12 168,000 years

APPENDIX I

A-1 $k_1 [CH_4][OH^{\cdot}]/k_2[O_2]$; yes, directly proportional

A-2 $k_1[O_3]/k_2 [H_2O]$ for O
   $(k_1/k_3)^{1/2} [O_3]^{1/2}$ for O_3

INDEX

absorption, of particulates, 99

absorption spectrum, of oxygen, 19, 20

Acadia National Park (Maine), smog in, 77

acaricides, target organisms of, 217

acetaldehyde, as air pollutant, 203

acetic acid, formulas for, 204

acetone, in tobacco smoke, 110

acetylcholine-destroying enzyme, insecticide effects on, 239

acid
    definition of, 439
    ecological effects of, 94–99
    weak and strong types of, 439

acid air, health hazards of, 103

acid-base chemistry
    of carbonate system in natural waters, 325–346
    equilibrium principles of, 439–441
    of water, 290

acid-base conjugate pair, 328, 440

acid dissociation constant, derivation of, 439–440

acid fog, 90, 98, 99

acid mine drainage, soluble iron compound formation from, 301–302

acid mist, 98, 99

acid rain, 90–99
    acids in, 91, 95, 136, 137, 338
    dry deposition effects of, 95, 97
    formation of, 91
    health hazards of, 90
    pH of, 91, 95, 96, 137

acid snow, 90

activated charcoal
    use for covering contaminated sediments, 391
    use for water purification, 302–303, 316

activation energies, of fusion reactions, 421

addition reactions, in troposphere, 116

adhesives, formaldehyde emission from, 107

adsorption, of particulates, 99

aeration, use for water purification and quality improvement, 302

aerosols
    as greenhouse gases, 178–180, 186
    particulates in, 99, 178, 370, 380

aerosol sprays
    chlorofluorocarbon use in, 53, 59, 60
    pollutant gas release from, 15, 87

Agent Orange, dioxin levels in, 247

agricultural crops
    air pollution effects on, 99
    pesticide use on, 217, 218

agricultural workers, deaths due to pesticide handling by, 240

AIDS, non-Hodgkins lymphoma with, 246

air
    heavy metal transport in, 347–348, 353

    kinetics of reactions in, 433

air chemistry
    of stratosphere, 14–73
    of troposphere, 8, 77–148

air conditioners
    CFC-12 use in, 52–53, 61, 177
    HCFC-22 use in, 57
    HFC-134a use in, 58
    pollutant gas release from, 15

air pollution
    from BTX hydrocarbons in gasoline, 400–401
    from carbon monoxide, 108–110
    from diesel and engine exhausts, 99, 275, 279–280, 396, 399
    effects of, on plants and trees, 98, 99
    from formaldehyde, 107–108
    ground-level air chemistry and, 77–148
    health effects of, 104–106, 106–112
    interconversion of concentrations of, 77–81
    minerals as cause of, 102
    from nitrogen dioxide, 108–110
    from ozone, 105
    from PAHs, 274, 275
    particulates in, 99–104
    pollutant levels in, 107
    from radon gas, 141–146
    role in ozone hole phenomenon, 14–15
    scientists' warnings concerning, 427–428
    smog as. *See* smog
    from toxic organic chemicals, 216
    from uncapped toxic landfills, 319

Air Quality Accord (1991), between United States and Canada, 94

air quality standards, for total suspended particulates (TSP), 101

air stripping, in removal of nitrogen from water, 316

air-water-rock system, carbon dioxide/carbonate reactions in, 326

Albuquerque (New Mexico), pollutant PAN from ethanol fuel use in, 407

alcohols
    functional group of, 202–203
    as gasoline additives, 400, 401, 403, 406, 407
    as vehicular fuels, 396, 401, 403–409, 414 energy density, 402, 403

aldehydes
    definition and formation of, 125–126, 128
    emission from alcohol fuels, 407, 409
    functional group of, 202
    role in eye irritation from smog, 128–129

aldicarb, LOD_{50} of, 240

aldrin
    biological conversion to epoxide, 234
    as cyclodiene insecticide, 233
    dieldrin formation from, 234

algae
    growth in polluted lakes, 288–289
      effects on alkalinity, 341–342
    from nitrate ion in wastewater, 311